GEOLOGY AND MINERALOGY RESEARCH DEVELOPMENTS

GRAPHITE

PROPERTIES, OCCURRENCES AND USES

Geology and Mineralogy Research Developments

Additional books in this series can be found on Nova's website under the Series tab.

Additional e-books in this series can be found on Nova's website under the e-book tab.

GEOLOGY AND MINERALOGY RESEARCH DEVELOPMENTS

GRAPHITE

PROPERTIES, OCCURRENCES AND USES

QUINTON C. CAMPBELL
EDITOR

nova publishers
New York

For permission to use material from this book please contact us:
Telephone 631-231-7269; Fax 631-231-8175
Web Site: http://www.novapublishers.com

Additional color graphics may be available in the e-book version of this book.

Library of Congress Cataloging-in-Publication Data

Graphite properties, occurrences and uses / editor, Quinton C. Campbell.
p. cm.
Includes index.
ISBN 978-1-62618-576-0 (hardcover)
1. Graphite. I. Campbell, Quinton C., 1951-
TA455.G7G74 2013
662'.92--dc23

2013010955

Published by Nova Science Publishers, Inc. † New York

CONTENTS

PREFACE

Graphite is the most stable form of carbon under standard conditions and is used in thermochemistry as the standard state for defining the heat of formation of carbon compounds. In this book, the authors discuss the advances in gasification of nuclear graphite; properties, occurrences and uses of graphite. Topics include the structural mimicry of carbon driven by ultrashort laser pulses; nanoscale sp2->sp3 conversion by visible light irradiation in graphite; graphite powder and related material as the principal component of carbon plastic electrodes; microstructural factors that govern the oxidative properties of graphite; application of graphite nano-particles in ultrafast fiber lasers; cluster deposition and implantation on/in graphite; thermal stability of fully lithiated graphite; and ethylene-octene copolymer/graphite composites.

Chapter 1 – This chapter reviews the gasification modes and fabrication methods of nuclear graphite used in Very High Temperature and High Temperature gas-Cooled Reactors (VHTRs and HTRs). It introduces chemical-reactions kinetics parameters and a model for the gasification of different grades of nuclear graphite. The chemical kinetics parameters are determined, using a multi-parameters optimization algorithm, from the reported experimental measurements of the total gasification rate and transient weight loss at different temperatures. The parameters for the nuclear graphite grades of IG-110 and NB-25, with super fine and medium size petroleum coke filler particles, are nearly the same, but different from those for NBG-18 and IG-430, with superfine and medium size coal tar pitch filler particles. Experimental measurements of the total gasification rate and transient weight loss for nuclear graphite grades of NBG-18, IG-430, NBG-25 and IG-110 successfully validate the chemical-reactions kinetics model for a wide range of temperatures and weight loss fractions up to 0.40. The model incorporates a recently developed correlation of Sherwood number for calculating the effective velocity of oxygen transport through the boundary layer in the diffusion-limited gasification mode at high temperatures. Results show that the initial specific surface area of free active sites is inversely proportional to the square root of the initial mass of graphite specimens in the experiments (≤ 50 g). However, the total gasification rate and the production rates of CO and CO_2 gasses increase with increasing the initial mass of graphite. Also presented are transient gasification analysis results for NBG-18 graphite in a helium coolant channel of a typical VHTR prismatic fuel element.

Chapter 2 – The transient and permanent structural mimicry of graphite, diamond and nanotubes in response to its moderate and strong excitation by femtosecond laser pulses was experimentally studied by means of diverse complementary experimental techniques. Sub-picosecond changes of the optical probe reflection, as well as pronounced picosecond hypersonic modulation of the probe reflectivity in the photoexcited graphite were related to its transient structural transformations and melting, respectively. The final, permanent structural modification were envisioned by micro-Raman spectroscopy as atomic-scale local structural transformations (point defect generation) at moderate photoexcitation, and as well-defined graphite and diamond phases at high photoexcitation of the graphite surface, which proceeds through formation of a hot, critical electron-ion surface plasma and TPa-level plasma-driven shock-wave compressing the material and expanding it off. Similarly, femtosecond laser-driven structural transformation toward graphite was revealed on surface and in bulk diamond. Finally, single- and multi-shot IR and visible femtosecond laser ablation of single-wall carbon nanotubes was envisioned at microscopic and mesoscopic structural levels. Overall, the observed rich variety of carbon allotropes, their mutual transformations and the intermediate structural carbon forms are illustrated and discussed in details.

Chapter 3 – Various recent experiments indicate us, that a graphite photo-stimulated with femtosecond laser pulse results in the formation of a quite new, but stable domain with sp^3 interlayer bonds. Here, the authors will theoretically review this mysterious non-equilibrium phase transition phenomenon, in terms of the lattice relaxation and the proliferation of photo-generated inter-layer charge transfer excitations. The excitation takes place in the visible region, and at the Franck-Condon state, the resulting electron-hole pair spanning neighboring two layers, is quite unstable, being easily dissipated into the 2d-semimetallic continuum electronic states, as positive and negative free carriers. However, by a small but finite probability, the electron and the hole become stable, as the lattice relaxation proceeds, being bound with each other, just like an exciton, through the inter-layer Coulomb attraction. This exciton state will be shown to self-localize, contracting the inter-layer distance and buckling the six membered ring of graphite, only around it. Thus, local inter-layer σ bonds are formed, or a tiny sp^3 nano-domain appears in the semi-metallic continuum. The Local Density Approximation (LDA) electronic structure analysis reveals that the electronic state of the new phase is an insulator domain immersed in a semimetal. In the next, in terms of molecular dynamics calculations the authors will explain proliferation of this domain and will show that the overall process of the transition becomes a multi-photon one. Furthermore using LCBOP potential and molecular dynamics, the authors will explain a temperature-induced ablation of the two bottom layers with the diaphite inside. Finally a possible application of diaphite as hydrogen adsorption surface will be described.

Chapter 4 – In this chapter, graphite in powdered form is presented as the material of choice for the preparation and use of the so-called *carbon paste electrodes* (CPEs) representing one of the most popular types of electrodes in electrochemical measurements for more than five decades. Starting with a short historical reminiscence, when highlighting also the importance of carbon materials in electrochemistry as such, powdered graphite is presented as particularly suitable material for an endless variety of electrodes, sensors, and detectors; the individual forms having found the widespread use in theoretical electrochemical studies, practical electro analysis, as well as in other instrumental measurements. Special attention is paid to the graphite powder with respect to its applicability in carbon paste mixtures and, subsequently, in the configuration of CPEs. In this respect, classification of traditional types of graphite powders is also given, completed with a survey of related and/or alternate carbonaceous materials whose applicability to the preparation of CPEs has undergone a real boom in the new millennium and, in fact, resulted in a new renaissance of the field. Furthermore, graphite powders (together with related materials) are discussed via physical, physicochemical, and principal electrochemical properties and of similar interest are also the corresponding graphite/carbon paste mixtures, including the experimental know-how on their laboratory preparation and way(s) of housing in specially designed holders and similar assemblies. At the end of this section, one special example is included, summarizing for the first time the recently performed studies on less well-known phenomena within the electrochemistry with CPEs – the ohmic resistance of carbon pastes in relation with their structure and behavior. Last but not least, the electrochemistry and electroanalysis with CPEs are briefly surveyed, documenting the versatility of these electrodes in modern instrumental analysis, their adaptability to the present day's trends, as well as great potential in the future, regardless whether one considers traditional graphite powder-based configurations or mixtures from new alternate materials.

Chapter 5 – Carbon is an element easily found in different polymorphic arrangements, existing in at least three main well-established forms (diamond, graphite and fullerenes), each of which can exist in a variety of materials with different electrochemical and mechanical properties; such characteristics lead to many advantages of carbon electrodes such as low cost, wide potential window, relatively inert electrochemistry, and electrocatalytic activity for a variety of redox reactions, specially to organic compounds. The most common allotrope are based on the graphite structure, consisting of ideally infinite sheets of graphene stacked like parallel sheets. In this arrange carbon atoms are all sp^2 hybridized, with an intraplanar and interplanar C-C bond lengths of 1.42 Å and 3.354 Å, respectively. Diamond is entirely sp^3 hybridized and tetrahedral, with a C-C bond length of 1.54 Å, its poor electrical properties requires the use of dopants (such boron) to provide sufficient conductivity for electrochemical purposes; geometrically, graphite and the newer forms of carbon have a strong family resemblance, as they are all built of sp^2-hybridised atoms. The two-dimensional structure of graphite is constructed by hexagons, and, as a result, diamond has a three-dimensional crystal structure (covalent network solid), while graphite consists of carbon layers (with covalent and metallic bonding within each layer) which are stacked in such a sequence different from that found in a hexagonal close packed crystal structure, being linked by a weak van der Walls interaction produced by a delocalized π-orbital. The anisotropy observed in graphite provides a good electrical and thermal conductor within the layers (due to the in-plane metallic bonding) and a poor electrical and thermal conductor perpendicular to the layers (due to the weak van der Waals forces between the layers), anisotropy in such case also allows these carbon layers to slide with respect to one another quite easily, making graphite a good solid lubricant, besides these graphite is able to undergo chemical reactions by allowing the reactant/intercalate to be located between the graphene layers, forming compounds (called intercalation compounds). Although carbon materials used in electrochemistry share some of the electronic properties of metals, their structures and chemistry differ dramatically from all metallic electrodes; being the development and application of carbon-based composites specially cited.

Chapter 6 – Graphite is widely utilized in high temperature applications under oxidative conditions. As such an understanding of the oxidative behaviour of graphite and hence the factors which govern these properties is essential. The exact kinetic mechanism remains a complex and disputed topic. However the effect of the microstructure on the oxidation is often neglected. The primary microstructural factors which influence the oxidation and graphite properties which lead to these are discussed. This includes the complex and varied microstructures found in real graphite materials from different origins. As well as the influence of both catalytically active and inert impurities on the development of the microstructure. This affords a comprehensive view of the influence of the microstructure on the oxidative properties of graphite. It is possible to represent the simplest of these behaviours qualitatively through simulation of the microstructural development. This indicates that the use of intricate and arbitrary kinetic reaction schemes is not necessary to accurately describe and sensibly compare the oxidative properties of different graphite materials.

Chapter 7 – Ultrafast optical pulse sources have impacted scientific research and industrial applications for many years, especially in the fields of optical communication, optical measurements, semiconductor fabrication and diagonostics, and medicine. Furthermore, ultrashort laser pulses with extremely wide band *spectra* also show potential associated with the development of future device technologies. In view of previous reports, the majority of ultrafast lasers are generated by mode-locking techniques. Most ultrafast lasers with sub-pspulse widths can be easily attained by using a passive mode-locking technique, due to the net gain window induced by the transient transparency in a saturable absorber with the fast carrier relaxation time. The passive mode-locking technique with a fast saturable absorber is more suitable to generate an ultrafast laser than the active mode-locking technique because the reaction time of the fast saturable absorber is faster than most optoelectronic devices. Graphene saturable absorbers have demonstrated to passively mode-lock versatile lasers in 2009. Graphene with a broadband nonlinear transmission feature makes it a superior saturable absorber for all kinds of lasers operating at a wide wavelength range, which replaces the typical saturable absorbers such as semiconductor saturable absorber mirrrors (SESAMs) and carbon nanotubes (CNTs). Since 2011, the graphite nano-particle has been found to possess similar optical properties as graphene, including an ultrafast carrier relaxation time, a large nonlinear modulation depth, a wideband wavelength tuning range and a high optical damage threshold. The graphite nano-particle has been used to mode-lock the erbium-doped fiber laser. It is noted that fabrications of graphite nano-particles are convenient such as with the direct polishing process. A complex chemical vapor deposition (CVD) process or chemical solution assistance is not needed. Moreover, the generated passively mode-locked pulse width with a graphite nano-particle is comparable to that which uses a grapheme saturable absorber. Except for its unique optical properties, the graphite nano-particle is electrically and thermally conductive due to the distribution of the orbital over the graphene sheet, which can freely move the electron on the plane. To date, graphite nano-particles also have important roles in a variety of fields including with batteries, conductive coatings, metallic alloys and nuclear reactors. In this chapter, the fabrications of a graphite nano-particle are discussed. Subsequently, the structural and optical properties of the graphite nano-particle are studied. The structural properties of the graphite nano-particle are investigated by using X-ray diffraction (XRD) and Raman scattering spectroscopy. The optical properties of the graphite nano-particle include linear absorption and nonlinear absorption. At last, the applications of graphite nano-particles as being the saturable absorber for the passively mode-locked fiber lasers are introduced.

Chapter 8 – Cluster ion beam technique is a versatile tool which can be used for controllable formation of nanosize objects on a surface as well as modification and processing of surfaces and shallow layers on an atomic scale. In this chapter an overview of research on cluster interaction with graphite is presented. One of the emphases is put on pinning of metal clusters on graphite with a possibility of following selective etching of graphene layers. The other main topic of concern is related to the phenomenon of cluster stopping and the development of scaling law for cluster implantation in graphite. Graphite is chosen for experiments because it is a good model material; it has an atomically smooth surface that makes it easy to resolve very small deposited clusters or damaged areas. Layered structure of graphite with strong covalent bonds in the graphene sheets and very week van der Waals interactions between them is an interesting type of crystalline arrangement for modelling of clusters implantation. Additionally, there is rapidly increasing research activities on graphene which promises a number of applications in nanoelectronics, plasmonics, sensing etc. Therefore, understanding of particle behaviour on soft landing or pinning on graphene is of considerable practical importance.

Chapter 9 – The safety-related performance of lithium ion batteries is a major concern in their potential applications. The safety of lithium ion batteries is closely related to the thermal stability of the cell materials, in which appears that the thermal stability of anode can be critical to the thermal runaway of lithium-ion batteries. Here is an overview of the studies on the thermal stability of fully lithiated natural graphite investigated using differential scanning calorimetry (DSC), which deal with the effect of particle size and carbon coating on thermal stability of lithiated graphite anode. The initial heat generation is due to the solid electrolyte interface (SEI) decomposition. The onset temperatures for exothermic reactions after initial SEI decomposition, which is associated with the release of intercalated lithium from graphite, appear to be affected by particle sizes and surface carbon coating. A thermal induced delithiation is facilitated by reduced diffusion path and higher surface area in smaller graphites, but alleviated by carbon coating. These structural changes in graphites during DSC scan are confirmed by ex situ X-ray diffraction analysis.

Chapter 10 – A series of ethylene-octene copolymer (EOC) composites have been prepared by melt-mixing with different weight ratios of expandable graphite filler (0-50% by weight). Electrical conductivity [both alternating current (AC) and direct current (DC)] and thermal conductivity studies were carried out. Effect of filler loading and frequency on electrical conductivity was studied. DC conductivity has increased from 1.51×10^{-13} Scm^{-1} to 1.17×10^{-1} Scm^{-1}. Percolation threshold by DC and also AC methods was observed at about 16 vol. % of the filler. Real part of permittivity was found to be decreasing with increase in frequency while conductivity was increasing. Thermal conductivity was also found to be increasing gradually from 0.196 to 0.676 $Wm^{-1}K^{-1}$ which is about 245% increase. Graphite not only increases the electrical and thermal conductivities but at and above 40 wt. %, also acts as a halogen-free, environmental friendly flame retardant. Horizontal burn rate was studied in detail with interesting results. Initially the addition of graphite increases the horizontal burn rate but above 35 wt. % of graphite the burning speed decreases. Shore-A hardness of EOC/graphite composites shows that even with high graphite loading, the hardness is increased from about 50 to 68 only so that the rubbery nature of the composite is not affected very much.

In: Graphite
Editor: Quinton C. Campbell

ISBN: 978-1-62618-576-0

Chapter 1

ADVANCES IN GASIFICATION OF NUCLEAR GRAPHITE: CHEMICAL REACTIONS, KINETICS MODEL VALIDATION AND ANALYSES RESULTS

Mohamed S. El-Genk[1,2,3,*] and Jean-Michel Tournier[1]

[1]Institute for Space and Nuclear Power Studies,
University of New Mexico, Albuquerque, NM, US
[2]Chemical and Nuclear Engineering Department,
University of New Mexico, Albuquerque, NM, US
[3]Mechanical Engineering Department,University of New Mexico,
Albuquerque, NM, US

ABSTRACT

This chapter reviews the gasification modes and fabrication methods of nuclear graphite used in Very High Temperature and High Temperature gas-Cooled Reactors (VHTRs and HTRs). It introduces chemical-reactions kinetics parameters and a model for the gasification of different grades of nuclear graphite. The chemical kinetics parameters are determined, using a multi-parameters optimization algorithm, from the reported experimental measurements of the total gasification rate and transient weight loss at different temperatures. The parameters for the nuclear graphite grades of IG-110 and NB-25, with super fine and medium size petroleum coke filler particles, are nearly the same, but different from those for NBG-18 and IG-430, with superfine and medium size coal tar pitch filler particles. Experimental measurements of the total gasification rate and transient weight loss for nuclear graphite grades of NBG-18, IG-430, NBG-25 and IG-110 successfully validate the chemical-reactions kinetics model for a wide range of temperatures and weight loss fractions up to 0.40. The model incorporates a recently developed correlation of Sherwood number for calculating the effective velocity of oxygen transport through the boundary layer in the diffusion-limited gasification mode at high temperatures. Results show that the initial specific surface area of free active sites is inversely proportional to the square root of the initial mass of graphite specimens in the experiments ($\leq$ 50 g). However, the total gasification rate and the production rates of CO

*Corresponding Author, Regents' Professor and Director, mgenk@unm.edu, 1-505-277-5442.

and CO_2 gasses increase with increasing the initial mass of graphite. Also presented are transient gasification analysis results for NBG-18 graphite in a helium coolant channel of a typical VHTR prismatic fuel element.

ABBREVIATIONS

A	Flow area of channel (m^2)
ASA	Active sites surface area (μmole)
C_{f,ε_i}	Active free sites with activation energies in bin i
C_{f,ε_j}	Active free sites with activation energies in bin j
CO	Carbon monoxide gas
CO2	Carbon dioxide gas
$(CO)_{\varepsilon_i}$	Surface stable oxide complexes in energy bin i
$C(O_2)_{\varepsilon_i}$	Un-dissociated oxygen complexes in energy bin i
D	Diameter of flow channel (m)
$D_{A,B}$	Binary diffusion coefficient for A and B gas mixture (m^2/s), Equation (9b)
D_{i-m}	Effective diffusion coefficient of gas species i in mixture (m^2/s), Equation (9a)
f(E)	Distribution function as a function of energy E (kJ/mole)
f(ε)	Distribution function as a function of energy ε (kJ/mole)
k	Rate Coefficient
k^o, k^*	Pre-exponential rate constants ($mole^{-n}.s^{-1}$)
k_m	Effective diffusion velocity of oxygen in boundary layer (m/s), Equation (8)
M_j	Molecular weight of gas species *j* (kg/mole)
L	Flow channel total length (m)
m	Mass ofgraphite specimen in experiments
$\dot{m}$	Gas mass flow rate (kg/s)
m_i''	Mass flux of specie i ($kg/m^2.s$)
N_a	Avogadro number, $N_a = 6.0225 \times 10^{23}$ ($mole^{-1}$)
$[\hat{O}_2]_b$	Oxygen concentration in bulk gas mixture ($mole/m^3$)
$[\hat{O}_2]_s$	Oxygen concentration at graphite surface ($mole/m^3$)
P_j	Partial pressure of gaseous species *j* (Pa)
P	Total pressure of gas mixture (Pa)
R	*Total* gasification rate (mg/hr)
$\dot{R}$	Consumption / production rate (mole/s)
R_2	Outer radius of nuclear graphite sleeve (m)
R_c	Radius of flow channel (m), (D/2)
Re	Reynolds number at bulk gas temperature
Re_{in}	Reynolds number at inlet air temperature, T_o
Re_s	Local Reynolds number at wall surface temperature, $\dot{m}D/(\mu_m A)$

R_g	Perfect gas constant (8.3144 J/mole.K)
Sc	Schmidt number at wall surface temperature, ν_m / D_{O2-m}
Sh	Sherwood number at wall surface temperature, $k_m D / D_{O2-m}$
S_m^o	Specific Active Sites (μmole/g)
t	Time (s)
t_{10}	Time to reach 10% local weight loss at channel entrance (s)
T	Temperature (K)
T_o	Initial temperature and air inlet temperature (K)
U	Average velocity of gas mixture (m/s)
X	Local weight loss fraction of graphite ($0 \leq X \leq 1$)
x_i	Molar fraction of species i in gas mixture
Y_i	Mass fraction of species i in gas mixture
$Y(X)$	Fractional change in weight loss
z	Axial location along flow channel (m)

Greek

δ	Thickness (mm)
ΔH	Enthalpy of formation (J/kg)
ε	Specific activation energy (J/mole)
$\bar{\varepsilon}$	Gaussian most-probable specific activation energy (J/mole)
μ	Gas viscosity (kg/m.s)
ν	Kinematic viscosity (m^2/s), $\nu = \mu / \rho$
ρ	Mass density (kg/m^3)
σ	Gaussian standard deviation (J/mole)
σ_j	Effective molecular diameter of gaseous species j (m)
$\Omega^{(1,1)*}$	Dimensionless collision integral for self-diffusion, Equation (9b)

Subscript/Superscript

a	Adsorption of oxygen onto active free sites to form un-dissociated complexes
b	Breakup of un-dissociated surface oxides to form stable complexes
c	Carbon
b	Bulk gas
CO	Carbon monoxide gas
CO_2	Carbon dioxide gas
D	Desorption of CO gas
He	Helium
i	Gaseous species i
in	Flow channel inlet
j	Gaseous species j

L	Laminar
m	Gas mixture
o	Initial, initial condition at t = 0
s	Inner surface of nuclear graphite sleeve
tot	Total
w	Wall
*	Desorption of CO_2 gas

INTRODUCTION

An air ingress into the massive nuclear graphite core of a Very-High Temperature or High-Temperature gas-cooled Reactor (VHTR and HTR), following a break in the pipes connecting the reactor vessel to the closed Brayton cycle turbo-machinery (Figure 1), is a design basis accident with potential safety implications. In such an accident, graphite gasification resulting in a ~ 10% weight loss of the core support columns in the lower plenum (Figure 1) could compromise their mechanical strength and structure integrity. Subsequent graphite gasification in the vertical, helium coolant channels of the core fuel elements by air natural convection could potentially release entrapped fission products in the graphite matrix of the coated fuel particles compacts (Figures 2a and 2b). The release of CO and CO_2 gaseous products is also a safety concern due to their self-ignition potential and polluting effect. The heat released by the exothermic chemical reactions producing these gases increases the graphite temperature, further accelerating its gasification and weight loss.

A typical HTR or VHTR (Figure 1) prismatic core comprises 102 hexagonal fuel assemblies, or elements, loaded with cylindrical fuel compacts of coated particles (Figures 1b and 2c) and stacked 8 m high in 3 concentric rings [1-4]. The hexagonal fuel elements in the core are 0.36 m flat-to-flat and 0.80 m tall (Figure 2c). The radial reflector of the annular reactor core consists of hexagonal graphite assemblies, in five inner and 2-3 outer rings (Figures 1 and 2d). Graphite reflector assemblies are also stacked (1.2 m) on top and and at the bottom (0.8 m) of the radially reflected core (Figure 1). The fuel compacts consist of Tristructural-isotropic (TRISO) particles dispersed in graphite [5]. The ~1 mm diameter particles are made of uranium oxy-carbide spherical kernels surrounded by porous graphite, PyC and SiC coatings (Figure 2a). The fuel compacts, typically 1.245 cm in diameter and 4.95 cm tall (Figure 2b), are loaded into vertical channels with a triangular lattice in the hexagonal fuel assemblies. On average, six fuel compacts surround a helium coolant flow channel (Figure 2c) that carries approximately the fission heat generated in two vertical stacks of fuel compacts in the core fuel elements during reactor operation. The helium coolant channels in the hexagonal fuel assemblies or elements (Figure 2c) are 8.0 m long and most are 1.5875 cm in diameter.

GRAPHITE GASIFICATION

The gasification of nuclear graphite is a complex process involving a multitude of chemical reactions whose rates depend on many variables. These include temperature, oxygen partial pressure, total pressure of the bulk gas mixture, chemical kinetics parameters, graphite grade (e.g., origin of binder and filler particles (petroleum or coal pitch) and the filler particles size (super-fine, fine or medium)) and the initial active surface area of free sites and stable complexes. Many models have been introduced for calculating the total gasification rate of different grades of nuclear graphite. They employ a semi-empirical approach of fitting experimental measurements (Figure 3).

Most reported models for calculating the total gasification of nuclear graphite use empirical Arrhenius or Langmuir-Hinshelwood expressions with apparent activation energies and pre-exponential rate constants, which are empirically determined from the reported experimental measurements at low and intermediate temperatures (Figures 3 and 4). These expressions are relatively simple and easy to integrate into computational fluid dynamics (CFD) and reactor thermal-hydraulics and safety analyses codes, but inheren- tly limited. The reported gasification models typically employ control-volume or finite-elements techniques to solve the multi-dimensional flow, mass transport and energy equations.

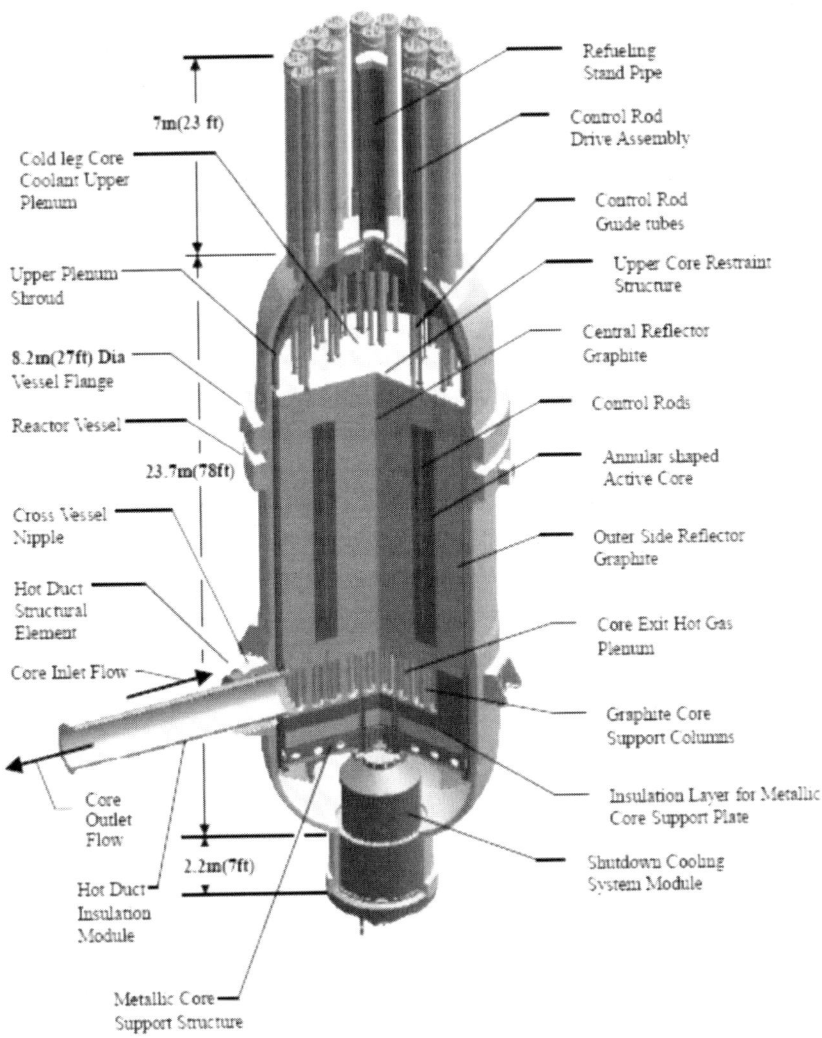

Figure 1. A Cut-Away View in a Prismatic Core Very High Temperature Reactor (VHTR) [1].

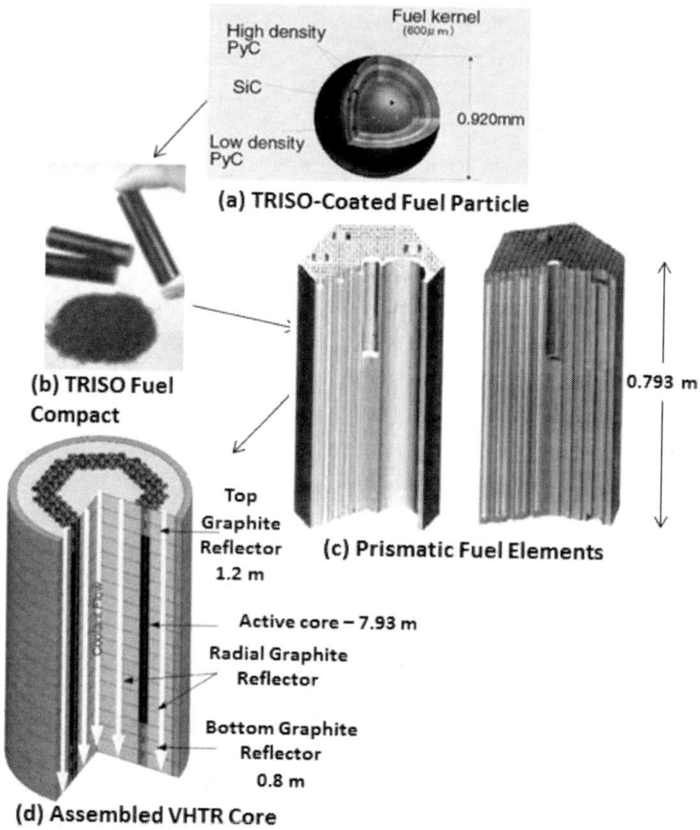

Figure 2. A TRISO Particle, Fuel Compacts, Prismatic Fuel Elements and a Cutaway Vertical View in a VHTR core. [6-8].

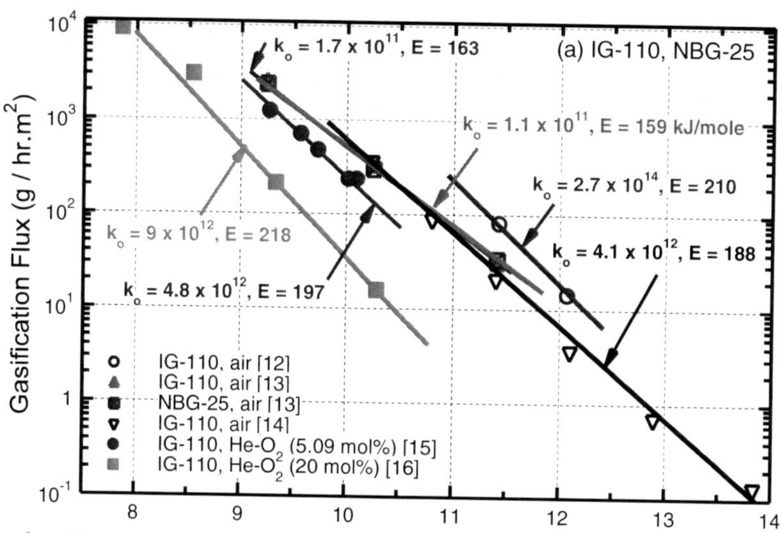

Figure 3. (Continued).

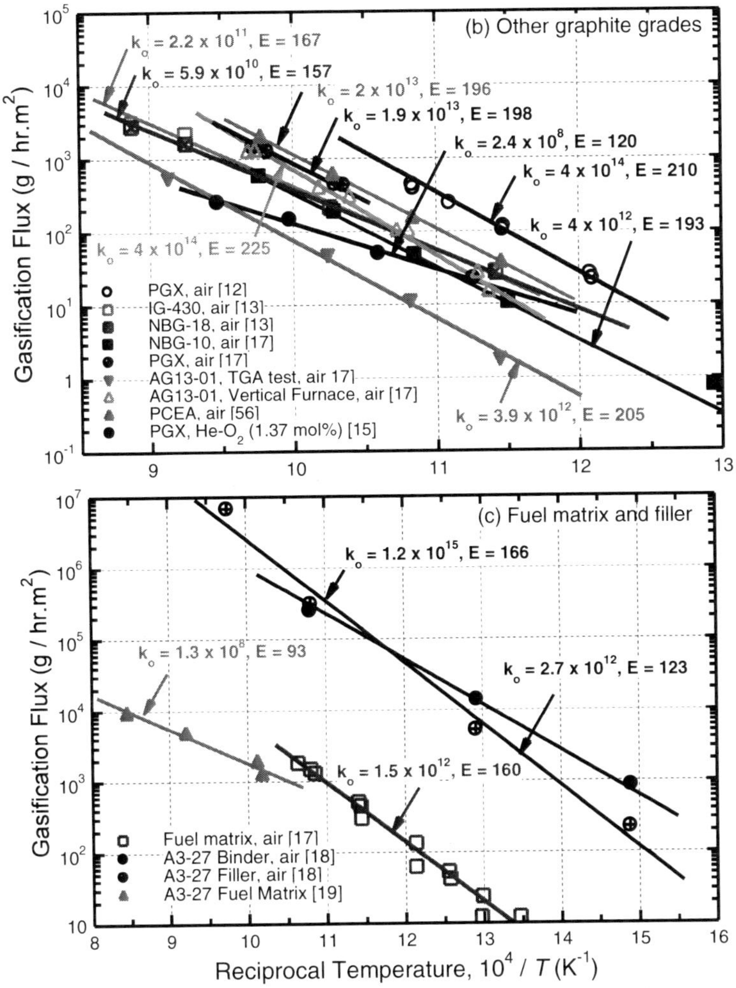

Figure 3. Reported Oxidation Data and Empirical Arrhenius Relations of the Total Gasification Flux for Various Grades of Nuclear Graphite.

The empirical correlations of graphite gasification are only valid in the range of temperatures and test conditions used in the experiments that generated the total gasification rate measurements used in the empirical fits (Figure 3). These correlations are not suitable for directly calculating the production rates of the CO and CO_2 gases during graphite gasification. The empirical values of apparent activation energy and pre-exponential rate constants for the empirical fits reported by various investigators are different, even for the same nuclear graphite grade, (Figure 3). This would introduce large differences in the safety analysis results of VHTRs and HTRs, following a massive air ingress accident. In addition, the empirically determined apparent activation energies are different in the different oxidation modes, each prevailing in a certain range of temperatures (Figure 4). Thus, the gasification rate predictions exhibit discontinuities at transition temperatures among the successive modes of graphite gasification (Figure 4).

The chemical-reactions kinetics approach introduced recently [9-11] avoids the inherent limitations of the empirical approach for calculating the gasification rate of nuclear graphite.

Experimental results for the gasification of different grades of nuclear graphite in atmospheric air over a wide range of temperatures successfully validated the chemical-reactions kinetics model by the authors [9-11].In this model, the chemical kinetics parameters for the prevailing gasification reactions depend in part on the grade of graphite, and the rate of gasification depends on the values of the specific activation energies and increases exponentially with temperature, subject to practical limitations (Figure 4). During gasification, the active surface area of free sites and stable complexes changes with weight loss as the size distribution of the volume and surface pores change. The extent of these changes depends on the temperature and hence, the prevailing model of gasification (Figure 4).

The chemical-reactions kinetics model [9-11] incorporates four primary reactions involved in the gasification of nuclear graphite, with different specific activation energies and pre-exponential coefficients. These reactions characterize the following successive processes: (a) adsorption of oxygen onto the surface active free sites; (b) simultaneous formation of C-O bonds and break up of C-C bonds to form stable surface complexes; and (c) desorption of CO and CO_2 gaseous. The values of the chemical-reactions kinetics parameters for these reactions depend on the grade of nuclear graphite. The initial specific surface area per unit mass of the free active site for adsorption of oxygen and the formation of stable oxide complexes decreases with increasing the initial mass of the graphite specimens in experiments [11].

The model uses Gaussian-like distributions of the specific activation energies for adsorption of oxygen and desorption of CO gas [9-11]. These and other chemical kinetics parameters and the initial active surface area of free sites are determined using a multi-parameters optimization algorithm. This determination is from the reported experimental measurements of the transient weight loss and total gasification rate for the different grades of nuclear graphite at different temperatures. The chemical kinetics model accounts for the change in the active surface area of free sites with weight loss. It calculates the total gasification rate and transient weight loss of graphite and the production rates of CO and CO_2 gases during graphite gasification. Employing the chemical-reactions kinetics model will predictably have important consequences for the safety analyses of VHTRs and HTRs. Although many parameters are measured in gasification experiments, others are difficult to measure directly but could be estimated reasonably accurately based on the reported measurements of the total gasification rate and transient weight loss [9-10,20]. Examples are the active surface area of free sites and its change with weight loss, the fractional coverage of free sites with surface complexes as a function of temperature and oxygen partial pressure, and the progressive transition in the total gasification rate among the principal modes with increasing temperatures (Figure 4).

Objectives

This book chapter reviews gasification modes and fabrication methods of nuclear graphite in VHTRs and HTRs. It also reviews the chemical-reactions kinetics approach developed by the authors [9-11] for modeling graphite gasification and the validation of the model results with experimental measurements for nuclear graphite grades of NBG-18, IG-430, NBG-25 and IG-110. These measurements cover a wide range of temperatures, traversing all three modes of gasification (Figure 4), and weight loss fraction up to 0.40. The chemical-reactions kinetics model incorporates a recently developed Sherwood number correlation for calculating the effective transport velocity of oxygen through the boundary layer in the diffusion-limited mode of gasification, Mode (c) in Figure 4. Also presented are the results of an analysis investigating the effect of the initial graphite mass on the total gasification rate and the production rates of CO and CO_2. In addition, the results of a parametric transient analysis of the gasification of NBG-18 nuclear graphite in a helium coolant channel of a typical VHTR prismatic fuel element are presented and discussed.

The following section presents and discusses the three primary modes for the gasification of nuclear graphite with increasing temperature; namely: (a) in-pores gasification, (b) in-pores diffusion-limited gasification, and (c) diffusion limited gasification (Figure 4).

Gasification Modes

The graphite's rate of gasification strongly depends on temperature and, hence the prevailing mode (Figure 4). Below 673 K, graphite gasification is negligible, but the rate increases exponentially with increasing temperature (Figures 3 and 4). There are three primary modes of graphite gasification, depending on the temperature [10,14,21,22], namely:

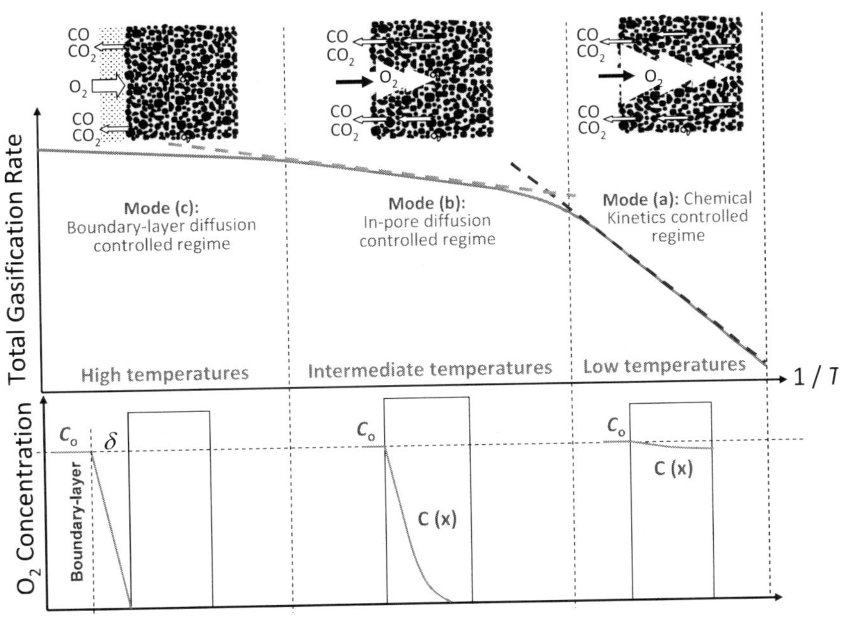

Figure 4. Prevailing Modes of Graphite Gasification with Temperature [9-11].

Mode (a)

At low temperatures (~ 673 – 800 K), in-pore gasification dominates and the rate depends solely on the chemical reactions kinetics. Increasing temperature limits the supply of oxygen from the bulk gas flow and its penetration into the open volume pores, transitioning gradually to the in-pores limited-diffusion mode of gasification (Mode (b)) at intermediate temperatures of 800 - 1123K (Figure 4). In Mode (a), gasification occurs entirely in the open volume pores. Although does not change the outside dimension, this mode weakens the graphite structure, and could compromise its integrity. The weight loss in the volume pores progressively increases their size distribution and access to previously closed pores, further weakening the mechanical strength of the graphite structure. Local weight loss in pores decreases with distance from the surface [17, 23-25]. The size distributions of volume pores in graphite also change with weight loss (Figures 5 and 6).

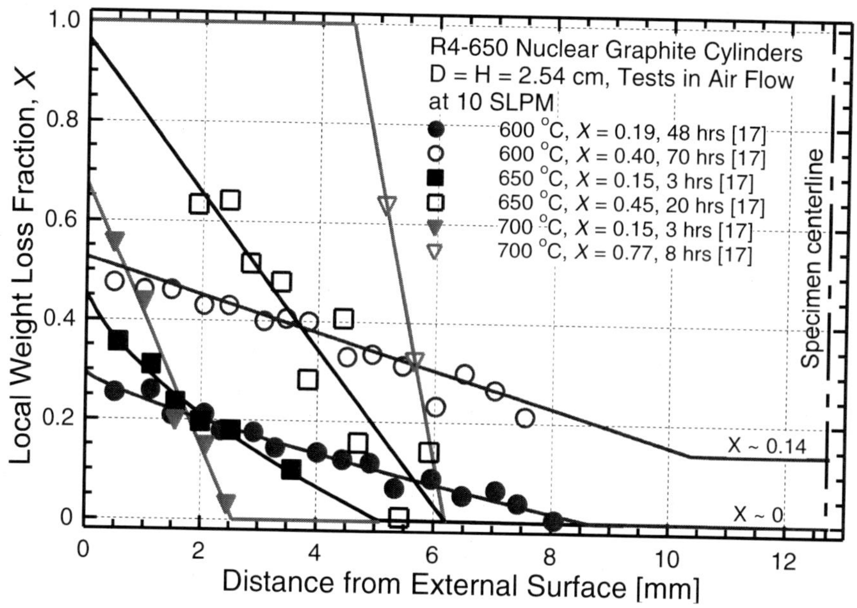

Figure 5. Measured Distribution of the Local Weight Loss Fraction for R4-650 Gra- phite in Gasification Experiments (Modes (a) and (b)); Data from [17].

Mode (b)

This mode could influence graphite gasification at temperatures as low as 800 K and extend to 1123 K, depending on the flow rate and Reynolds number of the bulk gas mixture, the initial mass of graphite and both the total and oxygen partial pressures. Although this mode of graphite gasification mostly occurs in open volume pores, oxygen diffusion into the pores becomes increasingly limited with increasing temperature (Figures 5 and 6). This is because of the depletion of oxygen within a smaller depth into the graphite and the counter-current diffusion of CO and CO_2 gases out of the pores (Figure 4). Despite the exponential increase in gasification with increasing temperature, the limited oxygen penetration into the volume pores in Mode (b) reduces the slope of the gasification rate with increasing temperature below that in Mode (a) (Figures 4 and 5). With a further increase in temperature, graphite gasification eventually ceases within the volume pores and shifts to the external surface (Figures 4-6), marking the transition to Mode (c).

Mode (c)

The transition occurs gradually from the in-pores diffusion limited Mode (b) to the diffusion-limited Mode (c) with increasing temperature. Thus, within a range of temperatures both modes co-exit, but with different proportions. In Mode (c), although the rate of the chemical-reaction kinetics is the highest, the transport of oxygen from the bulk gas, through the boundary layer, to the external surface limits graphite gasification. Mode (c) dominates at temperatures > 1123 K and unlike Mode (a) and Mode (b) it erodes the external surface, but does not affect the mechanical properties or compromise the structural integrity of graphite.

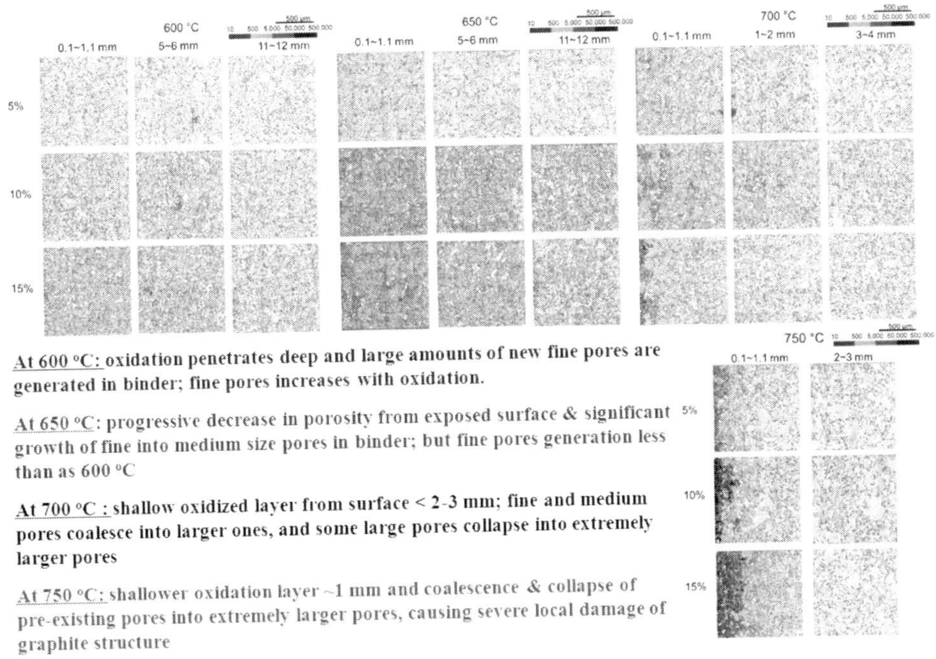

Figure 6. Micrographs Illustrating Changes in Volume Porosity in Gasification Experiments of IG-110 Graphite in Mode (a) and Mode (b) [25].

NUCLEAR GRAPHITE

Graphite is an aggregate of crystallite grains of almost identical properties and with sizes varying from 400 to 800 Å, depending on the grade of nuclear graphite [27]. The fundamental building block of the grains is the graphite crystal (Figure 7). The basic graphite crystal has a hexagonal unit cell structure with a predominant ABABA stacking of the planes of carbon atoms that have a large aspect ratio *c*/*a* (6.7/2.46) of 2.72 (Figure 7). The height of the graphite crystal, *c*, is twice the inter-layer spacing, *d*, of the atomic planes. The nearest distance between the carbon atoms in these planes is 1.42 Å. In a perfect graphite crystal, the basal planes are stacked with an inter-layer spacing d = 3.35 Å and a strong covalent bonding [28].

The weak van der Waal bonds between the basal planes of the crystal and the strong covalent bonding within these planes cause large differences in the surface energies. This results in much lower gasification rates from the basal planes than the edge and prismatic surfaces [26], referred to as "active surface." Thus, graphite gasification mostly depends on the active surface area (ASA) of free sites, *the fraction of the geometrical total surface area (TSA) that is chemically active*. The basal planes ideally terminate by arm-chair and zig-zag faces (Figure 8).

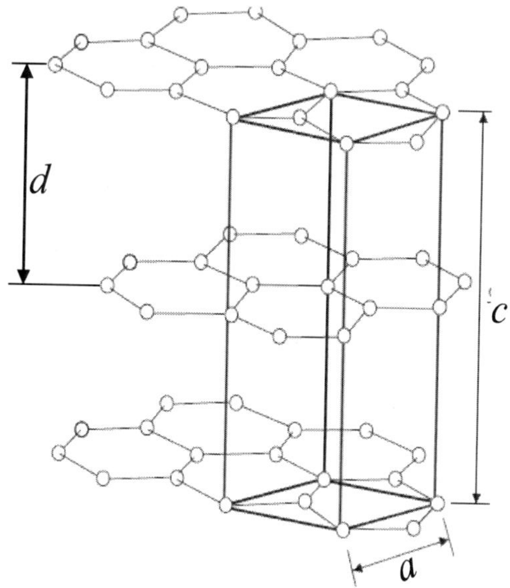

Figure 7. A Unit-Cell Graphite Crystal [27].

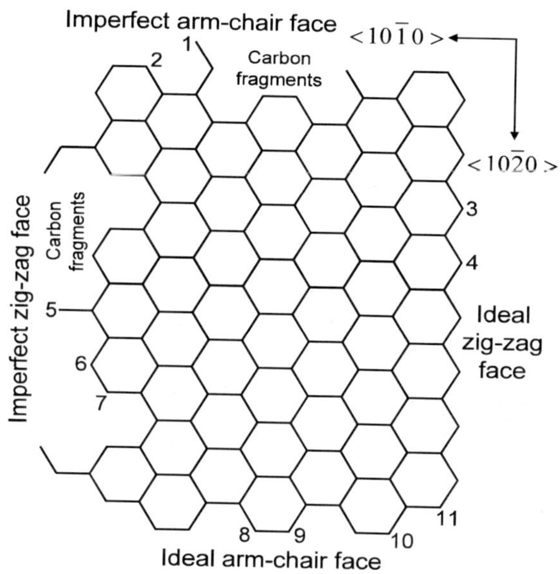

Figure 8. Edge Configurations of Graphite Basal Plane [26, 29].

The carbon fragments of one or 2 atoms protruding from the main condensed-benzene ring structure in the basal planes could contribute to forming active surface free sites and oxide complexes [26,29]. Determining the active surface area (ASA) of free sites and the fractions of the free sites and complexes on the surface as functions of temperature and the oxygen partial pressure is not straightforward [30]. The adsorption of oxygen atoms onto the free active sites depends on the reactivity of these sites, their affinity to surface compounds, the strengths of the covalent and van der Waal bonds, and the relative stability of the forming complexes.

The crystallite grains in a nuclear graphite grade are non-homogeneous composites of filler or coke particles and a binder. The random orientation of the filler particles or the grains provides nuclear graphite with its unique isotropic properties and structural stability during irradiation in nuclear reactors.

The micro-pores (~5% porosity) forming in the cokes filler particles during cool-down introduce micro-cracks due to the large anisotropy of the graphite crystals' thermal expansion and Young's modulus.

The cracked crystallites join and follow the general shape of the coke filler particle. The cracked crystallites enhance the shock resistance of graphite and allow large expansion in the *c* direction of the graphite crystal without inter-crystalline cracking [31]. These cracks and the internal volume pores form by the evolution of gases due to the combustion of impurities in the binder during heat treatment of nuclear graphite. They are caused by the anisotropic thermal shrinkage during cool-down, partially accommodating the irradiation-induced crystal growth. Propagating cracks are typically drawn toward volume pores in their vicinity [32].

The binder, a conglomerate of very small graphite crystals, does not graphitize well during the fabrication of nuclear graphite and has open and closed volume pores of different sizes [33,34]. The binder is typically stronger than the filler particles and consists of a superposition of "domains" extending > 100 μm with a common basal plane alignment, and of "mosaics," regions of small and randomly oriented pseudo-crystallites (< 10 μm) with a common orientation of the basal planes [32].

During an air ingress accident, the binder with open volume porosity and more impurities than the filler particles oxidizes more readily [35]. Depending on temperature, the preferential oxidation of the binder enlarges the volume pores and changes their size distribution (Figures 5 and 6), and could eventually degrade the structural strength of nuclear graphite. A bulk gasification of graphite resulting in a weight low of ~10% could decrease its mechanical strength in excess of 50% [36].

Based on the average size, the grains or filler particles in graphite are classified as ultra-fine, fine, medium or large. The average sizes of the ultra-fine and fine grains are ~ 2 and 10 μm, while in medium-grained nuclear graphite, the filler particles ranges from 100 μm to 4 mm.

The porous structure of nuclear graphite has open and closed volume pores of different sizes and an average volume porosity of 20% to 30% (Figures 5 and 6), compared to < 1% for a single graphite crystal. The internal surface area of porous graphite for potential gasification could be as much as 100 m^2/g.

The volume pores are classified into: large or macro-pores of an average size > 200 Å, transitional pores with sizes ranging from 20 – 200 Å, and small or micro-pores that are < 20 Å in average size (e.g., Figure 6). Early grades of nuclear graphite had an average density of ~1.6 g/cm^3 and a large volume porosity of ~ 29% [35-37]. In modern grades, in order to increase the Young's modulus and mechanical strength, the density has recently been increased to ~ 1.75 g/cm^3 (~ 23% porosity) and could reach ~ 1.81 g/cm^3 (~ 20% porosity) in the future.

The high graphitization temperature of nuclear graphite increases its thermal conductivity (~140 – 150 W/m.K at room temperature). The high purity graphite experiences low neutron activation [38] and would have a low rate of gasification during an air ingress accident involving a VHTR or HTR [31, 35, 36].

Nuclear Graphite Fabrication

Manufacturing of nuclear graphite involves partial re-crystallization of the carbonaceous materials of cokes or filler particles and the viscoplastic coal tar pitch binders at high temperatures [27].

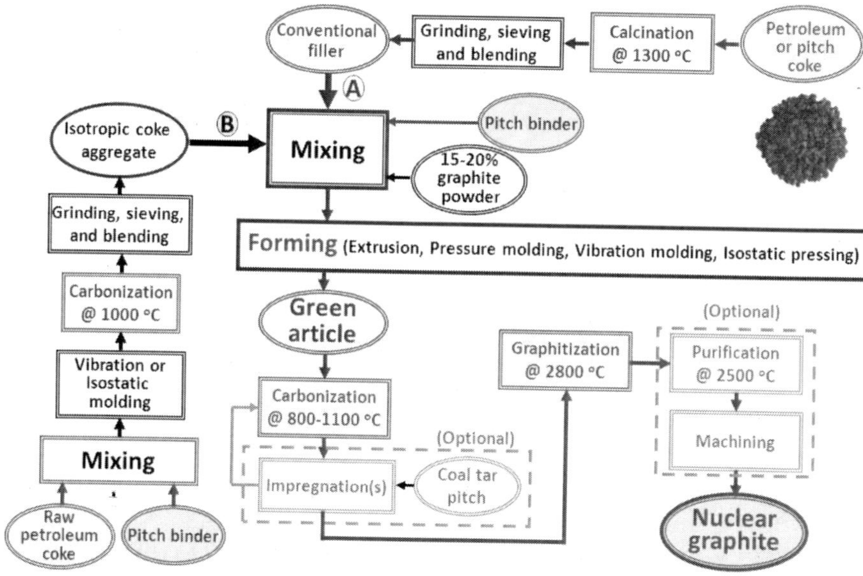

Figure 9. Manufacturing Processes of Nuclear Graphite [27, 31, 39, 41, 42].

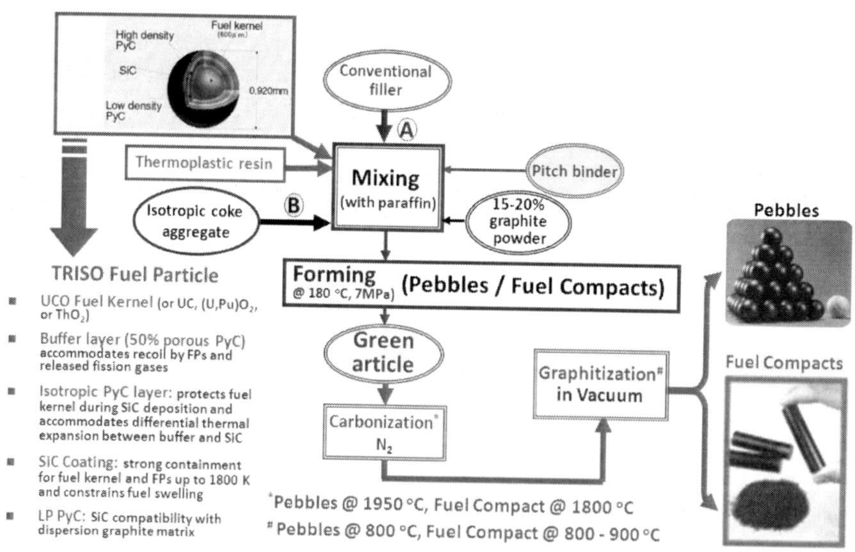

Figure 10. Fabrication of Fuel Pebbles and Fuel Compact [9, 44].

The cokes could be of a petroleum origin, coal by-product or naturally occurring pitch, varying considerably in microstructure, size and purity. The coke particles are heated up to 900 - 1300°C in order to densify and remove volatiles and impurities by combustion, preventing excessive dimensional changes in subsequent manufacturing processes [31,39]. The heat-treated coke particles are crushed, milled and sieved to the desired sizes (Path A in Figure 9), and subsequently mixed uniformly with the pitch binder to produce the "green article". This article could be formed into the desired shape or blocks, by extrusion, pressure molding, isostatic molding, vibration molding [40], or warm molding in the plastic temperature range of the binder resin (Figure 9). The randomly oriented and uniformly distributed filler grains within the binder produce macroscopically isotropic graphite.

Alternatively, secondary "precursor" cokes with isotropic structure can be produced commercially using inexpensive anisometric pitch [27, 41, 43]. One technique, illustrated in Figure 9, Path B, mixes the anisotropic precursor coke particles (grain size < 0.12 mm) with pitch binder then compresses the mixture by vibration molding to avoid preferential orientation of the grain. Following baking above 1000°C, the resulting graphite blocks are ground to a grain size < 1 mm [40, 41]. This technique produces secondary isotropic coke aggregates with high bulk density, excellent mechanical properties, low ash content and nearly spherical grains. The grains could be used, like conventional filler particles, in the fabrication of graphite.

The graphitization of the shaped green article takes place at high temperatures of 2600-3000°C (Figure 9). At temperatures > 1900°C the *a* and *c* dimensions of the graphite crystal (Figure 7), the graphitization and the thermal conductivity of the graphite all increase, but the mechanical strength, the modulus of elasticity and the potential rate of oxidation decrease. Above 2300°C, these parameters stabilize, and experience reversed trends at higher temperature (> 2600°C) due to carbon vaporization. The purity of nuclear graphite increases by reheating to ~2400°C in halogen gas flow (Figure 9).

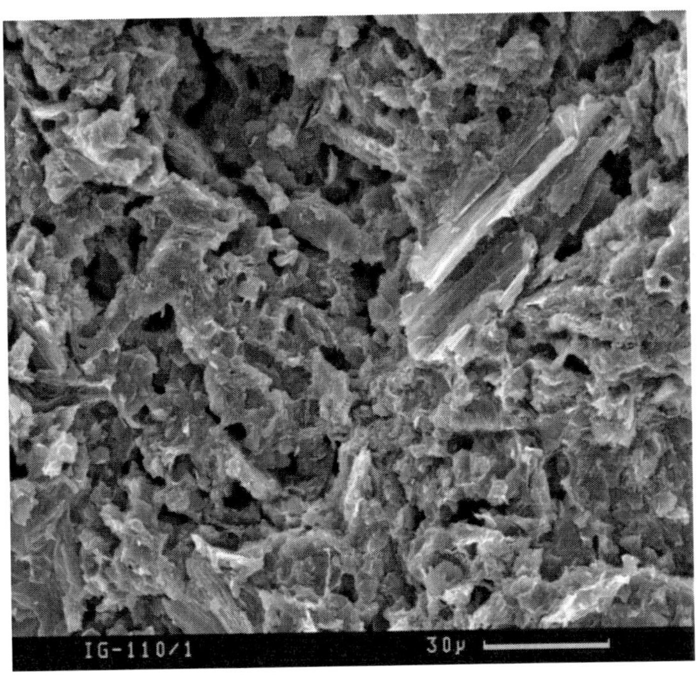

Figure 11. A Micrograph of IG-110 Nuclear Graphite showing Cracked Crystallite Structure, and Cracks and Fissures that Follow Coke Particles Shape [31].

The fuel elements in a VHTR or HTR are fabricated by thoroughly mixing the spherical Tristructural-isotropic (TRISO) fuel particles (Figure 2a) with the heat-treated graphite powder and a thermoplastic resin. The resulting mixture is shaped into fuel compacts or spherical pebbles [44, 45] (Figure 10). The cylindrical fuel compacts are loaded into vertical channels arranged in a triangular lattice within the hexagonal fuel elements of a prismatic core VHTR or HTR (Figures 2b and 2c).

(a) Unirradiated graphite

(b) Irradiated at 200°C to fluence of 2.7 x 10^{20} n/cm^2

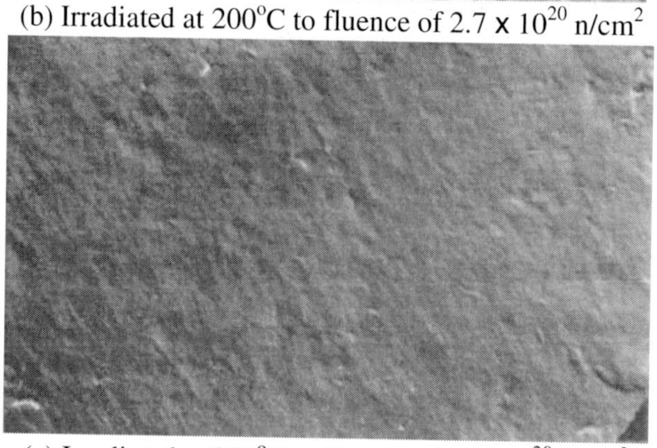

(c) Irradiated at 200°C to fluence of25 x 10^{20} n/cm^2

Figure 12. Electron Scoping Micrographs of IG-110 Crystallites show Irradiation-Induced Growth by Mrozowski Microcracks (5000x) [31].

Figure 11 presents a micrograph of IG-110 nuclear graphite showing its porous structure with cracked crystallites as well as micro-cracks and fissures that follow the shape of the coke filler particles [31]. Graphite gasification at low and intermediate temperatures in Modes (a) and (b) (Figures 4-6) would take place in the open cracks and volume pores of the binder. At higher temperatures of Mode (c), in-pore gasification practically ceases (Figure 4).

Neutron irradiation during reactor operation changes the graphite microstructure (Figure 12) and likely the active surface area of free sites and the gasification kinetics parameters for gasification. In addition to changing the active surface area of free sites and complexes, neutron activation of ^{12}C to ^{14}C isotope increases radioactivity of graphite, both during reactor operation and after shutdown.

Grades of Nuclear Graphite

The Fort St Vrain HTR used the near-isotropic H-451 nuclear graphite, fabricated by the Great Lakes Carbon Corporation, USA, using a mixture of coke particles [46-48]. The objective of using a mixture of different size coke filler particles, for example needle petroleum coke and more isotropic coke such as Gilsonite, is to achieve precise control and optimization of the graphite properties for high irradiation stability. The extruded, medium-grain H-451graphite (Table 1) for the moderator and core structure has an average mean coke particle size of 500 μm, with an aspect ratio of ~ 2.5 [49], and relatively large volume pores. For comparison, fines (< 100 μm grain size) and super-fine (< 20 μm grain size) cokes have aspect ratios of 3.1. H-451 graphite is no longer in production due to the lack of precursors, but often used as a reference for new graphite materials. It is structurally similar to NBG-10 and NBG-18 nuclear graphites (Table 1).

The use of fine coke particles and cold isostatic pressing or vibration molding to form the green article produces isotropic nuclear graphite with uniform properties. The fine-grains IG-110 nuclear graphite is nearly isotropic and contains much smaller petroleum coke particles than H-451 [50-52]. The filler particles in IG-110 are ~10 - 150 μm in size (or 20 μm mean), compared to 100 - 1500 μm (or 500 μm mean size) in H-451 [32], and the volume pores (16 μm mean size) are considerably smaller than in H-451 (42 μm) (Table 1).

The NBG-10, a medium-grain and nearly isotropic nuclear graphite, is manufactured by the SGL Carbon Company in France. It uses coal tar pitch cokes with a grain size < 1.6 mm and is formed by extrusion and doubly impregnated (Table 1) [48,53,54]. The PCEA nuclear graphite [48,55,56] is a candidate for the replaceable reflector and core blocks in the prismatic core, Next Generation Nuclear Plant (NGNP) in USA. It uses petroleum coke with grain sizes < 0.8 mm and is manufactured by extrusion [53,55]. Both NBG-10 and PCEA have a similar densification to that of H-451 nuclear graphite (Table 1).

Other types of nuclear graphite considered are NBG-17 and NBG-18 [53]. Both use pitch coke filler particles and are vibration molded. They exhibit nearly isotropic properties. The medium-grain (300 μm mean size) NBG-18 graphite has been considered for the core structure of the Pebble Bed Modular Reactor (PBMR), while NBG-17, a finer-grained version of NBG-18, is a candidate for the prismatic core fuel compacts and structure (Table 1). Finally, the IG-430 manufactured in Japan uses fine-grain, high-density coal pitch cokes. It is isostatically-molded and developed for the Japanese GTHTR-300 HTR design [57].

Table 1. Characteristics of Different Grades of Nuclear Graphite

Graphite Grade	**PGA**	**GR-1**	**H-451**	**ASR-2RS**	**AXF-8Q1**	**PCEA**	**NBG-10**	**NBG-25**	**NBG-17**	**NBG-18**	**IG-110**	**IG-430**
Reference	[59]	[42]	[32,62]	[41]	[60]	[53,56]	[53,54]	[53]	[53]	[13,53,56]	[41,42,56]	[59]
Reactor	Magnox		Ft St. Vrain, GT-MHR	HTR-500	Aero-space, furnace	NGNP	PBMR, NGNP	PBMR	NGNP	PBMR, NGNP	HTTR, NGNP	GTHTR-300
Manufacturer	UKAEA		SGL Carbon	GHT KFA GmbH	POCO Graphite	Graftech Int. Holdings	SGL Carbon GmbH	SGL C GmbH	SGL C GmbH	SGL Carbon GmbH	Toyo Tanso Ltd	Toyo Tanso Ltd
Country	UK	Russia	USA	Germany	USA	EC	France	Germany	Germany	Germany	Japan	Japan
Filler Cokes type	PC	PC	PC			PC	CC	PC	CC	CC	PC	CC
Filler Cokes grain size (μm)	coarse, 1100		medium, 500	< 1000	fine, < 20	medium, < 800	medium ,< 1600	medium, 300	fine	med., 300	fine, 20 (10-250)	fine
Forming process	E	M	E			E	E	V	V	V	I	I
Number of impregnations	1	none		2			2					
Graphitizing temperature (°C)		2800		> 2800	2800							
Ash content (ppm)		< 0.05w%		40	5		140	< 11		< 10	80	
Density (g / cm^3)	1.75	1.73	1.76	1.87	1.8		1.81	1.82		1.85	1.75	1.88
Volume porosity, ε	0.228	0.236	0.223	0.174	0.205		0.201			0.183	0.227	0.170
CTE x 10^6 (K^{-1}) [//,⊥] at 30°C	1.9, 3.9	4.8, 4.6	3.5, 4.5	3.9, 4.12	7.1	3.5, 3.71	4.2, 4.3				4, 3.6	

Graphite Grade	**PGA**	**GR-1**	**H-451**	**ASR-2RS**	**AXF-8Q1**	**PCEA**	**NBG-10**	**NBG-25**	**NBG-17**	**NBG-18**	**IG-110**	**IG-430**
Reference	[59]	[42]	[32,62]	[41]	[60]	[53,56]	[53,54]	[53]	[53]	[13,53, 56]	[41,42, 56]	[59]
Reactor	Magnox		Ft St. Vrain, GT-MHR	HTR-500	Aero-space, furnace	NGNP	PBMR, NGNP	PBMR	NGNP	PBMR, NGNP	HTTR, NGNP	GTHTR-300
Manufacturer	UKAEA		SGL Carbon	GHT KFA GmbH	POCO Graphite	Graftech Int. Holdings	SGL Carbon GmbH	SGL C GmbH	SGL C GmbH	SGL Carbon GmbH	Toyo Tanso Ltd	Toyo Tanso Ltd
Country	UK	Russia	USA	Germany	USA	EC	France	Germany	Germany	Germany	Japan	Japan
CTE x 10^6 (K^{-1}) [//,⊥]] at 400°C		6.8, 6.7			7.1	4.0, 4.24						
Anisotropy factor ((α⊥ / α//)	2.05	0.96	1.28	1.05		1.06	1.02				0.9	
Conductivity (W / m.K) [k//, k⊥]	209, 126	86, 90	150, 135	146, 142	120	143, 146	161, 157				124, 138	140 mean
Resistivity ((μΩ.m) [ρ//, ρ⊥]	6, 10	15, 14			16	7.7, 7.9	8.4, 8.6				< 13	
Young's modulus (GPa) [E//, E⊥]	12, 5.6	7.5, 7.7	11, 9.6	10.5, 10	12	11.8, 11.3	10.3, 9.9				9.4, 10	10.8 mean
Tensile strength (MPa) [//,⊥]		15, 18	15, 13	19, 18.1	33.1	23.5, 21.6	18, 21				24.9, 24	37 mean

Table 1. (Continued)

Graphite Grade	**PGA**	**GR-1**	**H-451**	**ASR-2RS**	**AXF-8Q1**	**PCEA**	**NBG-10**	**NBG-25**	**NBG-17**	**NBG-18**	**IG-110**	**IG-430**
Reference	[59]	[42]	[32,62]	[41]	[60]	[53,56]	[53,54]	[53]	[53]	[13,53, 56]	[41,42, 56]	[59]
Reactor	Magnox		Ft St. Vrain, GT-MHR	HTR-500	Aero-space, furnace	NGNP	PBMR, NGNP	PBMR	NGNP	PBMR, NGNP	HTTR, NGNP	GTHTR-300
Manufacturer	UKAEA		SGL Carbon	GHT KFA GmbH	POCO Graphite	Graftech Int. Holdings	SGL Carbon GmbH	SGL C GmbH	SGL C GmbH	SGL Carbon GmbH	Toyo Tanso Ltd	Toyo Tanso Ltd
Country	UK	Russia	USA	Germany	USA	EC	France	Germany	Germany	Germany	Japan	Japan
Compress. strength (MPa) [//,⊥]		70,84	60,60	79.8, 80	110.3		70, 60				73.4, 69.6	90 mean
Bending strength (MPa)[//,⊥]		28, 40	20, 24	28.5, 28	96.5		30, 27				34.7, 32.8	

(Legend: E = extrusion; I = isostatic molding; M = molding; PC = petroleum coke; CC = coal tar pitch coke; V = vibration molding).

Table 2. Experimental Conditions and Reported Apparent Activation Energies and Pre-exponential Coefficients for Graphite Gasification (Figure 3)

Nuclear Graphite Material	Impurities (ppm)	Specimen density, shape and dimensions(mm)	Temperature (°C)	Test gas	Total pressure	O_2 mole fraction	Flow rate or velocity	Pre-expcoefficient(g / hr.m^2)	Activation Energy (kJ/mole)	Reference
Ultra-pure graphite	< 5 ppm	1.65 g/cc, sphere, 12.7	550 - 850	air	1 atm	21	0.13 - 1 cm/s	1.9×10^{12}	188	[62]
PyC - HPG		2.24 g/cc, plate, 15 x 12 x 1	327 - 727	air	1 atm	21	No flow	5.0×10^{7}	130	[61]
AXF-5Q		1.85 g/cc, plate, 15 x 12 x 1	327 - 727	air	1 atm	21	No flow	2.8×10^{10}	164	[61]
EK98		1.86 g/cc, plate, 15 x 12 x 1	327 - 727	air	1 atm	21	No flow	1.9×10^{10}	164	[61]
AG 13-01	(TGA)	Right Cylinders, 25.4	600 - 750	air	1 atm	21	10 SLPM	3.9×10^{12}	205	[17]
AG 13-01	(furnace)	Right Cylinders, 25.4	600 - 750	air	1 atm	21	10 SLPM	4.0×10^{14}	225	[17]
IG-110	**< 20 ppm**	**Cylinders, 21 x 30**	**540 - 630**	**He-O_2**	-	**2.5 - 20**	**40 SLPM**	$\mathbf{9.0 \times 10^{12}}$	**218 ± 4**	**[16]**
IG-110	**30 ppm**	**Cylinders, 20 x 30**	**450 - 650**	**air**	**1 atm**	**5 - 21**	**5 SLPM**	$\mathbf{2.7 \times 10^{14}}$	**210**	**[12]**
IG-110		**Cylinders, 8.4 x 19.05**	**450 - 750**	**air**	**1 atm**	**21**	**0.5 SLPM**	$\mathbf{4.1 \times 10^{12}}$	**188**	**[14]**
IG-110	**< 10 ppm**	**Right Cylinders, 25.4**	**600 - 800**	**air**	**1 atm**	**21**	**10 SLPM**	$\mathbf{1.1 \times 10^{11}}$	**159**	**[13]**
IG-110	**< 100 ppm**	**Sleeve, D = 19, 22, H = 135**	**700 - 970**	**He-O_2**	**1.2 bar**	**5.09**	**6.6, 20 m/s**	$\mathbf{4.8 \times 10^{12}}$	**197**	**[15]**
IG-11	1900 ppm	Cylinders, 20 x 30	450 - 650	air	1 atm	5 - 21	5 SLPM	4.8×10^{11}	155	[12]

Table 2. (Continued)

Nuclear Graphite Material	Impurities (ppm)	Specimen density, shape and dimensions(mm)	Temperature (°C)	Test gas	Total pressure	O_2 mole fraction	Flow rate or velocity	Pre-exp coefficient (g / hr.m^2)	Activation Energy (kJ/mole)	Reference
IG-11	479 ppm	Right Cylinders,10.0	600 - 800	air	1 atm	21	0.02 SLPM	4.5×10^{11}	158.6	[21]
R4-650		Right Cylinders, 25.4	500 - 750	air	1 atm	21	10 SLPM	4.0×10^{13}	202	[17]
PGX (gas purified)	1.77 g/cc	Right Cylinders, 25.4	500 - 750	air	1 atm	21	10 SLPM	1.9×10^{13}	198	17]
PGX	1000 ppm	Cylinders, 20 x 30	450 - 650	air	1 atm	5 - 21	5 SLPM	4.0×10^{14}	210	[12]
PGX	< 7000	Sleeve, D = 19, 22, H = 135	600 - 1500	$He-O_2$	1.2 bar	1.37	6.63 m/s	2.4×10^{8}	120	[15]
NBG-10		Right Cylinders, 25.4	500 - 750	air	1 atm	21	10	4.0×10^{12}	193	[17]
H-451 (K022)		Cylinders, 8.382 x 19.05	500 - 900	air	1 atm	21	0.1 SLPM	1.0×10^{10}	159	[63]
H-451 (K018)		Cylinders, 8.382 x 19.05	500 - 900	air	1 atm	21	0.1 SLPM	1.0×10^{10}	159	[63]
A3-27 - filler	< 60 ppm		400 - 750	air	1 atm	21	< 1 cm/s	1.2×10^{15}	166	[18]
A3-27 - binder	< 60 ppm		400 - 750	air	1 atm	21	< 1 cm/s	2.7×10^{12}	123	[18]
A3-3 fuel matrix	< 60 ppm	Sleeve, D = 10, 20, H = 30	650 - 950	$He-O_2$	1.5 bar	1, 5, 10	-	-	128	[19]
A3-27 fuel matrix	< 60 ppm	Sleeve, D = 10, 20, H = 30	650 - 950			1 - 20	-	1.3×10^{8}	93	[19]
Fuel matrix		Right Cylinders, 25.4	500 - 650	air	1 atm	21	5 - 10	1.5×10^{12}	160	[17]
NBG-25	**< 11 ppm**	**Right Cylinders, 25.4**	**600 - 800**	**air**	**1 atm**	**21**	**10 SLPM**	$\mathbf{1.7 \times 10^{11}}$	**163**	**[13])**

Nuclear Graphite Material	**Impurities (ppm)**	**Specimen density, shape and dimensions(mm)**	**Temperature (°C)**	**Test gas**	**Total pressure**	**O_2 mole fraction**	**Flow rate or velocity**	**Pre-exp coefficient ($g / hr.m^2$)**	**Activation Energy (kJ/mole)**	**Reference**
NBG-18	**< 10 ppm**	**Right Cylinders, 25.4**	**600 - 800**	air	**1 atm**	**21**	**10 SLPM**	**5.9×10^{10}**	**157**	**[13]**
IG-430	**< 10 ppm**	**Right Cylinders, 25.4**	**600 - 800**	air	**1 atm**	**21**	**10 SLPM**	**2.2×10^{11}**	**167**	**[13]**
V483T	340 ppm	Sleeve, D = 10, 20, H = 30	650 - 950			1 - 20	-	1.6×10^{10}	132	[19]
ASR-1RS	200 ppm	Sleeve, D = 10, 20, H = 30	650 - 950			1 - 20	-	-	132	[19]
PCEA	**< 11 ppm**	Right Cylinders, 25.4 and 12.7 x 25.4	600 - 750	air	1 atm	21	**10 SLPM**	2.0×10^{13}	196	[56]

The microstructure, density, volume porosity and reactivity of nuclear graphite for oxidation depend on the origin and size of the coke particles, the molding processes used and the heat treatment of the green article. These variables affect the alignment and size of the crystallites and the size distribution of the volume pores forming during heat treatment. The larger are the crystallites the lower is the Active Surface Area (ASA) and the smaller is the average diameter of the volume pores. Increasing the heat treatment temperature and time at temperature, both increase the crystallites size, reducing graphite reactivity for oxidation [58]. Preferential oxidation of the binder occurs in coarse and fine-grain graphites [17,18,23,64]. The binder is relatively disordered and more chemically active than the filler coke particles. The volume pores in the binder increase the effective surface area for oxidation and the impurities and impregnates, if present, are a catalyst for accelerating oxidation. Degradation in the mechanical properties of graphite have been reported due to a preferential oxidation of the binder, a propagation of cracks at binder-grains interface, and increase in volume porosity [17,18,23,36,38,64].

Table 2 provides a summary of the conditions used in the reported gasification experiments for the different grades of nuclear graphite and lists the empirically determined values of the apparent activation energies and pre-exponential coefficients from the best fit of total gasification rate measurements (Figures 3a-3c).

Worth noting are the differences in the reported values of the apparent activation energy and the pre-exponential coefficient by various investigators, even for the same grade of nuclear graphite grade and temperature range in the experiments. The alternative approach of the chemical-reactions kinetics for estimating the rates of the total gasification and production of the CO and CO_2 in all three modes in Figure 4 [9,10] is briefly reviewed later in this chapter. Presented next is a review of the gasification kinetics of graphite.

Gasification Kinetics of Nuclear Graphite

The gasification of nuclear graphite involves a rapid chemisorption of oxygen onto the surface-active free sites in the crystal basal planes and the formation of stable complexes (Figures 8 and13). The production rates of CO and CO_2 gases and the graphite's total gasification rate depend on the surface fractions of the free active sites and complexes, which change with weight loss. As Figure 13 shows, surface complexes include acid anhydride, alcohol, carboxylic acid, ethers, ketene, lactones and quinones with oxygen/carbon ratios of 0.5 (ether) to 2 (carboxylic acid) [65-68].

The formation of stable hydrogen-containing aldehydes, carboxylic acids and alcohols typically occurs at the edges of the basal planes (Figure 13). For example, a two-atom carbon fragment (1) in Figure 8 can form a hexagonal δ-lactone ring with an embedded oxygen atom between carbon atoms (1) and (2). The carbonyl complexes of ketones, aldehydes, carboxylic acids and acid anhydrides would form on the conventional zig-zag sites (3) and (4), or on the corner sites (10) and (11) (Figures 8 and 13). The dangling carbon atom (5) in Figure 8 could stimulate the formation of γ-lactone or pentagonal ether (Figure 13), with an embedded oxygen atom between sites (5) and (6) in Figure 8. Alternatively, carbon site (5) can form ketene or alcohol with a hydroxyl (OH) functional group. The formation of quinone and semi-quinone groups can also occur on the ideal arm-chair sites (8) and (9) or on sites (6) and (7) of an imperfect zig-zag face (Figure 8).

Many techniques have been used to investigate the role the surface active sites and stable complexes in graphite gasification and the production of CO and CO_2 gases [65-70]. It is widely accepted that carbon atoms at the prismatic and pyramidal faces of the graphite crystallites (edge atoms, Figure 8), and at defective centers in the basal planes are the most reactive [26, 58]. There is also some evidence that graphite oxidation may proceed into the bulk of the lattice, as pits along screw cores and cores of non-basal edge dislocations, often catalyzed by impurity atoms.

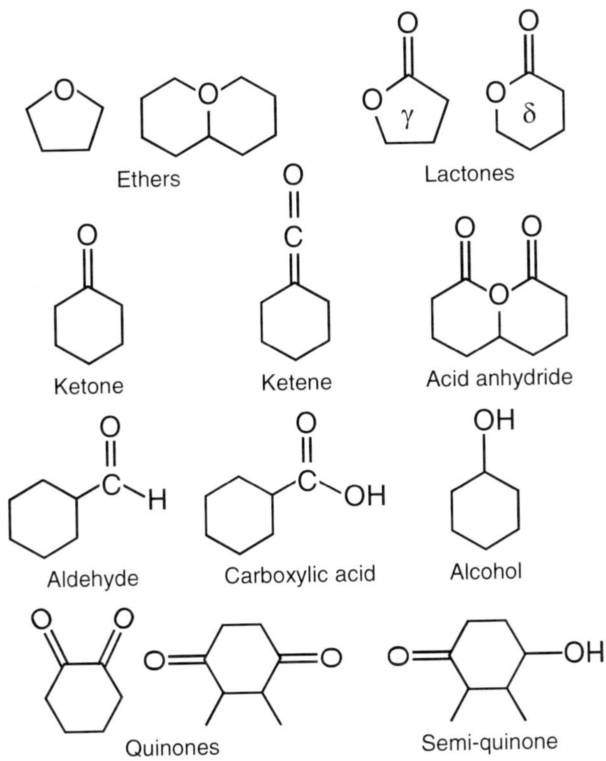

Figure 13. Organic Compounds Forming by Oxygen Adsorption onto Graphite.

The active surface area (ASA) is measured directly as a function of weight loss in temperature-controlled adsorption and desorption experiments [71-75]. In addition to the ASA, the values of the specific activation energies for adsorption of oxygen and desorption of CO and their Gaussian-like distributions and the specific activation energy for the desorption of CO_2 gas can also be measured directly or deduced from the reported total gasification rate and transient weight loss measurements in the experiments [9,10].

Researchers at the University of Sydney, Australia, have studied chemisorption of oxygen on out-gassed Spherocarb, an amorphous macroporous synthetic carbon of high purity, over a wide range of conditions (temperatures from 420 to 920 K and oxygen partial pressures of 0.01 to 8×10^4 Pa [66, 76, 77]. They identified two distinct mechanisms for forming surface complexes: (1) Type-A forms by adsorption of oxygen and is prevalent without gasification at both low temperature and oxygen partial pressure and for a short exposure time; and (2) Type-B forms with the gasification of one carbon atom for every chemisorbed oxygen atom. The formation of the surface complexes does not reach equilibrium in laboratory gasification experiments, even after several weeks. This is apparently due to the substantial heterogeneity of the surface sites, as indicated by Laine et al. (1963).

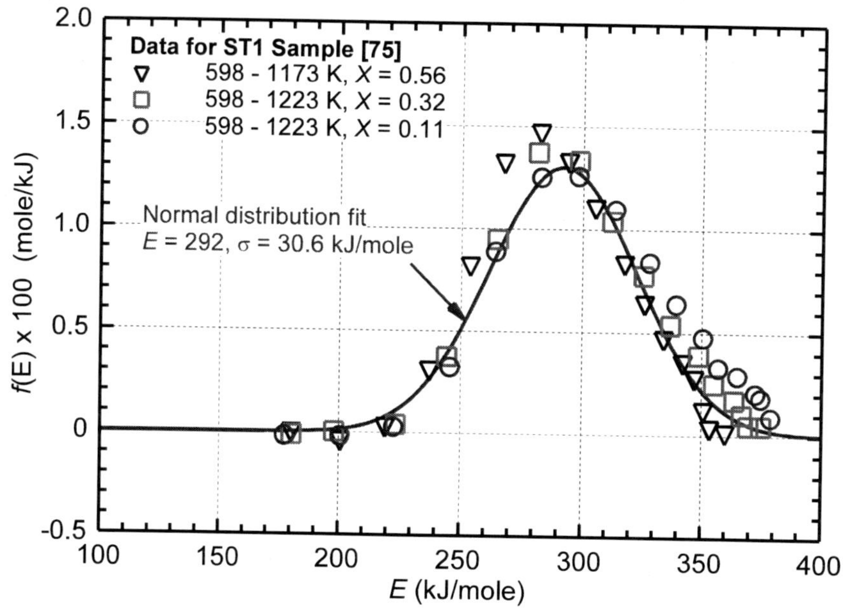

Figure 14. Desorption Energy Distribution for ST1 Uncatalyzed Soot from STD Measurements at Different Temperatures and weight loss, *X* [75].

Recent studies [65, 71, 75-79] have shown that the specific activation energies for desorption of oxygen and desorption of CO gas have Gaussian-like distributions (e.g., Figure 14). This is attributed to the wide varieties of the geometrical arrangements for the edge carbon atoms (Figure 8) and the diversity in forming oxygen-carbon surface complexes (Figure 13). The Gaussian-like distributions of the specific activation energies for adsorption of oxygen and desorption of CO have been measured experimentally using a controlled linear change in the temperature of the graphite test specimens with time [65, 71, 75-79].

Du et al. (1990 and 1991) performed experiments with argon flow to investigate the distribution of the desorption energy for uncatalyzed particles of ash-free soot (ash < 32 ppm by atom). The soot particles, with an average diameter of 42 nm, were collected in a pre-mixed ethylene flame (C/O = 0.98). The obtained distribution, the same for 3 different temperature ramps in the experiments, was well fitted by a Gaussian-like function with most probable activation energy, $\overline{E}$ = 292 kJ/mole and a standard deviation, σ = 30.6 kJ/mole (Figure 14). Experimental results involving different types of graphite have shown that the distributions of the specific activation energies are reproducible. For a given grade of nuclear graphite, they are independent of temperature and weight loss.

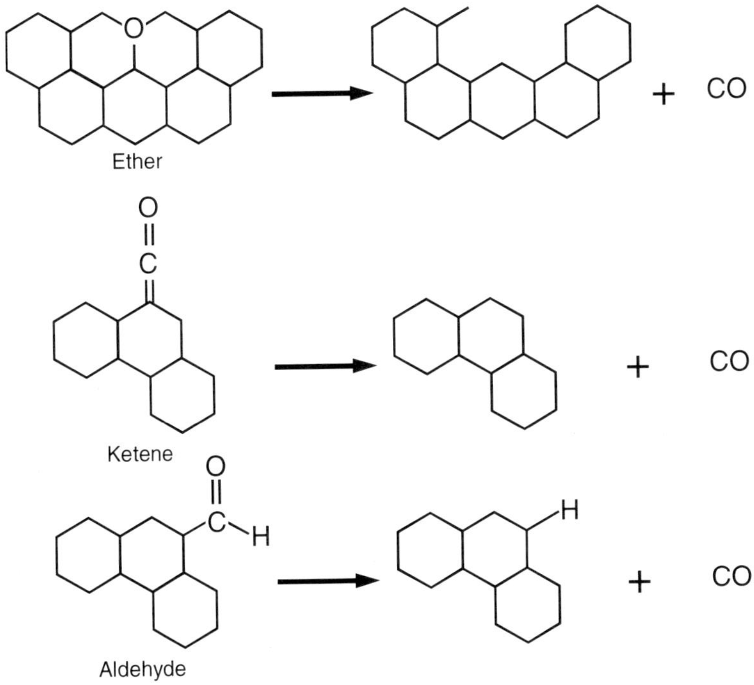

Figure 15. Direct Routes for CO Production by Desorption of Oxygen Complexes.

During graphite gasification in Modes (a) and (b), the ASA and the specific active surface area, S_m, change with weight loss. The ASA increases initially with weight loss to a peak, and then decreases with further increase in weight loss. The peak ASA and the corresponding weight loss fraction of 0.3 - 0.4 depends on the grade of graphite. Gasification in Modes (a) and (b) enlarges the open volume pores, increasing the accessible surface area for oxidation. As oxidation proceeds, some of the volume pores eventually coalesce into larger pores (Figures 5 and 6), decreasing the internal surface area for oxidation and hence the ASA [14, 21, 22]. This is somewhat mitigated by the possible access to originally closed volume pores.

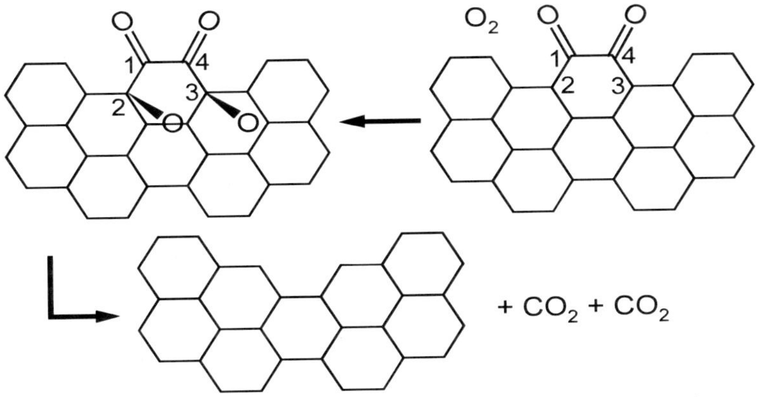

(a) From Quinone Complexes on Arm-Chair Face of Basal Plane

Figure 16. (Continued).

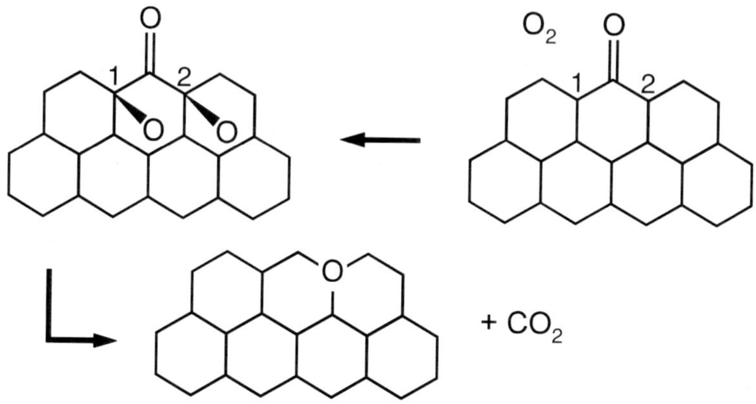

(b) From Semi-Quinone Complex on Zig-Zag Face of Basal Plane

Figure 16. Routes for the Production of CO_2 Gas during Graphite Oxidation.

Graphite gasification proceeds in a series of adsorption and desorption processes [58, 66, 79-81]. These are:(a) Diffusion of oxygen to the graphite surface and into open volume pores; (b) Adsorption of oxygen atoms onto the surface-active free sites and forming stable complexes and C-O and C-H bonds and the breaking of the C-C bonds; and(c) Desorption of CO and CO_2 gases and transport to bulk gas mixture.

The oxygen atoms attach to the surface active sites and form intermediate complexes (Figures 5, 8 and 13). Different carbon sites have different stability and reactivity, so are the surface complexes. Many elementary steps for the formation and desorption of CO and CO_2 gases have been proposed. The direct route for the production of CO is by thermal desorption of stable complexes, possibly ethers and carbonyl groups (Figure 15). Due to the stability of these complexes, the production of CO is not observed at low temperatures (< 500°C), but increases exponentially with increasing temperature. At high temperatures > 800°C, CO is the major product of graphite gasification.

Unlike CO, the CO_2 production during graphite gasification occurs at low temperatures, below those at which surface complexes decompose. As Figure 16 illustrates, a postulated primary venue for the production of CO_2 gas during gasification is the weakening of the graphite microstructure by the bonding of an oxygen atom onto a saturated carbon atom adjacent to an edge carbon atom already bonded to an oxygen atom [67]. Figures 16a and 16b illustrate this process for arm-chair and zig-zag faces of graphite planes supporting quinone and semi-quinone complexes, respectively.

In Figure 16a, carbon atoms (1) and (4), in the arm-chair face of the basal plane, support quinone groups. Thus, the carbon sites (2) and (3) carry net negative charges, and are likely targets for oxygen chemisorption because of the oxygen's strong electro-negativity. The newly adsorbed oxygen atoms at sites (2) and (3) are off-plane, as indicated by the solid triangular arrows. Oxygen chemisorption weakens the C-C bonds (1)–(2) and (3)–(4), which become much easier to break than the C-O quinone bonds. The high reactivity permits desorption of the semi-quinone radicals at low temperature, releasing CO_2.

Figure 16b shows a semi-quinone complex on the zig-zag face of the basal plane. The incoming oxygen molecule chemisorbs at the electronegative carbon sites (1) and (2), weakening the bonds to the semi-quinone group. A CO_2 molecule desorbs, possibly leaving the excess oxygen atom embedded in an ether bridge configuration. Note that the pathways in

Figures 15 and 16 are not inclusive, but only illustrations of plausible processes for the production of CO and CO_2 gases during graphite gasification. The next section describes the chemical-reactions kinetics model developed by the authors for calculating the production rates of CO and CO_2 gases and the total gasification rate for different grades of nuclear graphite [9-11, 20].

CHEMICAL-REACTIONS KINETICS MODEL

The gasification model recently developed by the authors [9-11] employs the chemical kinetics parameters and pre-exponential coefficients of four elementary reactions. These are the chemisorption of oxygen molecules to form oxygen radicals at active sites; the dissociation and adsorption of oxygen radicals to form stable complexes; desorption of stable complexes to produce CO gas; and the desorption of CO_2 gas. The eleven chemical kinetics parameters in the input to the model include:

(a) The values and Gaussian-like distributions of the specific activation energies and pre-exponential coefficients for adsorption of oxygen and desorption of CO (6);
(b) The specific activation energies and the pre-exponential coefficients for forming stable (CO) complexes and the desorption of CO_2 gas (4); and
(c) The initial active surface area of free sites, ASA (1)

For each grade of nuclear graphite investigated (Table 3), we used a multi-parameters optimization algorithm to obtain these chemical-reactions kinetics parameters from the reported measurements of the total gasification rate and transient weigh loss in experiments conducted by different investigators [10]. Employing the chemical-reactions kinetics model [9, 10], instead of the empirical approach (Figures 3a-3c), predictably will have important consequences for the safety analyses of VHTRs and HTGRs in the unlikely event of a massive air ingress accident.

The chemical-reactions kinetics model calculates the total gasification rate and the transient weight loss of graphite and the production rates of CO and CO_2 gases as functions of the graphite grade; the total pressure, composition and flow rate of the bulk gas mixture; the oxygen partial pressure and the temperature, geometry and dimensions of the graphite structure. The model [9,10] also predicts the change in the active surface area of free sites (ASA) with weight loss. However, it does not account for the migration, diffusion and interaction of surface species [80], or the thermal annealing of free active sites into non-reactive compounds [58, 69, 82]. The four gasification processes incorporated into the model are characterized by the following chemical reactions [9, 10]:

(a) Adsorption of oxygen molecules to form un-dissociated $C(O_2)$ complexes:

$$C_{f,\varepsilon_i} + O_2 \overset{k_{a,i}}{\rightarrow} C(O_2)_{\varepsilon_i} \, . \tag{1a}$$

The rate constants for these reactions are:

$$k_{a,i} = k_a^o \cdot e^{-\varepsilon_{i,a}/(R_g T)}. \tag{1b}$$

(b) Breakup of un-dissociated $C(O_2)$ radicals to form (CO) complexes:

$$C_{f,\varepsilon_j} + C(O_2)_{\varepsilon_i} \overset{k_b}{\rightarrow} (CO)_{\varepsilon_i} + (CO)_{\varepsilon_j}. \tag{2a}$$

The rate constants for these reactions are:

$$k_b = k_b^o \cdot e^{-\varepsilon_b/(R_g T)}. \tag{2b}$$

(c) Desorption of (CO) complexes to produce CO gas:

$$(CO)_{\varepsilon_j} \overset{k_{d,j}}{\rightarrow} CO + C_{f,\varepsilon_j}. \tag{3a}$$

The rate constants for these reactions are:

$$k_{d,j} = k_d^o \cdot e^{-\varepsilon_{j,d}/(R_g T)}. \tag{3b}$$

(d) Desorption of CO_2 gas:

$$C_{f,\varepsilon_i} + (CO)_{\varepsilon_j} + O_2 \overset{k_{CO_2}}{\rightarrow} CO_2 + (CO)_{\varepsilon_j} + C_{f,\varepsilon_i}. \tag{4a}$$

The rate constants for these reactions are:

$$k_{CO_2} = k_d^* \cdot e^{-\varepsilon_d^*/(R_g T)}. \tag{4b}$$

Although desorption of a CO_2 molecule frees an active site, the reactions given in Equation (4a) consume only an oxygen atom to remove a carbon atom from the surface. Desorption of CO_2 gas is limited by the availability of both surface free sites and (CO) surface complexes in close proximity (Equation (4a)).

The specific activation energy for desorption of a CO_2 molecule (Equation (4a)), ε_d^*, is considerably lower than that for the desorption of a CO molecule (Equation (3a)). The specific activation energies for adsorption of oxygen onto the surface-active free sites (Equation (1a)) have a Gaussian-like distribution. The examples in Figures 17a-17c indicate the values of the most-probable specific energies for oxygen adsorption, $\bar{\varepsilon}_a$, and the standard deviations, σ_a of their Gaussian-like distributions [9,10].

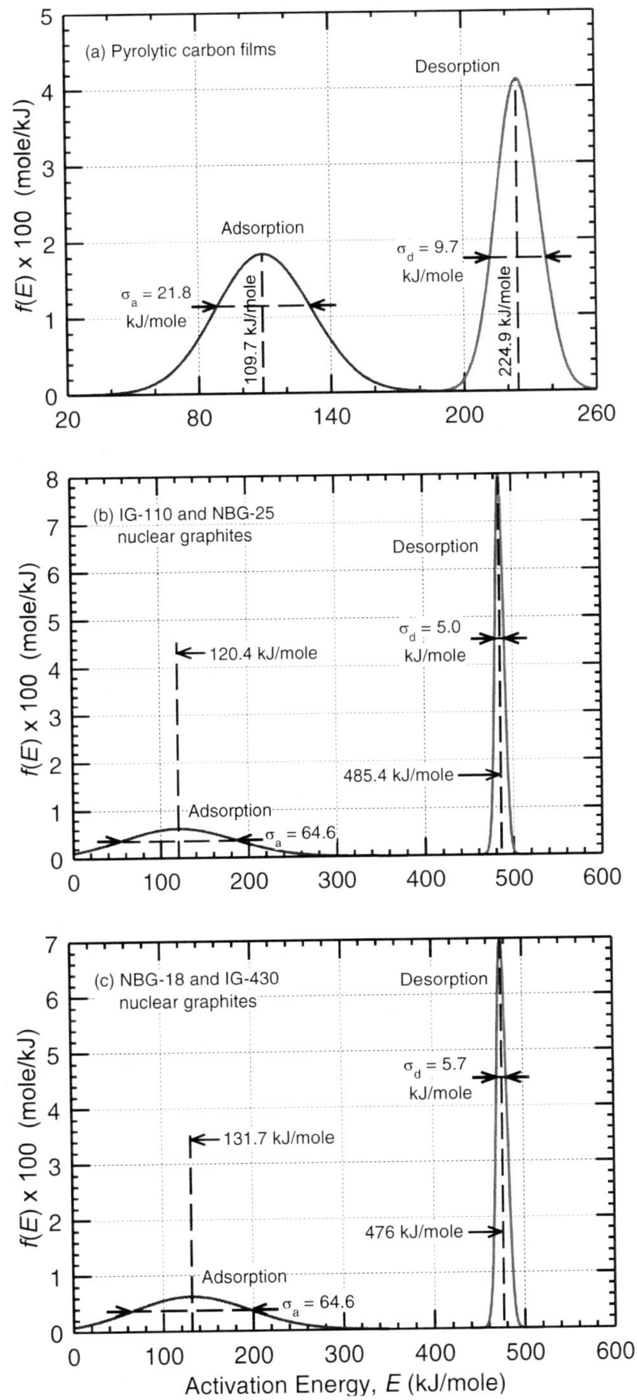

Figure 17. Specific Activation Energies for Adsorption of Oxygen and Desorption of CO Gas for Thin-Film Pyrolytic Graphite and Various Grades of Nuclear Graphite [9-11, 20].

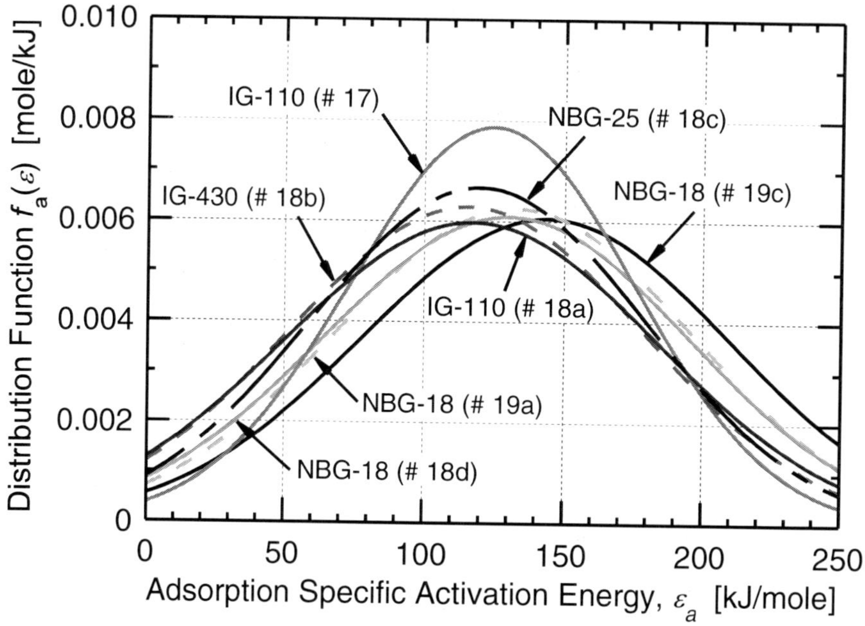

Figure 18. Gaussian-like Distributions of Specific Activation Energies of Adsorption of Oxygen for Different Grades of Nuclear Graphite (*Numbers in Parentheses Refer to Specimens in Table 3*) [20].

The specific activation energies for desorption of CO gas molecules also have a Gaussian-like distribution with a most-probable specific energy, and a standard deviation $\sigma_d < \sigma_a$. The constituent equations for calculating the rates of the four elementary reactions in the chemical-reaction kinetics model (Equations (1a) – (4a)) are detailed elsewhere [9, 10] and not repeated here.

The determined chemical-reactions kinetics parameters for IG-110, IG-430, NBG-18, and NBG-25 nuclear graphites are given in Figures 18 and 19 and listed in Tables 3 and 4 [9-11,20]. In addition to the ASAo, the input to the chemical-reactions kinetics model [9-11] for the gasification of nuclear graphite includes 10 parameters. These are: (a) the most-probable specific activation energies for adsorption of oxygen and desorption of CO ($\bar{\varepsilon}_a$ and $\bar{\varepsilon}_d$) and the standard deviations (σ_a and σ_d) characterizing their Gaussian distributions; (b) the specific activation energies and for elementary reactions in Equations (2a) and (4a); and (c) the pre-exponential rate coefficients, k_a^o, k_b^o, and for the rate constants of the four chemical reactions given by Equations (1b), (2b), (3b) and (4b).

The values of the chemical kinetics parameters [9-11,20] are specific to the grade of nuclear graphite, while the initial Active Surface Area of free sites(ASA) depends on the grade of graphite and the size of the specimens used in gasification experiments (Tables 3 and 4). The change in the value of ASA with weight loss in the experiments indicates the reactivity of the graphite and depends on the graphite's microstructure and the amounts of reactive impurities present. The ASA changes with weight loss due to the widening and random overlap of reacting surfaces (or coalescence) within the volume pores during the graphite gasification in Modes (a) and (b) [14, 83].

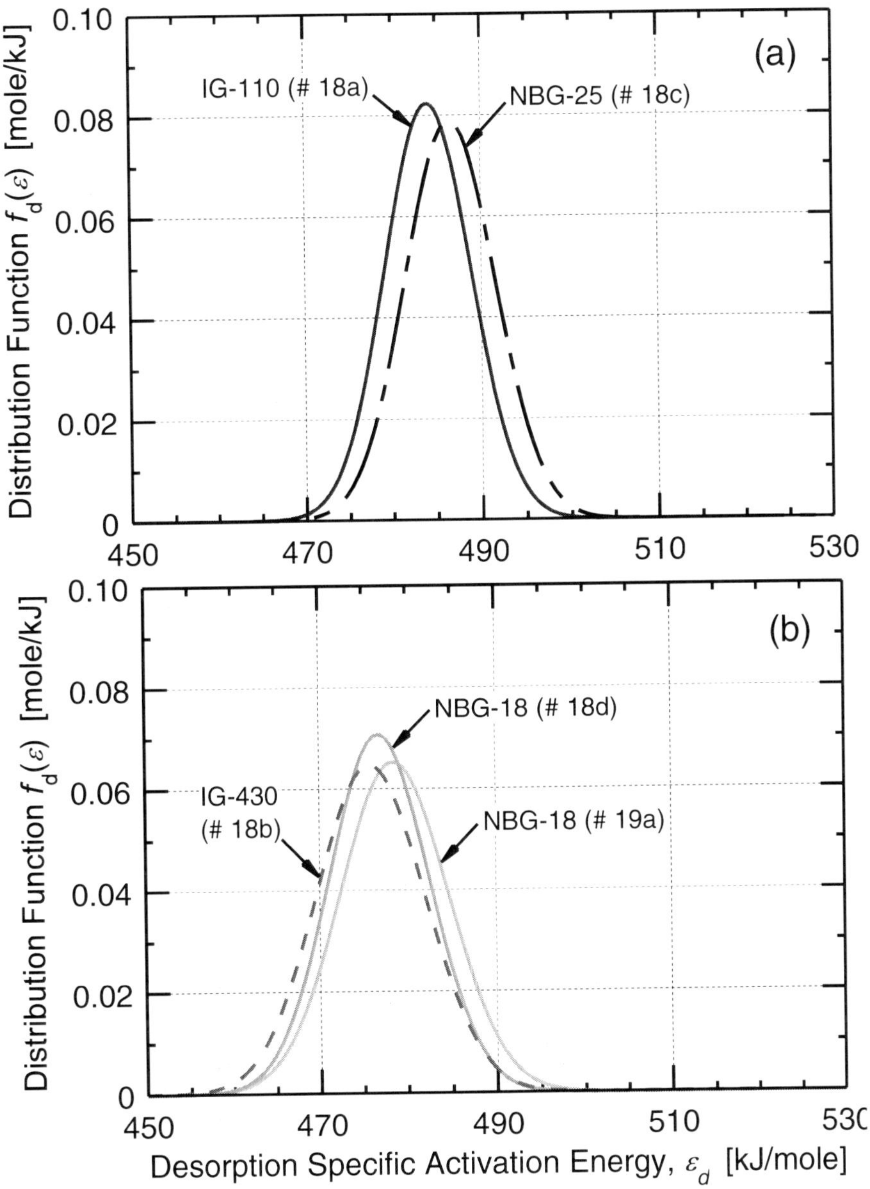

Figure 19. Gaussian-like Distributions of Specific Activation Energies for Desorption of CO for Grades of Graphite. (a) IG-110 and NBG-25; (b) IG-430 and NBG-18 (*Numbers in Parentheses Refer to Specimens in Table 3*) [20].

Su and Perlmutter [83] have examined the initial pore structure of coal chars using a combination of mercury porosimetry, CO_2 adsorption at 195 and 273 K, and pycnometry techniques. They compared the values of the structure parameter Ψ deduced from their measurements with those used for fitting the oxidation data in dry air at 653 to 773 K. The agreement of the Ψ values validated the Random Cylinders Model [83]. However, this model docs not accounts for the opening of initially closed volume pores and is not applicable at volume porosities > 60% and weight losses > 40%, when a breakup of the graphite is likely to occur.

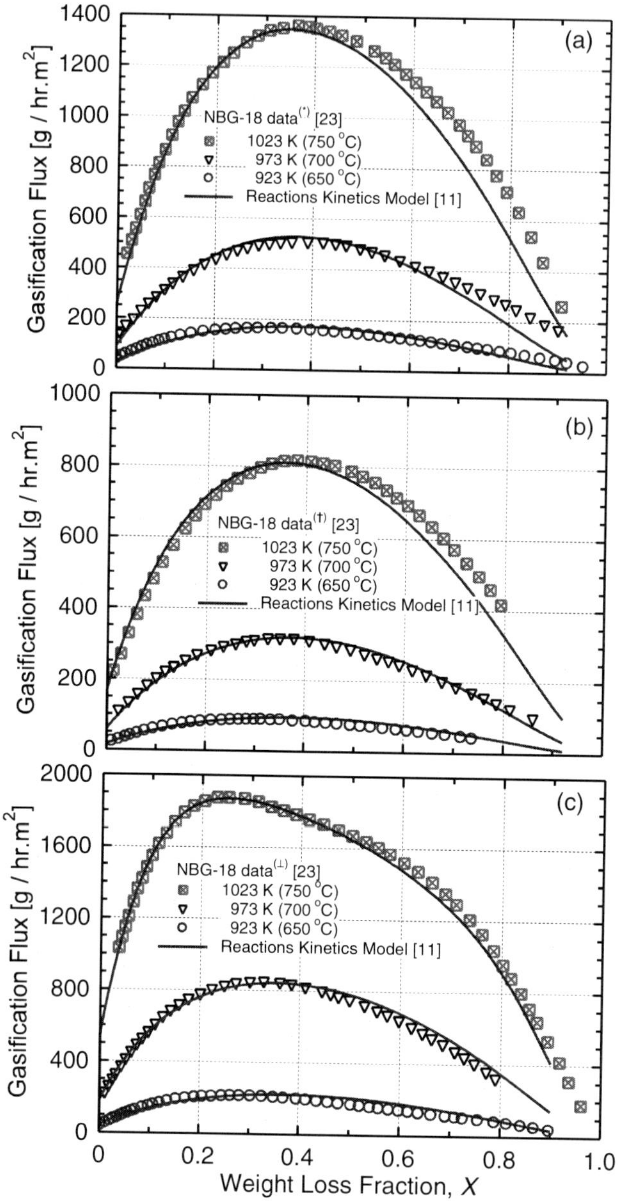

Figure 20. Comparisons of Model Results [11] with Reported Measurements for NBG-18 Nuclear Graphite [23]: *[(a) Specimen #19a (*); (b) Specimen #19b (†); and (c) Specimen # 19c (⊥) in Table 3]*.

The chemical-reactions kinetics model for graphite gasification [9-11] employs the approach proposed by Su and Perlmutter [83] to determine the change in the ASA with weight loss. The fractional change in the ASA with weight loss fraction, *X*, during graphite gasification is expressed as [83]:

$$Y(X) = (ASA/ASA_o) = (1-X)\sqrt{1-\Psi \ln(1-X)} \,. \tag{5}$$

The ASA initially increases with increasing weight loss to a maximum, then decreases with further increase in weight loss, as the widening pores begin to coalesce.

Results of the gasification experiments with nuclear graphite grades of NBG-18 [23], K018, K022 and IG-110 [14] have confirmed that the maximum gasification rate typically occurs at a weight loss of ~ 35%. Table 3 compares the obtained values of Ψ, ASA_o and S_m^o [10-11] for the IG-110, IG-430, NBG-18 and NBG-25 graphite specimens used in the gasification experiments [13,14,20, 23].

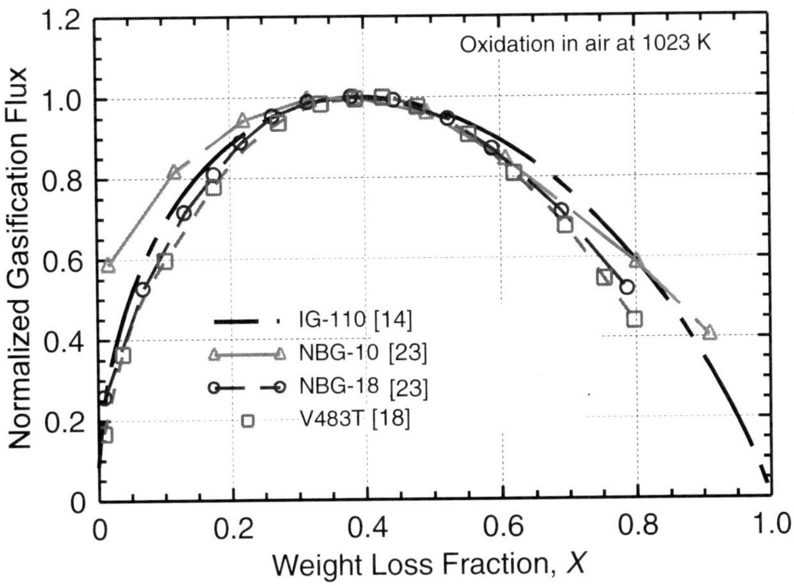

Figure 21. Normalized Gasification Flux for Different Grades of Nuclear Graphite.

The model of Su and Perlmutter [83] for predicting the reaction surface area in terms of the pores initial structural parameter, Ψ, is valid only in the absence of new pores generation and in Modes (a) and (b). It has been confirmed experimentally that for nuclear graphite $\Psi >$ 50 and $0.381 \leq X_{max} < 0.393$. This model's calculations [83] and the reported measurements of the total gasification rates for NBG-18 graphite specimens #19a and #19b [23] in Table 3 peak at weight loss fraction, $X \simeq 0.35$ (Figures 20a and 20b). The total gasification rate at 1023 K for the most reactive specimen peaks earlier at weight loss of ~ 0.25, apparently due to the limited oxygen diffusion through the boundary layer (Mode (c)). The total gasification rate then decreases slowly with further increase in weight loss, due to the decrease in the external surface area of the specimen in the experiments (Figure 20c). These results are representative of the variation in the active surface area of free sites (ASA) for these specimens with increased weight loss. In the experimental gasification data of Hinssen et al. [23] for the dime size NBG-18 graphite specimens, the determined structural parameter Ψ = 35 [10,11], compared to a value of 80 for the larger size specimens of the IG-110, IG-430, NBG-18 and NBG-25 nuclear graphite grades in Table 3 [10,11,13,14].

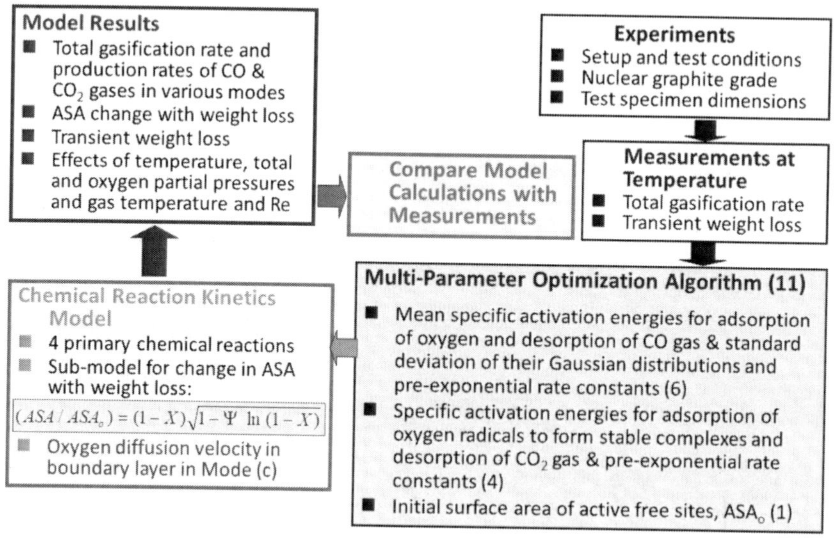

Figure 22. Building Blocks of the Chemical Reactions Kinetics Model for Gasification of Nuclear Graphite [9-11].

Figure 21 plots the normalized gasification mass fluxes reported in the experiments [14, 18, 23] with specimens of different grades of nuclear graphite at 1023 K versus the weight loss fraction. At this temperature, graphite gasification is in the in-pores limited diffusion Mode (b)(Figure 4). The experimental data in Figure 21 show the normalized gasification flux peaking at a weight loss fraction of 0.40 The collapse and coalescence of the open pores at high weight loss fractions decrease the normalized gasification flux, approaching zero when the entire test specimen is oxidized. In Figure 21, the largest weight loss fraction achieved was ~ 0.9 in the experiments of Hinssen et al. [23] using dime size specimens (Table 3) and 0.8 in the experiments of Fuller and Okho [14] (197) and Kühn et al. [18], which used larger size graphite specimens (Table 3).

The chemical-reactions kinetics model [9-11] would not be valid for weight loss fractions > 0.40 and/or volume porosities > 60%, even though the comparisons in Figures 20a – 20c with reported measurements are generally good over the full range of weight loss fractions in the experiments. At weight loss fraction, X ~ 0.10, nuclear graphite could loss structural strength and experience fragmentation at X> 0.40. This may explain why measured gasification rates at X > 0.40 are higher than calculated by the chemical reactions kinetics model [9-11] (Figure 20).

Figure 22 shows the building blocks of the chemical-reactions kinetics model developed by the authors [9-11]. It predicts the rates of the total gasification and the production of CO and CO_2 gases as functions of temperature and oxygen partial pressure, and the flow rate and the composition of the bulk gas mixture, in all three Modes in Figure 4. In Mode (a) and Mode (b), the chemical-reactions kinetics and the temperature control the rate of graphite gasification, while in Mode (c) at high temperatures the oxygen diffusion through the boundary layer limits the effective rate of gasification. In this diffusion-limited gasification Mode (Figure 4), the rate of oxygen consumption is expressed as [10,11,20]:

$$\dot{R}_{O_2} = (A_s\sqrt{1-X})\, k_m \times \left([\hat{O}_2]_b - [\hat{O}_2]_w\right). \tag{6}$$

In this expression, the term $\sqrt{1-X}$ accounts for the decrease in the geometrical surface area with weight loss fraction, X, for the cylindrical specimens of graphite in the gasification experiments. The terms and are the average oxygen concentrations in the bulk gas mixture and at the graphite's external surface, respectively, and k_m is the oxygen diffusion velocity through the boundary layer. The consumption rate of oxygen on the left-hand-side of Equation (6) is the sum of those of the elementary reactions in the chemical kinetics model for graphite gasification [10,11,20]. These are of the adsorption of oxygen atoms onto surface active sites to form un-dissociated [C(O_2)]complexes and of the desorption of CO_2 gas, such as:

$$\dot{R}_{O_2} = \sum_i \dot{R}_{a,i} + \sum_{ij} \dot{R}_{CO_2,ij} \,. \tag{7}$$

In Equation (6), the oxygen diffusion velocity in the boundary layer is expressed in terms of the flow Sherwood number, Sh, as [10,11,20]:

$$k_m = \text{Sh} \times D_{O_2,m} / D. \tag{8}$$

The values of Sh and $D_{O_2,m}$ depend on the flow conditions, properties, and composition of the bulk gas mixture. The effective diffusion coefficient of oxygen in the boundary layer, $D_{O_2,m}$, is calculated using the first-approximation of the molecular theory of gases at low pressure [84]. For a 3-component gas mixture (O_2, N_2 and CO), the effective diffusion coefficient is given as:

$$D_{O_2,m} = \frac{1 - x_{O_2}}{x_{N_2} / D_{O_2,N_2} + x_{CO} / D_{O_2,CO}} \,. \tag{9a}$$

The sum of the molar fractions of the various species in bulk gas flow is unity (i.e., $x_{O_2} + x_{N_2} + x_{CO} = 1$), and the binary diffusion coefficients are calculated as [84]:

$$D_{A,B} = \frac{3}{8\sqrt{2\pi}} \sqrt{\frac{1}{M_A} + \frac{1}{M_B}} \left(\frac{(R_g T)^{1.5}}{N_a \, P_{tot} \, \sigma_{AB}^2 \, \Omega^{(1,1)*}} \right). \tag{9b}$$

The force constants σ_A = 3.433 x 10^{-10} m and ε_A / k = 113 K are for the oxygen molecules, σ_B = 3.681 x 10^{-10} m and ε_B / k = 91.5 K are for the nitrogen molecules and σ_B = 3.59 x 10^{-10} m and ε_B / k = 110 K are for CO gas molecules. The collision integral is a function of the dimensionless temperature (kT/ε_{AB}), and the interaction force constants, σ_{AB} and ε_{AB} are given as [84]:

$$\sigma_{AB} = (\sigma_A + \sigma_B)/2, \text{ and } \varepsilon_{AB} = \sqrt{\varepsilon_A \varepsilon_B} \,. \tag{10}$$

A Sherwood number correlation recently developed by the authors [85] for laminar flow condition ($0.006 \leq Re \leq 1{,}000$) is used to calculate the oxygen diffusion velocity in the boundary layer in Mode (c) of graphite gasification (Equation 8). This correlation is given as [85]:

$$Sh_L = [0.27 + 0.60 Re^{0.47}]\, Sc^{0.33} \left(\mu_b / \mu_w\right)^{0.14}. \quad (11)$$

As shown in Figure 23, this correlation is consistent with the high-temperature gasification rate measurements using different size specimens of nuclear graphite grades of NBG-18, NBG-25, IG-11, IG-110 and IG-430 in atmospheric air at $0.08 \leq Re \leq 30$ (Table 3). Figures 24a-24c compare the predictions of the chemical-reactions kinetics model [10,11,20] with the reported experimental measurement by Chi and Kim [13] for the gasification of NBG-18 nuclear graphite. The model uses the determined chemical kinetics and structural parameters listed in Table 3 [10, 11].

The chemical-reactions kinetics model calculations are in good agreement with the experimental measurements (Figures 24a, 24b, 24c). In Figure 24c, note the continuous Arrhenius curve that traverses the three Modes of graphite gasification (Figure 4) with increasing temperature in the experiments. In the diffusion-limited Mode (c) at high temperatures, the effective diffusion velocity of oxygen in the boundary layer is calculated using the correlation given by Equations (8) and (11). The properties in these equations are evaluated at the bulk temperature of the multi-species gas mixture. For ambient-air gasification experiments, the species in the bulk gas mixture could include oxygen, nitrogen, CO, CO_2 and helium. Table 4 lists the recommended [20] chemical-reactions kinetics parameters for the gasification of nuclear graphite grades: IG-110, IG-430, NBG-18 and NBG-25.

The chemical kinetics parameters in Table 4 as well as the Gaussian-like distributions of the specific activation energies for adsorption of oxygen and desorption of the CO gas are obtained from the reported gasification rate and transient weight loss measurements in the experiments [13, 14, 23]. The determined distributions are quite similar and the values of the most-probable specific activation energy ($\bar{\varepsilon}_a$) and the characteristic standard deviation (σ_a) are close (Figures 18 and 19).

The recommended values [20] of the specific activation energies for adsorption of oxygen ($\bar{\varepsilon}_a$) and desorption of CO gas ($\bar{\varepsilon}_d$) for nuclear graphite grades of IG-110 and NBG-25 are $\bar{\varepsilon}_a = 120.4 \pm 4$ kJ/mole and $\bar{\varepsilon}_d = 485.4 \pm 3$ kJ/mole (Table 4).They are slightly different from those recommended for IG-430 and NBG-18: $\bar{\varepsilon}_a = 131.7 \pm 16$ kJ/mole and $\bar{\varepsilon}_d = 476 \pm 24$ kJ/mole (Table 4). For all four grades of nuclear graphite in Table 4, the standard deviations of the Gaussian distributions for the specific energy for the adsorption of oxygen are nearly the same, $\sigma_a = 64.6 \pm 5$ kJ/mole. On the other hand, for IG-110 and NBG-25, the recommended standard deviation of the Gaussian distribution for the specific activation energy for desorption of CO gas is $\sigma_d = 5.0 \pm 0.7$ kJ/mole, compared to $\sigma_d = 5.7 \pm 0.4$ kJ/mole for IG-430 and NBG-18 [20]. The recommended values of the specific activation energy for the breakup of the un-dissociated $C(O_2)$ radicals to form stable (CO) complexes

(ε_b) are $\varepsilon_b = 204.7 \pm 7$ kJ/mole for IG-110 and NBG-25, and $\varepsilon_b = 195 \pm 7$ kJ/mole for IG-430 and NBG-18 (Table 4).

The recommended value of the specific activation energy for desorption of CO_2 during the gasification of nuclear graphite grades of IG-110 and NBG-25 is $\varepsilon_d^* = 158.5 \pm 1.2$ kJ/mole, compared to a slightly lower value of 153.4 ± 3.5 kJ/mole for the nuclear graphite grades of IG-430 and NBG-18 (Table 4).

The recommended value of the rate constant for oxygen desorption, $k_a^o = 13{,}000 \pm 200$ mole^{-1}.s^{-1} for the gasification nuclear graphite grades of IG-110 and NBG-25, is slightly larger than that recommended for IG-430 and NBG-18, $k_a^o = 11{,}730 \pm 1{,}100$ mole^{-1}.s^{-1} (Table 4).

Similar results are obtained of the chemical kinetics rate constant for the breakup of un-dissociated $C(O_2)$ oxygen radicals to form stable (CO) complexes. The recommended rate constant [20] for the nuclear graphite grades IG-110 and NBG-25 is $k_b^o = (7.46 \pm 0.1) \times 10^{14}$ mole^{-1}.s^{-1}, and $(10.1 \pm 0.7) \times 10^{14}$ mole^{-1}.s^{-1} for graphite grades of IG-430 and NBG-18. The recommended value of the pre-exponential rate coefficient for desorption of CO_2 gas, $k_d^* = (4.89 \pm 0.4) \times 10^{12}$ mole^{-2}.s^{-1} for nuclear graphite grades IG-110 and NBG-25 is 17% larger than that recommended for the IG-430 and NBG-18 grades, $k_d^* = (4.16 \pm 0.4) \times 10^{12}$ mole^{-2}.s^{-1} (Table 4).

Finally, the recommended chemical kinetics rate constant for desorption of CO during graphite gasification is $k_d^o = 4.3 \times 10^{26}$ s^{-1} for nuclear graphite grades of IG-110 and NBG-25, compared to $5.6 \pm 0.5 \times 10^{26}$ s^{-1} for IG-430 and NBG-18.

In summary, the determined chemical kinetics parameters for the four elementary chemical reactions for the gasification of nuclear graphite are:

(a) The values and Gaussian distributions of the specific activation energies for the adsorption of oxygen and desorption of CO gas;
(b) The specific activation energy for the desorption of CO_2 gas,
(c) The values of pre-exponential rate coefficients for the adsorption of oxygen and desorption of CO and CO_2 gases.

These parameters and the initial active surface area of free sites (ASA_o) are determined using a multi-parameters optimization algorithm [9-11], based on the reported measurements of the total gasification rate and transient weight loss in experiments performed by various investigators (Table 3 and 4). The chemical-reactions kinetics parameters for the IG-110 and NBG-25 nuclear graphite, with petroleum filler particles, are similar, but different from those of NBG-18 and IG-430 with coal tar pitch filler particles (Table 4).

Table 3. Determined Chemical Kinetics and Micro-Structural Parameters for Different Grades of Nuclear Graphite [10-11, 20]

Parameters	Fuller andOkoh [14]	Chi and Kim [13]				Hinssen*et al.* [23]		
Graphite type	IG-110	IG-110	IG-430	NBG-25	NBG-18	NBG-18(*)	NBG-18(†)	NBG-18(⊥)
Density (g/cm^3)	1.66	1.75	1.82	1.82	1.85	1.85	1.85	1.85
Mass (g)	1.756	22.52	23.42	23.42	23.81	0.25	0.25	0.25
Specimen	#17	#18a	#18b	#18c	#18d	#19a	#19b	#19c
Specimen dimensions	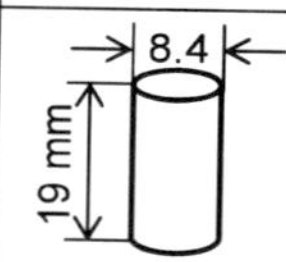	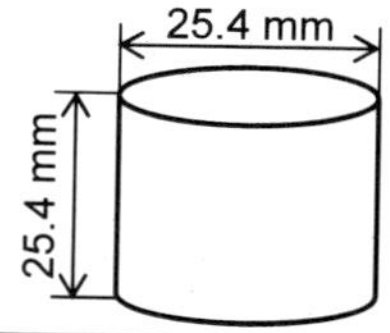				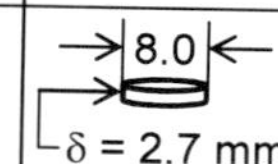		
Microstructural Parameters								
Ψ	80(c)	80(a)	80(a)	80(a)	80(a)	35(a)	35(a)	35(a)
ASA_o (μmole)	8.91(d,e)	29.28(b,e)	15.75(b,e)	29.81(b,e)	14.29(b,e)	4.24(b,e)	2.91(b,e)	5.41(b,e)
S_m^o (μmole/g)	5.07	1.30	0.673	1.27	0.60	16.95	11.65	21.62

Oxidation Kinetics Parameters								
$\bar{\varepsilon}_a$ (kJ/mole)	124.6	117.0	115.0	119.7	130.0	132.6	126.8	143.8
σ_a (kJ/mole)	50.70	64.35	63.25	59.85	65.49	63.80	39.41	65.87
k_a^o ($mole^{-1}.s^{-1}$)	13,199.	13,014.	12,842.	12,776.	11,853.	11,325.	11,843.	10,810.
k_b^o ($mole^{-1}.s^{-1}$)	8.844×10^{14}	7.568×10^{14}	10.60×10^{14}	7.351×10^{14}	9.558×10^{14}	9.754×10^{14}	12.50×10^{14}	9.802×10^{14}
ε_b (kJ/mole)	203.1	205.3	199.5	205.7	196.3	196.2	194.3	191.8
$\bar{\varepsilon}_d$ (kJ/mole)	416.3	484.5	475.6	486.3	476.8	498.5	478.5	503.6
σ_d (kJ/mole)	5.645	4.96	6.153	5.118	5.647	5.669	6.114	5.353
k_d^o (s^{-1})	7.08×10^{26}	4.234×10^{26}	5.462×10^{26}	4.333×10^{26}	5.673×10^{26}	5.864×10^{26}	6.052×10^{26}	5.333×10^{26}
k_d^* ($mole^{-2}.s^{-1}$)	4.477×10^{12}	5.112×10^{12}	3.712×10^{12}	5.091×10^{12}	4.217×10^{12}	4.298×10^{12}	3.964×10^{12}	4.597×10^{12}
ε_d^* (kJ/mole)	159.1	158.4	149.4	158.0	152.0	154.9	168.8	157.5

(a) Based on reported transient weight loss measurements; (b) Based on reported measurements of the transient weight loss and total gasification rate; (c) Assumed as in (a) for same IG-110 graphite type; (d) Based on reported total gasification rate measurements versus weight loss up to 86%; (e) From multi-parameter optimization algorithm [9-11]; and (*), (†)and(⊥) are specimens cut from different cylinders [23].

Table 4. Recommended Chemical-Reactions Kinetics Parameters for Different Grades of Nuclear Graphite [20]

Parameter	IG-110 and NBG-25	IG-430 and NBG-18
$\bar{\varepsilon}_a$, σ_a (kJ/mole) k_a^o (mole^{-1}.s^{-1})	120.4 $\pm$ 4, 64.6 $\pm$ 5 13,000 $\pm$ 200	131.7 $\pm$ 16, 64.6 $\pm$ 5 11,730 $\pm$ 1,100.
k_b^o (mole^{-1}.s^{-1}) ε_b (kJ/mole)	(7.46 $\pm$ 0.1) x 10^{14} 204.7 $\pm$ 7	(10.1 $\pm$ 0.7) x 10^{14} 195.0 $\pm$ 7
$\bar{\varepsilon}_d$, σ_d (kJ/mole) k_d^o (s^{-1})	485.4 $\pm$ 3, 5.00 $\pm$ 0.7 (4.3 $\pm$ 0.1) x 10^{26}	476.0 $\pm$ 24, 5.70 $\pm$ 0.4 (5.6 $\pm$ 0.5) x 10^{26}
k_d^* (mole^{-2}.s^{-1}) ε_d^* (kJ/mole)	(4.89 $\pm$ 0.4) x 10^{12} 158.5 $\pm$ 1.2	(4.16 $\pm$ 0.4) x 10^{12} 153.4 $\pm$ 3.5

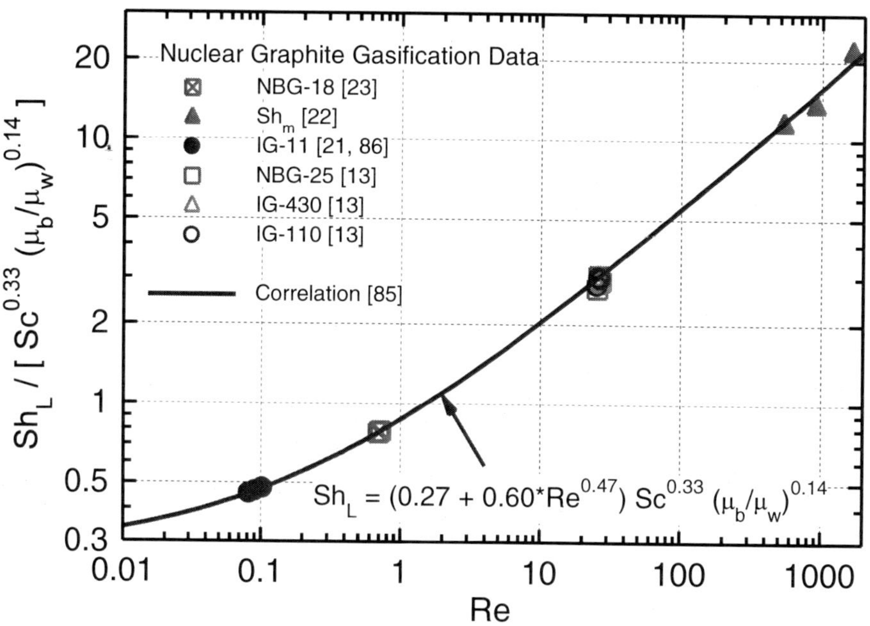

Figure 23. Comparison of Sherwood Number Correlation [85] with Graphite Gasification Measurements in Experiments.

The origin, not the size, of the filler particles appears to affect the reactivity of nuclear graphite. This is attributed to the differences in the value of ASA_o and of the corresponding specific surface area of the active free sites, (μmole/g). The values of these quantities for calculating the gasification rate of nuclear graphite also depend on the mass or volume of the graphite block or test specimens used in the gasification experiments [13,14,23]. The cutting plane of the test specimens also appears to affect the values of ASA_o and S_m^o, and hence the total gasification rate, R (Figures 25 and 26).

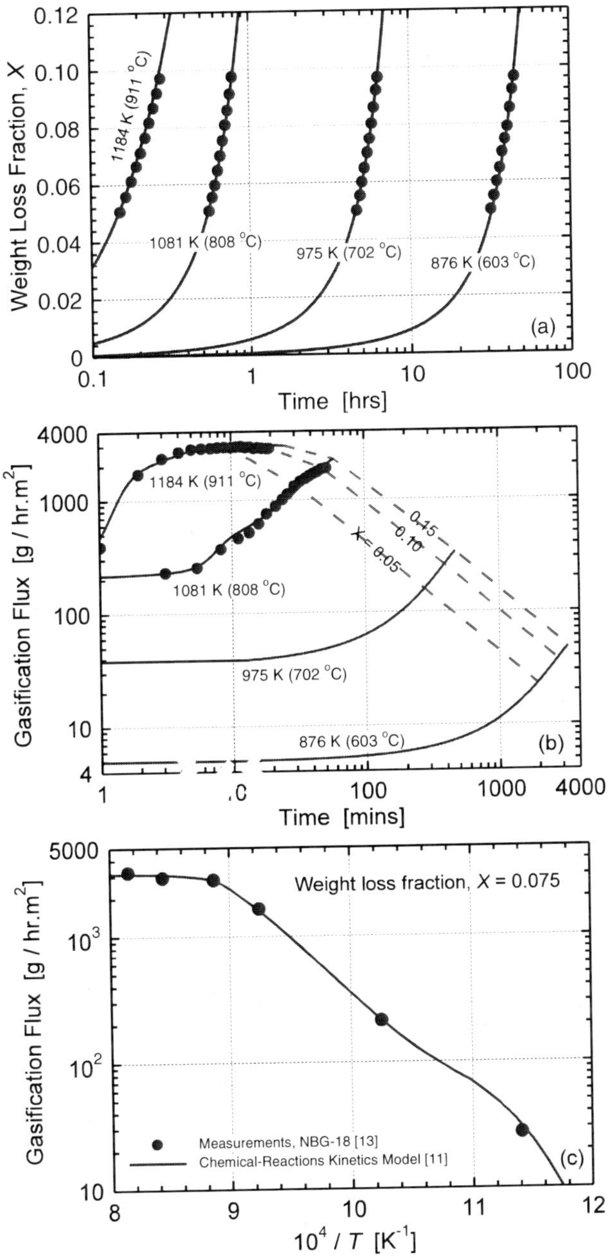

Figure 24. Comparison of Chemical Reactions Kinetics Model [11] with Gasification Measurements for NBG-18 [13]: (a) Transient Weight Loss; (b) Transient Gasification Flux; (c) Total Gasification Flux at $X = 0.075$.

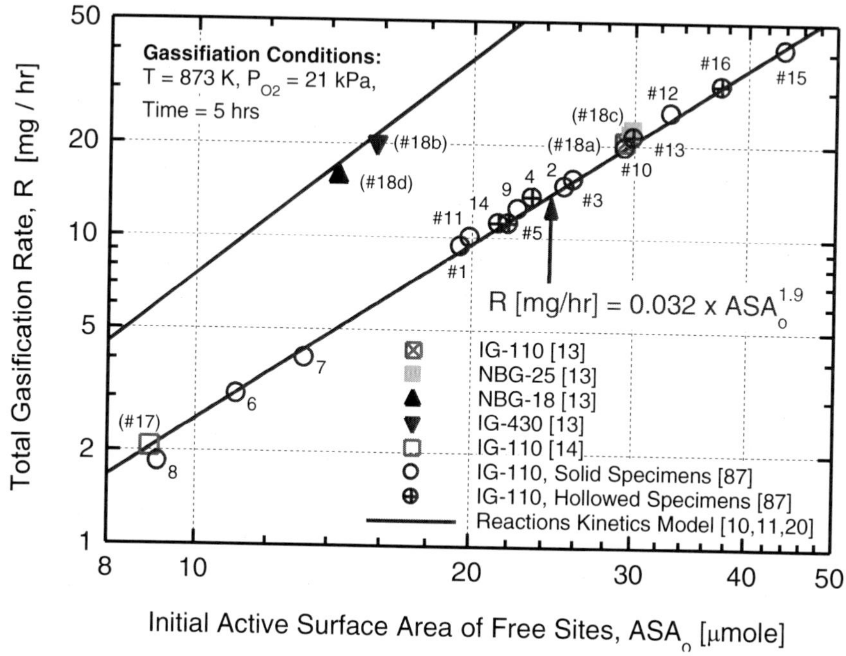

Figure 25. Effect of ASA$_o$ on the Total Gasification Rate for Different Grades of Graphite at 873 K (Numbers in Parentheses are of Specimens in Table 3, and Other Numbers are of Graphite Specimens in Figure 27) [20].

Effect of Initial Graphite Mass

Kim, Lee and No [87] have performed gasification experiments using sixteen IG-110 nuclear graphite specimens of different sizes and configurations and an initial mass ranging from 3.5 g (#6 in Figure 27) to 54.7 g (#15 in Figure 27). These specimens are shown in Figure 27 (#1 to #16), along with those used by Chi and Kim [13] (#18), Fuller and Okoh [14] (#17) and Hinssen *et al.* [23] (#19) in their gasification experiments (Table 3). The reported measurements of the total gasification rate by these investigators are used to investigate the effect of the initial mass of test specimens, m_o, on S_m^o.

In their gasification experiments, Kim, Lee and No [87] tested their IG-110 graphite specimens at 873 K (600°C) in a flowing dry atmospheric air. They reported the measured total gasification rates after 5 hours in the experiments. The final weight loss fraction achieved at those conditions was < 0.008. At such low temperature in the experiments, graphite gasification was in Mode (a) (Figure 4), solely controlled by the chemical kinetics of the oxidation reactions (Equation (1) – (4)).

Although the reported total gasification rate measurements by Kim, Lee and No [87] are limited, the oxidation-kinetics model developed by the authors [9-11,20] is used to infer the ASA$_o$ values for each of the IG-110 specimens used in the experiments (Figures 26 and 27). The model used the kinetics parameters and structural parameter $\Psi = 80$recommended in this work for IG-110 (Table 4). The determined ASA$_o$ values match the model's calculations of the total gasification rates to the reported measurements, after 5 hours at 873 K in the experiments. Figure 25 plots the reported total gasification rate measurements versus the determined ASA_o values for the different IG-110 specimens used in the experiments of Kim, Lee and No [87] (Figure 27). Figure 25 also includes the determined ASA_o values using the chemical-reactions kinetics model [9-11,20] for the IG-110 specimens used in the gasification experiments of Fuller and Okoh [14] and Chi and Kim [13], and the NBG-25, NBG-18 and IG-430 graphite specimens in the experiments of Chi and Kim [13].

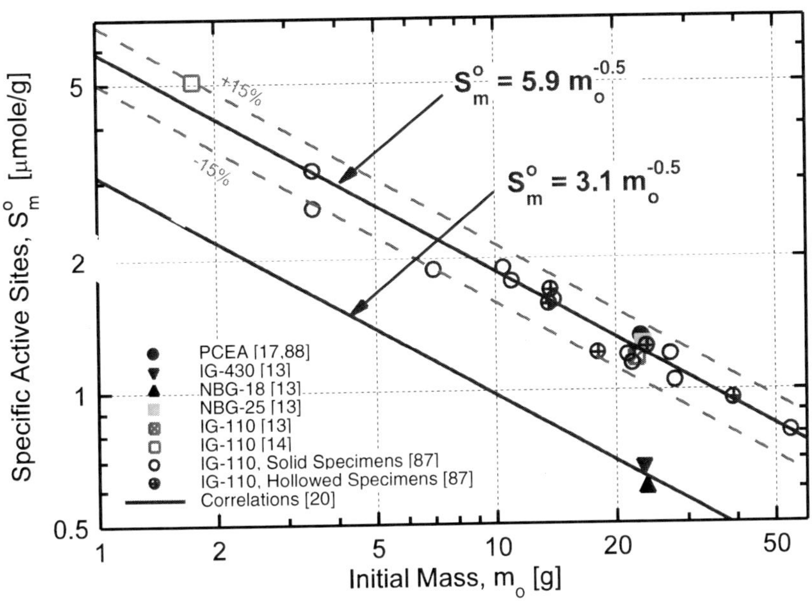

Figure 26. Specific Area of Free Active Sites, as a Function of Initial Mass of Graphite Specimens in the Experiments (Table 3 and Figure 27) [20].

The results presented in Figure 25 show the total gasification rate monotonically increases proportional to ASA_o to the power 1.9. This exponent reflects the dominance of CO_2 production (Equation (4a)) in the total gasification rate in the experiments (Mode (a) in Figure 4)). Figure 25 also shows that the dependence of the total gasification rate on ASA_o is consistent for all different shapes, configurations and sizes of the graphite specimens used in the experiments [13, 14, 87].

It is worth noting, however, that the results for the NBG-25 specimen #18c in the experiments of Chi and Kim (2008) (Table 3) are consistent with those of the IG-110 specimens. On the other hand, the results for the NBG-18 and IG-430 graphite specimens in the experiments of Chi and Kim [13] (#18d and #18b in Table 3) fall on a different curve. This supports the proposed classification of the nuclear graphite grades in terms of their gasification reactivity into two distinct groups: (a) IG-110 and NBG-25, and (b) NBG-18 and IG-430 (Figure 25 and Table 4).

The calculated change in the ASA with weight loss in the gasification Modes (a) and (b) (Figure 4) is for the values of the structural parameter, Ψ in Table 3. This parameter is included in the chemical-reactions kinetics model [9-11, 20] that incorporates the Random Cylinders approach of Su and Perlmutter [83].

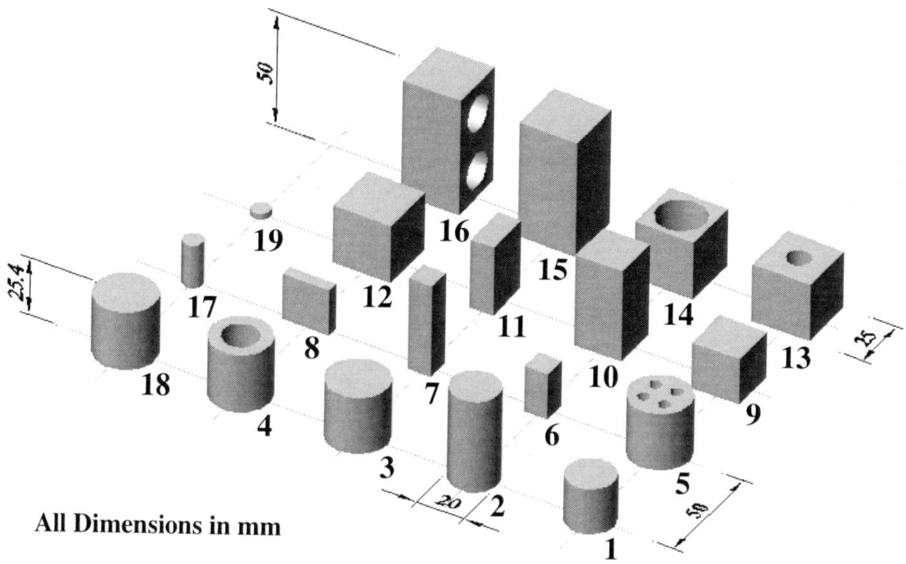

Figure 27. Graphite Specimens Used in the Gasification Experiments [10, 11, 13, 14, 23,87].

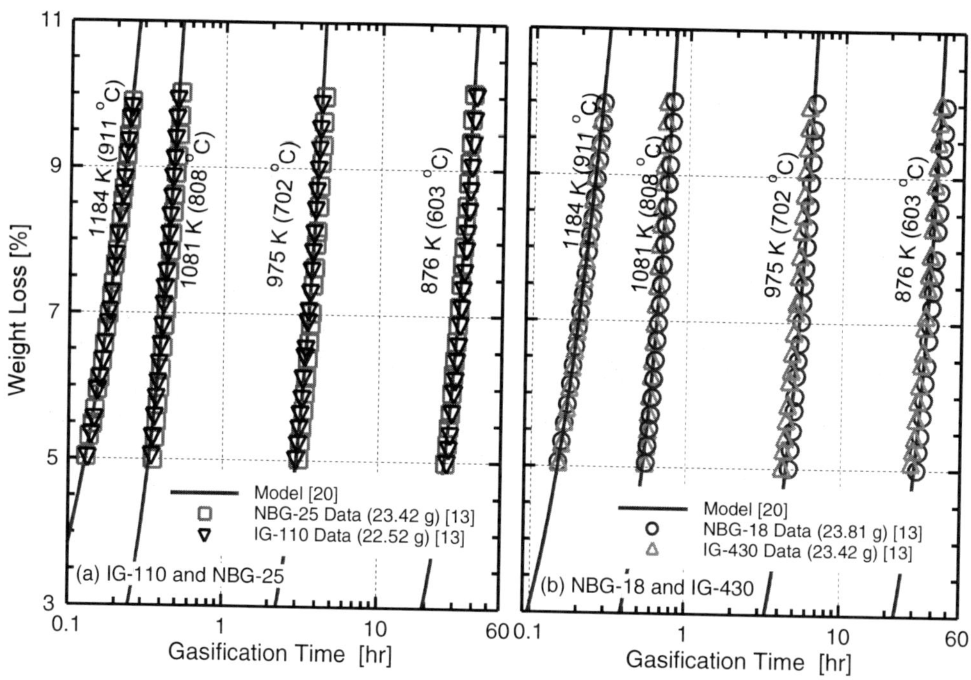

Figure 28. Comparison of Model Calculations [20] with Weight loss Measurements in Gasification Experiments [13] of Different Grades of Nuclear Graphite.

The determined values [9-11,20] from the gasification rate measurements of Chi and Kim [13] and Fuller and Okoh [14] are presented in Figure 26 for the different grades of graphite as functions of the initial mass of the specimens in the gasification experiments, m_o. This Figure also presents the determined values for PCEA nuclear graphite, based on the reported measurements of the total gasification rate and transient weight loss in recent experiments by Contescu et al. [17] and Contescu [90]. The graphite specimens used in these experiments are right cylinders 25.4 x 25.4 mm with an initial mass of ~ 23 g. The results for the NBG-18 specimens' #19a, 19b and 19c of Hinssen *et al.* [23] in Table 3 are not included in Figure 25 due to their extremely small size (0.25 g) and large variance in reactivity.

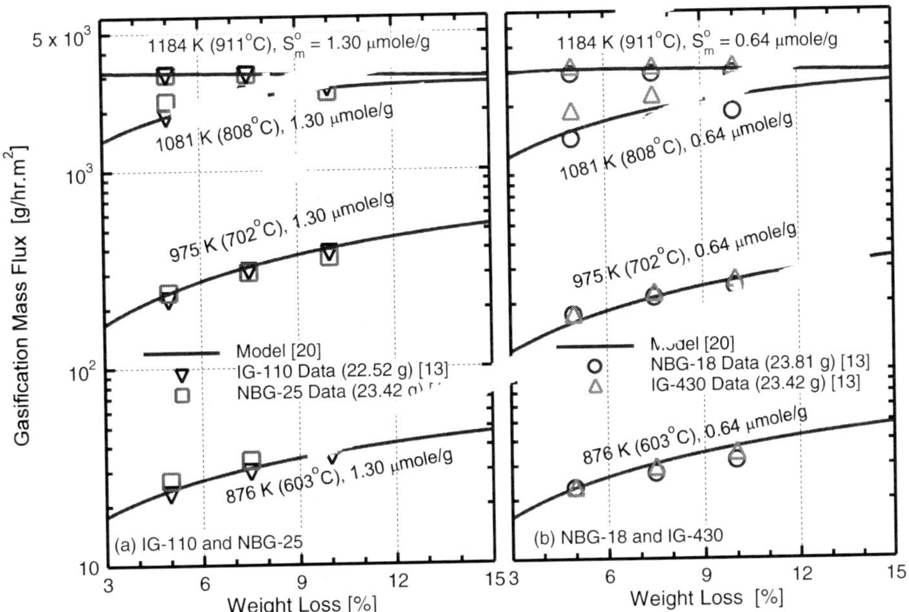

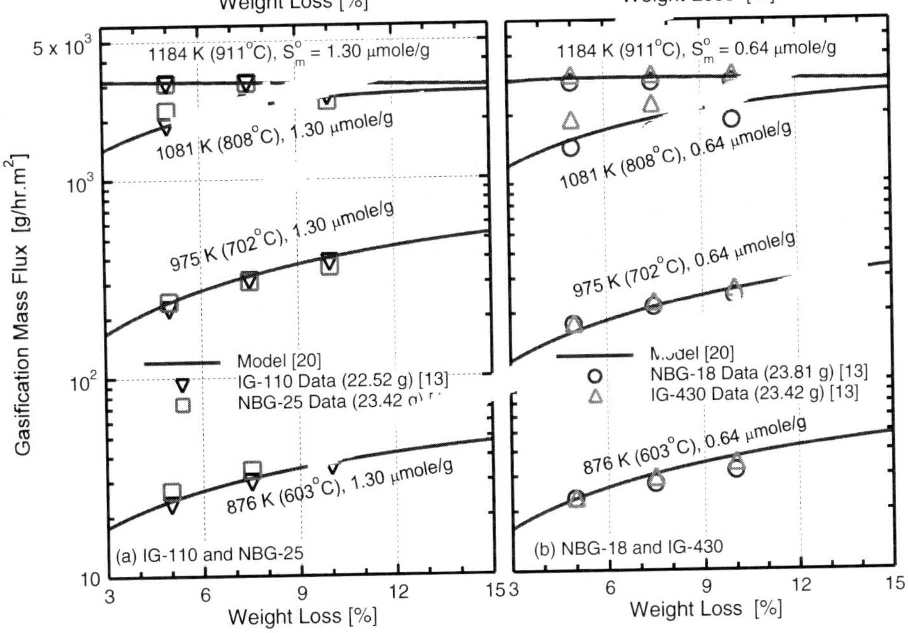

Figure 29. (Continued).

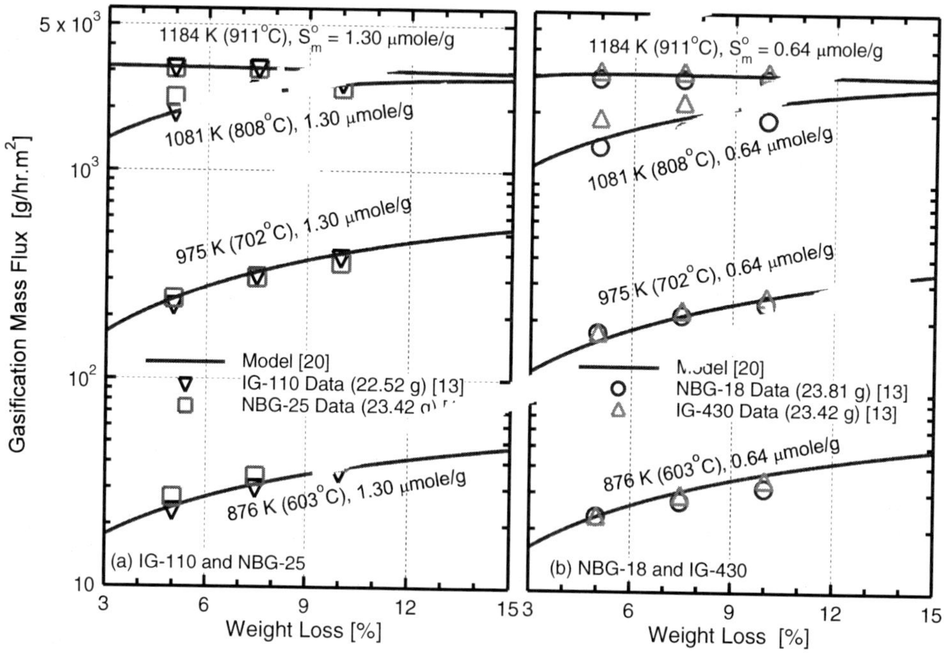

Figure 29. Comparison of Model Calculations [9, 10, 20] with Reported Experimental Measurements [13] of Gasification Mass Flux at Different Temperatures.

Figure 26 plots the obtained S_m^o values for the different grades of nuclear graphite versus the initial mass, m_o, of the graphite specimens, spaning from 1.7 to 55 grams. The dependence of on m_o is not the same for all grades of nuclear graphite. The identical-size right cylinders of IG-110 and NBG-25 (specimen #18a and #18c) in the experiments by Chi and Kim [13] have nearly identical values of 1.30 and 1.273 μmole/g (Table 3 and Figure 26). Similarly, the IG-430 and NBG-18 specimens' #18b and #18d in Table 3, of the same size and shape, have close values of 0.673 and 0.60 μmole/g, which are much smaller than for IG-110 and NBG-25 (Figure 26). The determined value for PCEA nuclear graphite is consistent with those for the IG-110 and NBG-25 specimens of the almost the same initial mass, suggesting similar gasification reactivity.

For the nuclear graphite grades of IG-110, NBG-25 and PCEA, S_m^o can be expressed in terms of the initial mass of the graphite specimens, m_o in grams, [20] as (Figure 26):

$$S_m^o\,(\mu mole/g) = 5.9\, m_o^{-0.5}. \tag{12}$$

This correlation is consistent to within ± 15% of the determined S_m^o values [20] from the reported measurements of the total gasification rate and transient weight in the experiments [13, 14, 17, 87, 90]. The initial specific active surface area of free sites, S_m^o, is inversely proportional to the square root of the initial mass, m_o. Conversely, the initial active surface area of free sites, $ASA_o = S_m^o \; x \; m_o$, increases proportional to the square root of the initial mass. Pending further qualification with gasification measurements for graphite specimens of larger masses, the correlation in Equation (12), along with the recommended values of the chemical-reactions kinetics parameters in Tables 3 and 4 may be used in future safety analyses of VHTRs and HTRs.

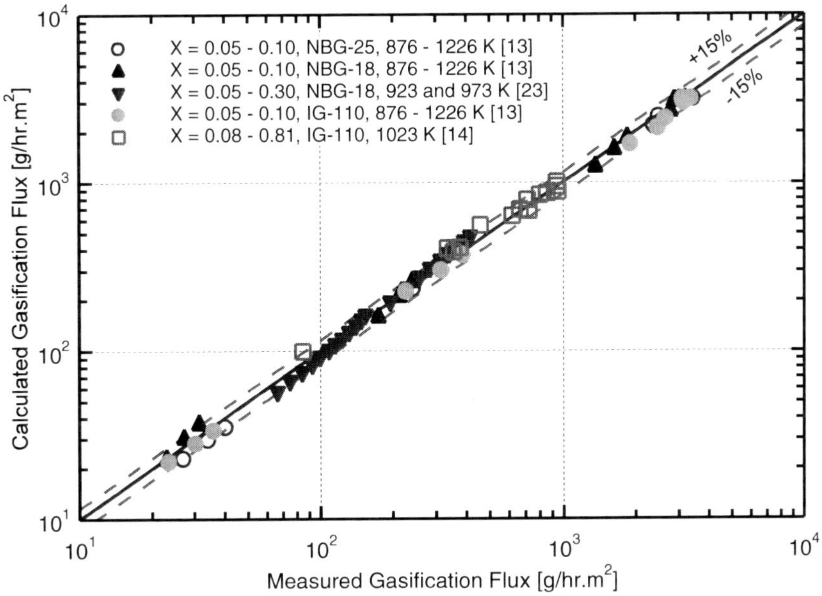

Figure 30. Comparison of All Experimental Measurements [13, 14, 23] of Gasification Flux with Calculations ofthe Chemical-Reactions Kinetics Model [9, 10].

For nuclear graphite grades of NBG-18 and IG-430, we proposed the following expression for S_m^o in terms of the initial mass [20], again pending further qualification with additional experimental data (Figure26):

$$S_m^o\ (\mu mole/g) = 3.1\, m_o^{-0.5}. \tag{13}$$

Model Validation

This section validates the chemical-reactions kinetics model [9-11,20] with reported measurements of the total gasification rate and transient weight loss for different grades of nuclear graphite in experiments performed at different temperatures [13,14,23]. The temperatures in the experiments cover all modes of graphite gasification in Figure 4. The chemical-reactions kinetics parameters in Table 4 and Equations (12) and (13) are used in the graphite gasification model [9-11,20] to calculate the transient weight loss (Figure 28) and total gasification rates (Figures 28 - 29) in the experiments [13]. Figure 28a compares the model estimates with the experimental measurements of weight loss [13] for IG-110 and NBG-25 graphite specimens. Figure 28b compares the reported measurements of weight loss for nuclear graphite grades of NBG-18 and IG-430 [13] with the model calculations [20].

The accuracy and effectiveness of the model [10, 11, 20] are also demonstrated in Figures 29a and 29b. These Figures compare the model's calculations of the total gasification flux versus weight loss with the reported experimental measurements by Chi and Kim (2008). The model calculations are in agreement with all measurements for IG-110 (Chi and Kim, 2008; Fuller and Okoh, 1997), NBG-25 (Chi and Kim, 2008), IG-430 (Chi and Kim, 2008) and NBG-18 (Chi and Kim, 2008; Hinssen *et al.*, 2008) to within $\pm$ 15% (Figure 30).The comparisons in Figures 28 and 29 span a wide range of temperatures (from 876 to 1226 K), covering all modes of graphite gasification (Figure 4), and weight loss fractions from 0.05 to 0.81 (Figure 30).

The good agreement with experimental measurements confirms the effectiveness and the accuracy of the chemical-reactions kinetics model (Figure 22). It also confirms the applicability of the recommended chemical kinetics parameters in Table (4) and the proposed expressions for the initial specific active area of free sites (Equations (12) and (13)) for future safety analysis of HTRs and VHTRs.

The calculated Arrhenius curves of the total gasification rate for nuclear graphite grades of IG-110 and NBG-18 in Figure 31a and for graphite grades of NBG-18 and IG-430 in Figure 31b are compared to the reported experimental measurements [10, 11]. The curves calculated by the chemical-reactions kinetics model [10, 11, 20] for weight loss fractions, *X*, of 0.05 and 0.10 are in good agreement with measurements [13].

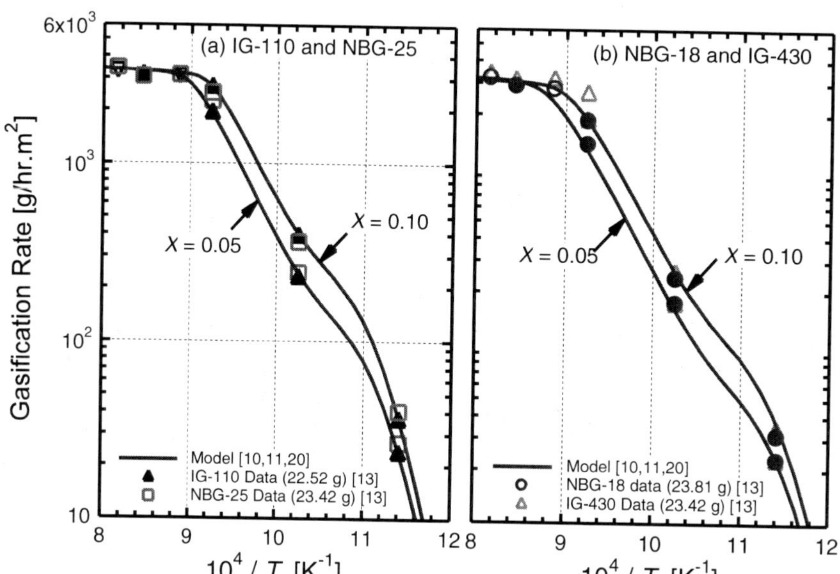

Figure 31. Comparisons of Calculated Arrhenius Curves with Reported Measurements of Total Gasification Flux for Different Grades of Nuclear Graphite.

The continuous Arrhenius curves in Figures 31a and 31b covering all three modes of graphite gasification (Figure 4) show a slight "hump" as the graphite gasification approaches the transitions from Mode (a) at low temperatures to Mode (b) at intermediate temperatures (Figure 4). These humps correspond to the peak production rates of CO_2 gas in the experiments. The model [20] calculates the peak rate to occur at ~ 935 K for nuclear graphite grades of IG-100 and NBG-18 (Figure 32a), and at a slightly lower temperature of ~ 925 K for NBG-18 and IG-430 (Figure 32b).

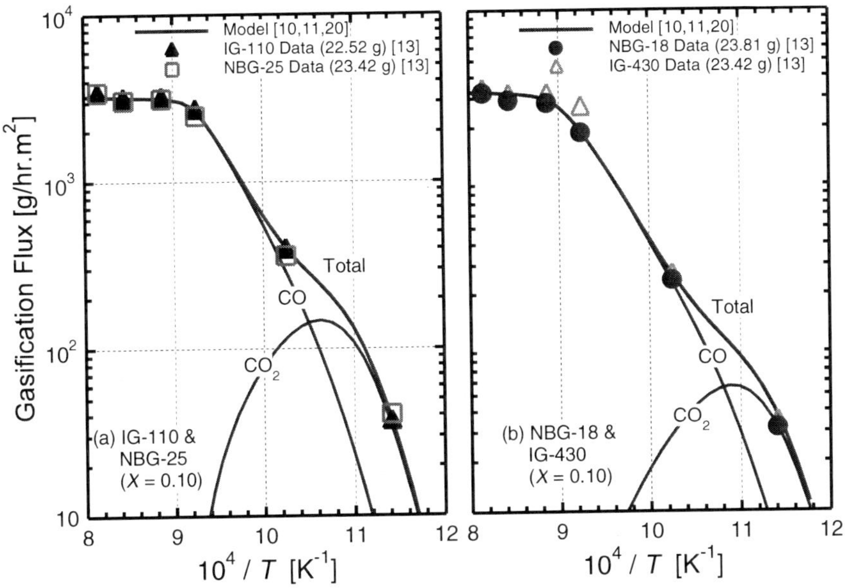

Figure 32. Calculated Production Fluxes of CO and CO_2 Gases and Comparisons of Total Gasification Flux with Measurements for Different Graphite Grades.

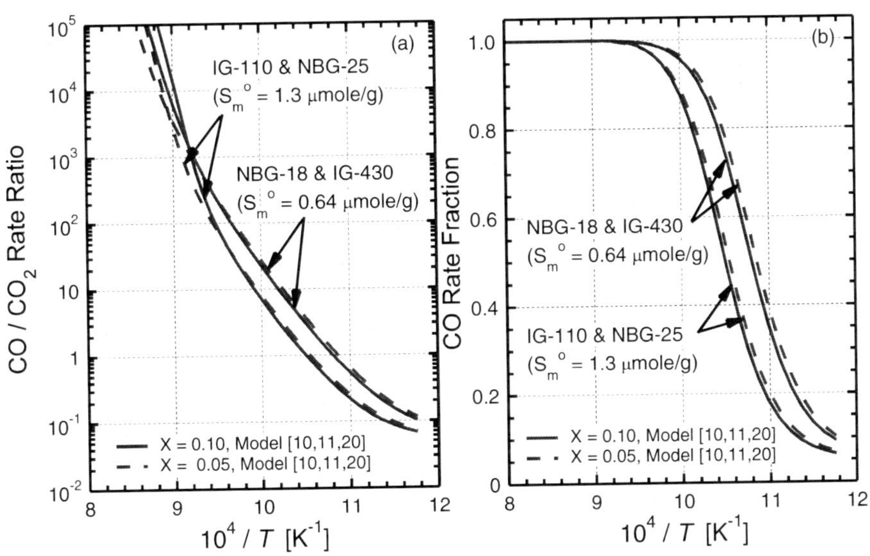

Figure 33. Calculated CO/CO_2 Production Rate Ratios and CO Fractions during the Gasification of the Different Grades of Nuclear Graphite.

The results presented in Figures 28-32 show that the gasification rates for the nuclear graphite grades with petroleum coke particles (IG-110 and NBG-25 in Tables1 and 4) are practically identical (Figures 31a and 32a), and so are those for the nuclear graphite grades with coal tar pitch coke particles (IG-430 and NBG-18 in Tables 1 and 4). The results in these Figures demonstrate that, in addition to the total gasification rate and transient weight loss results, the chemical kinetics model [9,10,20] calculates the production rates of CO and CO_2 gasses and their relative contributions to the total gasification rate as functions of temperature (Figures 32 and 33). These Figures for a weight loss fraction $X = 0.10$ confirms that the total gasification mass flux is the sum of the production fluxes of CO and CO_2 gases. The latter are discussed next.

Production Rates of CO and CO_2 Gases

In Mode (a) of graphite gasification (Figure 4) at low temperatures (< ~ 885 - 900 K), while both CO and CO_2 gases are produced, the latter contributes the most to the total gasification (Figures 32a and 32b). In Mode (b) (Figure 4) at intermediate temperatures (up to 1030 - 1060 K), graphite gasification is increasingly dominated by the production of CO gas, whose production rate increases, while that of CO_2 gas decreases, with increasing temperature. In the diffusion-limited Mode (c) (Figure 4) at high temperatures > 1030 - 1060 K, graphite gasification is entirely due to the production of CO gas (Figures 32a and 32b).

The calculated flux ratios of the CO and CO_2 gases during the gasification of NBG-18 and IG-430 at weight loss fractions of 0.05 and 0.10 [10, 11, 20] are compared in Figure 33a to those calculated for NBG-18 and IG-430 nuclear graphite. The calculated CO fractions of the total gasification flux are compared in Figure 33b. The ratios of the production fluxes of CO and CO_2 during the gasification of nuclear graphite grades of NBG-18 and IG-430 are identical, but consistently higher than for IG-110 and NBG-25.

The temperature corresponding to a CO/CO_2 production ratio of unity during the gasification of NBG-18 and IG-430 is ~ 935 K, slightly lower than that for IG-110 and NBG-25 (~ 950 K). These temperatures are ~ 10 - 15 K higher than those corresponding to the peak production fluxes [10, 11, 20] of CO_2 gas during the gasification of the various grades of nuclear graphite (Figures 32a and 32b). In the diffusion-limited Mode (c) at higher temperatures (> ~ 1123 K), the ratios of the production fluxes of CO/CO_2 for all grades of nuclear graphite are comparable and exceed 5×10^2 (Figure 33a).

In Mode (b) at intermediate temperatures, the fractions of the total gasification fluxes for NBG-18 and IG-430 in the form of CO gas increase with increasing temperature from ~ 30% to 0.98% and for IG-110 and NBG-25 they increase from ~ 18% to 96% (Figure 33b). The results presented in Figures 32 and 33 are an example of the potential usefulness of the chemical-reactions kinetics approach [9-11, 20] for future safety analysis of VHTRs or HTRs in the unlikely event of a massive ingress accident.

As indicated earlier in Figures 25 and 26, the initial mass of nuclear graphite, regardless of shape (Figure 27), affects both the initial active surface area, ASA_o, and the initial specific active area of free sites, S_m^o, and hence the reactivity of graphite for gasification (Equations (12) and (13)). These correlations, developed based on reported total gasification rate measurements for relatively small graphite specimens (≤ 55 grams), are used next in a parametric analysis. The analysis calculates the continuous Arrhenius curves for the gasification of large right cylinders of IG-10 and NBG-18nuclear graphites, with initial masses up to 100 kg in atmospheric airflow.

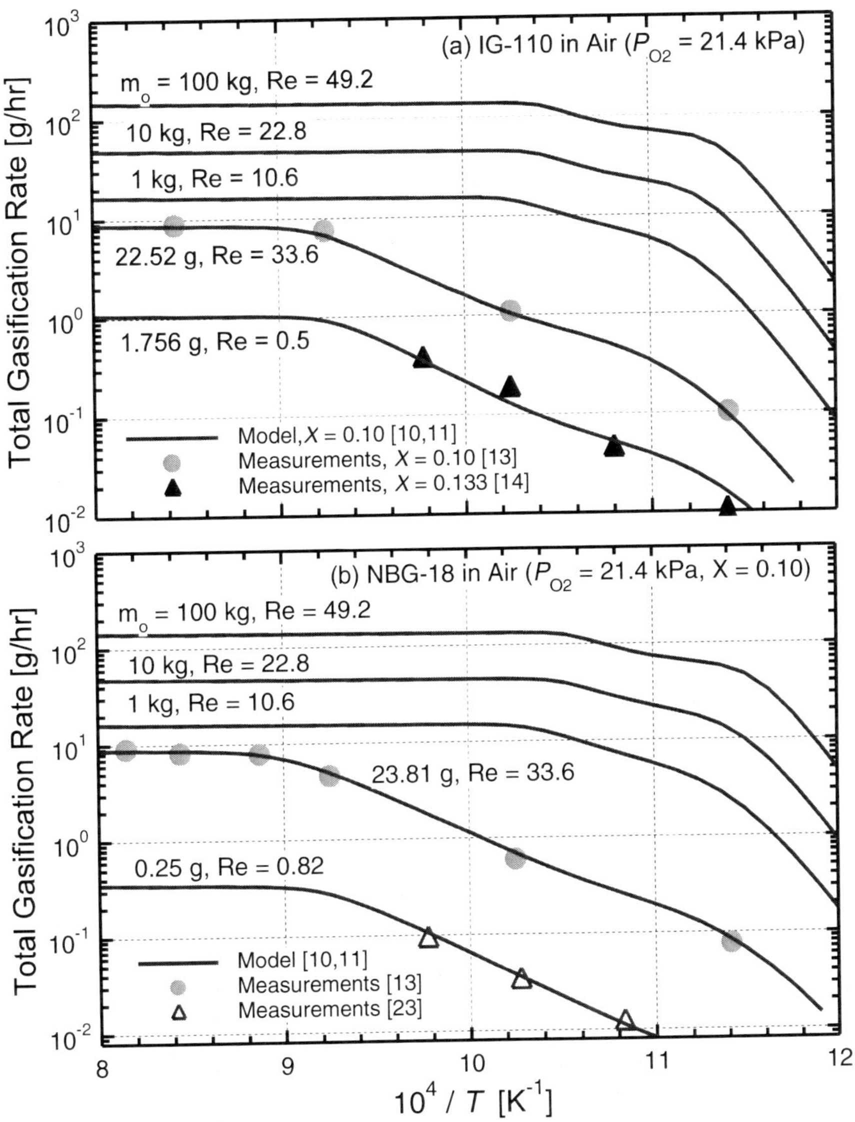

Figure 34. Calculated Arrhenius Curves of the Total Gasification Rate for Different Initial Mass of Graphite in Flowing Atmospheric Air: (a) IG-110; (b) NBG-18.

Effect of Initial Graphite Mass

A parametric analysis is performed using the chemical-reaction kinetics model [10, 11, 20] to examine the effects of temperature and the initial mass on the gasification of IG-110 and NBG-18 nuclear graphite. The analysis is for right cylinders of graphite weighting 1, 10 and 100 kg. The analysis assumes the top and bottom surfaces of the cylinders are shielded, thus gasification only occurs at the lateral surface exposed to slow ambient airflow at 1 cm/s. At an air temperature of 800 K, the corresponding Reynolds numbers for the airflow, based on the diameter of the graphite cylinders, are 10.6, 22.8 and 49.2.The values of the Sherwood number [85] for calculating the transport velocity of oxygen through the boundary layer in the diffusion-limited gasification Mode (c) [85] are 1.72, 2.35 and 3.31, respectively.

The performed gasification analysis for nuclear graphite grades of IG-110 and NBG-18 used the recommended values of the chemical kinetics parameters listed in Table 4, and Equations (12) and (13) for the specific area of active sites, S_m^o. Figures 34a and 34b present the calculated Arrhenius curves of the total gasification rate for IG-110 and NBG-18 at an oxygen partial pressure of 21.4 kPa (ambient air) and a weight loss fraction, $X = 0.10$. The temperatures in these Figures (830 K to 1250 K) cover all three modes of graphite gasification in Figure 4 and the calculated total gasification rates span more than 3 orders of magnitude.

Figure 34a shows that the temperature at which graphite gasification transitions to the diffusion limited Mode (c) for the large graphite cylinders decreases form ~ 980 K for the 1 kg initial mass to ~ 960 K for the 100 kg initial mass. These temperatures for IG-110 are slightly (~ 10 – 15 K) higher than those for the same masses of NBG-18 graphite (Figure 34b), but much lower for the initial masses (≤ 23.81 g) used in the reported gasification experiments [13, 14, 23].

For the large graphite cylinders in Figures 34a and 34b, at temperatures > 950 - 960 K, graphite gasification is diffusion-limited (Mode (c) in Figure 4), and the total gasification rate is weakly dependent on temperature. Instead, it depends on the oxygen diffusion velocity through the boundary layer (Equation (8)) and the lateral surface area of the graphite cylinders. As a result, the total gasification rate increases proportionally to the diameter of the graphite cylinders raised to the power ~ 1.40 (Figures 34a and 34b). The IG-110 right cylinders with initial masses of 1, 10 and 100 kg have diameters of 9, 19.38 and 41.75 cm, respectively. The NBG-18 cylinders of the same initial masses have slightly smaller diameters of 8.83, 19 and 41 cm, due to the higher density of NBG-18 (1.85 versus 1.75 g/cm^3 for IG-110).

At temperatures < 950 K (Modes (a)and(b) in Figure 4), the total gasification rate of the large graphite cylinders, 1-100 kg initial mass, is chemical-kinetics limited and increases exponentially with temperature (Figures 34a and 34b). Note that in the in-pores limited diffusion Mode (b) at intermediate temperatures, the slope of the Arrhenius curve is lower than in Mode (a) and decreases gradually with increasing temperature (Figures 34a and 34b). The large graphite cylinders, the temperatures at the transitions from Mode (a) (chemical-kinetics limited gasification) to Mode (b) (in-pores diffusion limited gasification) are also much lower than for the small graphite specimens used in the experiments [13,14,23] (Figures 34a and 34b).

The total gasification rate of the large nuclear graphite cylinders in Figures 34a and 34b is the sum of the production rates of the CO and CO_2 gases. At temperatures < ~ 890 - 900 K, gasification is controlled by the chemical kinetics (Mode (a)) and the rate is dominated by the production of CO_2 gas.

At these temperatures, the production rate of CO gas is lower than that of CO_2 because of the higher specific energy for desorption of CO. The production rate of CO gas, however, increases rapidly with increasing temperature into Mode (b). By contrast, the production rate of CO_2 in Mode (a) is initially higher than that of CO. It increases with increasing temperature, peaks near the transition from Mode (a) to Mode (b), where it is comparable to that of CO, and then drops with further increase in temperature.

For the specimens of IG-110 nuclear graphite used in gasification experiments of Chi and Kim [13], the diffusion-limited Mode (c) dominates at much higher temperatures (> 1073 K) than for the larger mass graphite cylinders for which the rates are much higher. Similar results are shown in Figure 34b for the small NBG-18 specimens in the experiments of Chi and Kim (2008) (#18d in Table 4, weighting 23.81 g) and of Hinssen et al. [23] (#19 in Table 3, weighting only 0.25 g). Again, the comparison in Figure 34b shows good agreement between the model's calculations [9, 11, 20] and the total gasification rate measurements in the experiments [13, 14, 23].

The next section presents and discusses the results of a parametric analysis investigating the gasification of NBG-18 nuclear graphite by atmospheric-air flow in a typical helium coolant channel of a VHTR or HTR fuel element (Figure 2c). The analysis quantifies the effects of the air inlet temperature and Reynolds number and the initial temperature of the graphite on the local gasification flux and weight loss fraction with distance into the channel. Calculated are the graphite local surface temperature, gasification flux and production fluxes of CO and CO_2 and the total graphite loss until reaching a weight loss fraction of 0.10 at the inlet of the channel.

Graphite Gasification in a VHTR Core Channel

For the safety analysis of a VHTR or HTR, in the unlikely event of massive air ingress, it is important to estimate accurately the gasification rate and weight loss fraction of the graphite of the support columns in the lower plenum and in the helium coolant channels in the reactor core.

This complex exercise involves coupling a 3-D Computational Fluid Dynamics (CFD) code for the reactor's lower plenum and coolant channels to the chemical-reactions kinetics model [10, 11, 20]. Although capable of calculating the local and total gasification rates, the local and total production rates of CO and CO_2 gases and the transient local weight loss fractions, such a coupling would be computationally taxing and time intensive.

An alternate approach, for speeding up the safety analysis and reducing the computation time without compromising accuracy of the results, comprises two steps. First, compile the results of graphite gasification for a wide range of possible conditions using the chemical-reactions kinetics model [10, 11, 20] into readout tables or arrays.

Second, couple the readout tables using a User Interface to the CFD code of choice for performing the safety analysis. The readout tables list the molar fluxes of CO and CO_2 gases as functions of the graphite temperature; local oxygen partial pressure; total pressure; the composition, temperature and flow conditions of the bulk gas mixture and the local weight loss fraction [89].

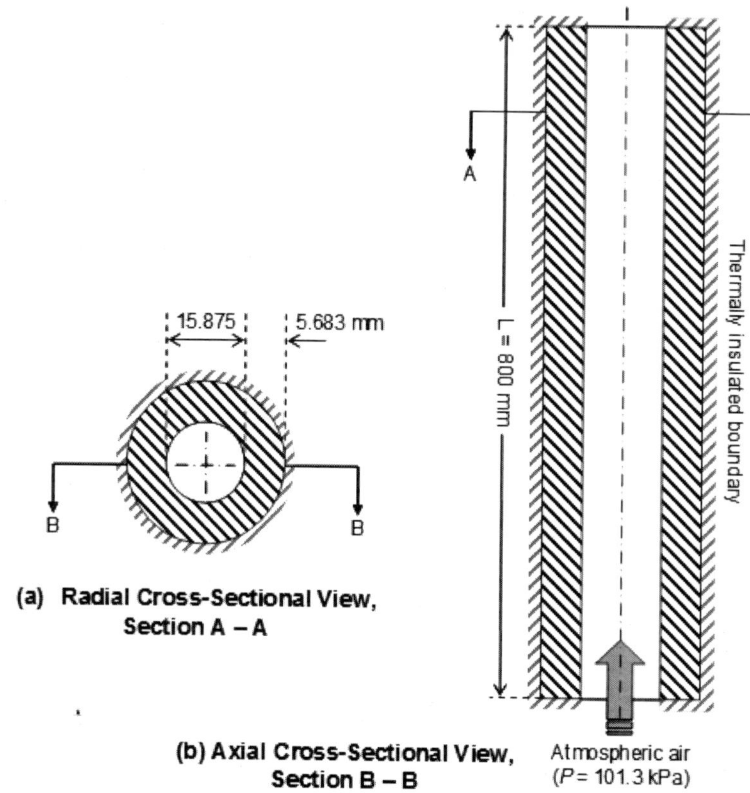

Figure 35. Cross-Sectional Views of an Equivalent Flow Channel in a VHTR Fuel Assembly [89].

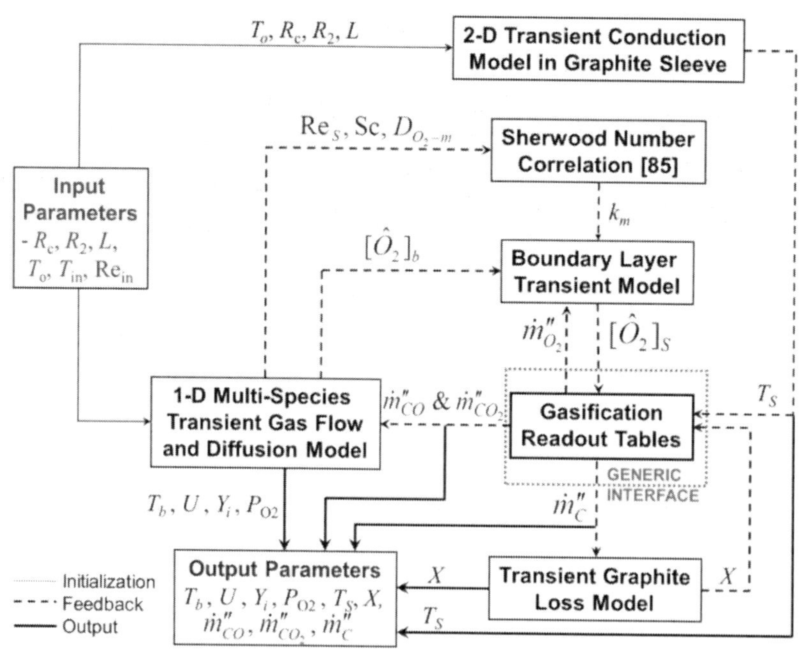

Figure 36. A Block-Diagram of the Integrated Gasification Model of Graphite in a Typical Flow Channel of a VHTR or HTR Fuel Assembly. [89].

The fidelity of implementing the proposed approach for the safety analysis of VHTRs and HTRs has been successfully demonstrated recently [89] in a transient gasification analysis of NBG-18 nuclear graphite in a typical flow channel of a VHTR or HTR fuel assembly (Figure 2c). Figure 35 shows cross-sectional views of the flow channel surrounded by a 5.683 mm thick NBG-18 graphite sleeve that preserves the amount of graphite for a single channel in a prismatic VHTR fuel assembly. For symmetry, the performed transient gasification analysis [89] assumes thermally insulated outer, top and bottom surfaces of the graphite sleeve in Figure 35. The channel height (L = 0.80 m) is that of a typical prismatic fuel assembly.

The performed gasification analysis [89] assumes that the graphite sleeve of the flow channel (Figure 35) is initially at a uniform temperature, T_o, and the channel is filled with helium gas at atmospheric pressure and a temperature T_o. At the start of the transient gasification analysis atmospheric air enters the channel at T_o and Re_{in}. The analysis varies T_o from 800 to 1100 K, uses Re_{in} = 5, 10 and 20 and continues until reaching a weight loss fraction, X, of 0.10 at channel entrance. The time, $t = t_{10}$, at which this condition is reached is calculated. The range of temperatures is the analysis covers all 3 modes of graphite gasification (Figure 4).

The air inlet Reynolds numbers considered in the analysis correspond to inlet velocities < 20 cm/s, covering a wide range of possibilities in natural convection phase of a massive air ingress accident in a VHTR or HTR. Owing to the lack of specific knowledge of the expected air inlet temperature and flow velocity in such an accident, the performed parametric analysis is none the less a useful exercise [89]. It investigates the impact of various scenarios on the progression of graphite gasification and weight loss in a typical VHTR channel.

A primary objective of the transient gasification analysis [89] is also to demonstrate the fidelity as well as the effectiveness of the developed generic interface for coupling the readout tables of graphite gasification to a transient flow and multi-species diffusion model in the channel (Figure 35). This model couples the bulk gas flow to heat conduction in graphite sleeves and accounts for the released energy in the formation of CO and CO_2 during graphite gasification. Actual safety analysis of a VHTR or HTR would replace this model with a CFD code of choice.

Figure 36 presents a schematic of the building blocks of the integrated model applied to the transient gasification analysis of NBG-18 graphite in the flow channel (Figure 35). The governing equations of the multi-species transient flow and diffusion model, including the heat conduction in the graphite sleeves are detailed elsewhere [89]. The constituents of the integrated gasification model in Figure 36 are the multi-species transient gas flow and diffusion model, the transient conduction model in the graphite sleeves, the readout tables for graphite gasification, a generic coupling interface, a boundary layer transient model for graphite gasification in the diffusion limited Mode (c) (Figure 4), and a transient graphite loss model.

Analysis Results

Figure 37 compares the calculated axial profiles of the graphite local weight loss in the coolant channel in Figure 35 at different times during the gasification transient when Re_{in} = 20 and T_o = 900 K.

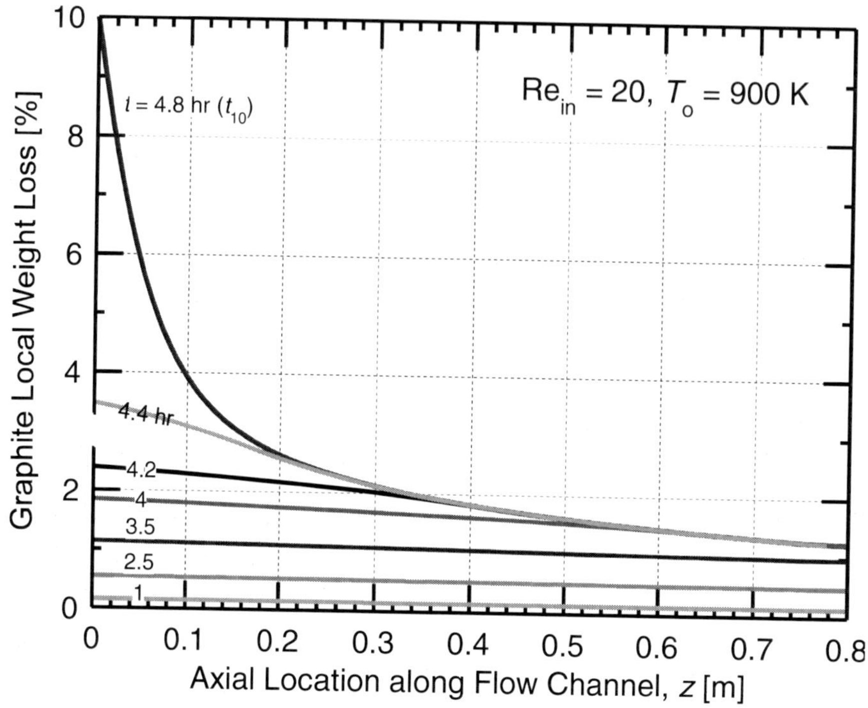

Figure 37. Graphite Local Weight Loss Along Flow Channel in Figure 35, at Different Times of the Gasification Transient [89].

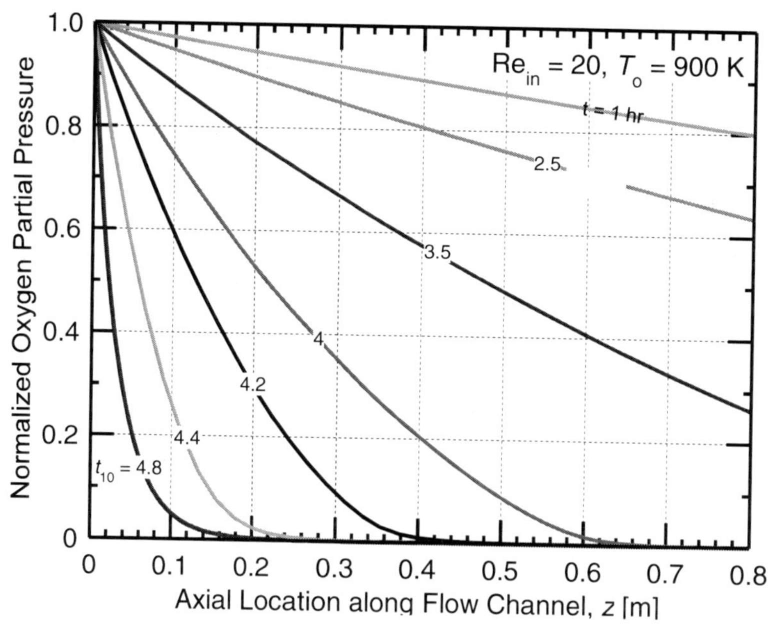

Figure 38. Oxygen Partial Pressure in the Bulk Gas Mixture along the Channel [89].

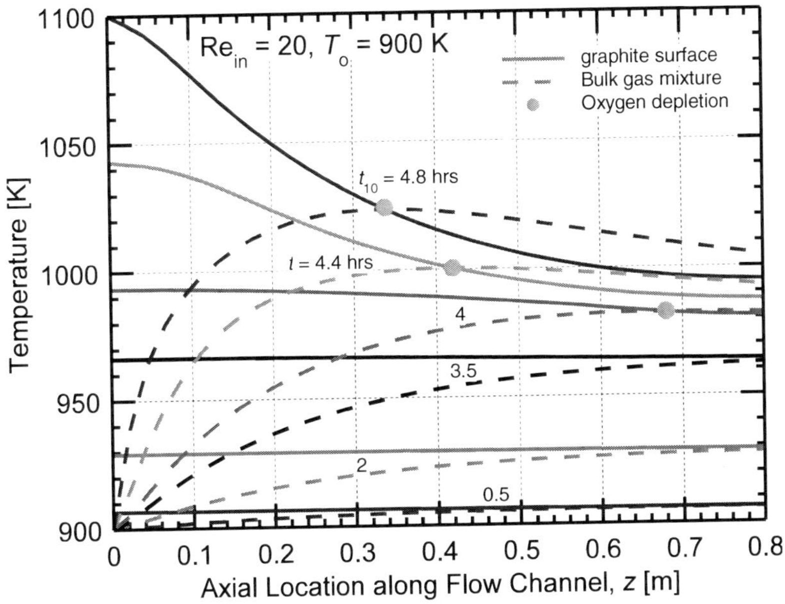

Figure 39. Axial Distributions of the Graphite and Bulk Gas Temperatures [89].

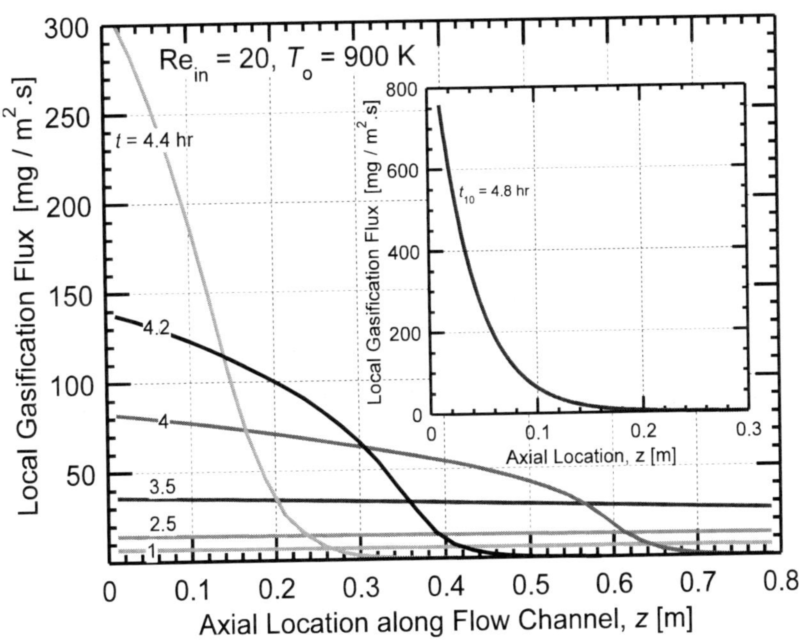

Figure 40. Axial Distributions of Local Gasification Flux in Flow Channel [89].

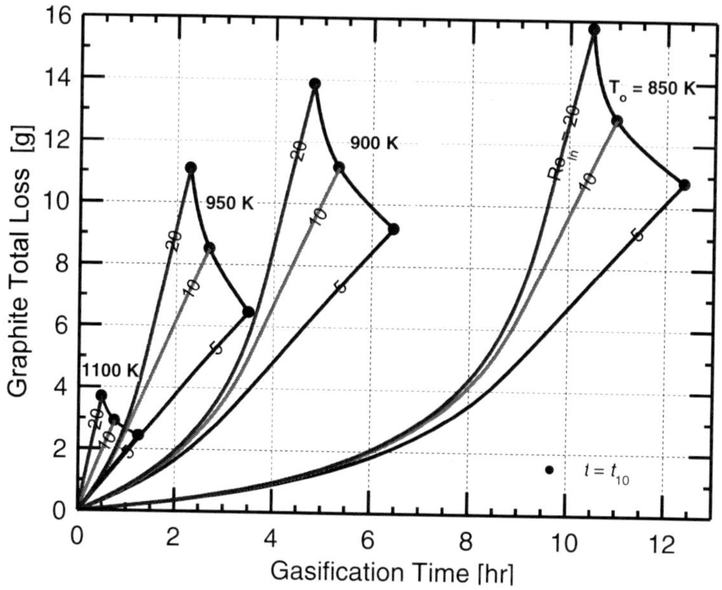

Figure 41. Effects of T_o and Re_{in} on Total Graphite Loss in Flow Channel [89].

During the first 3.5 hrs of the gasification transient in the channel (Figure 35), the local gasification rate is relatively low and graphite local weight loss is small (~ 1%) and nearly uniform along the channel. This is because at these conditions oxygen, although gradually consumed with distance into the channel, is still readily available in the bulk gas flow. As shown in Figure 38, the oxygen mole fraction in the bulk gas flow exiting the channel is 27% of its value at the channel entrance (0.2115).

In the analysis, the graphite sleeve temperature (Figure 35) increases also uniformly along the channel (Figure 39). Such an increase is due the released energies associated with the production of CO and CO_2 gases by the exothermic reactions:

$$2C + O_2 \rightarrow 2\,CO + \Delta H = 110.5 \text{ kJ/mole}, \quad (14)$$

$$C + O_2 \rightarrow CO_2 + \Delta H = 393.5 \text{ kJ/mole}. \quad (15)$$

In the gasification Mode (a) at low temperatures (Figure 4), the heat of formation of CO_2 is the major contributor to the total energy release, while that associated with the production of CO dominates at higher temperatures in Mode (c). In Mode (b), at intermediate temperatures, the reactions in Equations (14) and (15) both contribute, but with different proportions, to the total energy release during graphite gasification.

The axial distribution of the graphite local weight loss along the flow channel becomes increasingly non-uniform with time in the gasification transient (Figure 37). In addition, the values of the local gasification rate as well as the graphite local weigh loss increase with time. The steady increase in the local gasification rate with time along the channel is due to a number of causes, namely:

(a) The increase in graphite sleeve surface temperature by the release of the heats of formation of CO and CO_2 gases;

(b) The increase in Active Surface Area (ASA) of free sites with graphite weight loss, increasing CO gas production and to a lesser extent that of CO_2 [10]; and
(c) The decrease in local oxygen partial pressure in the bulk gas flow along the channel decreases the local gasification rate. The contribution of this effect is much smaller than those of (a) and (b), but becomes increasingly important with time in the gasification transient (Figure 39).

At larger times of the simulated transient, $t > 3.5$ hrs, the local gasification rate of graphite along the channel becomes non-uniform (Figure 37). The local weight loss and surface temperature increase faster near the entrance and decrease rapidly with distance into the channel, due to the decrease in the oxygen partial pressure in the bulk gas flow (Figure 38). After 4 and 4.2 hrs into the gasification transient, the oxygen in the bulk gas flow is fully consumed at axial distances of ~ 70 and 50 cm into the channel, respectively. This ceases graphite gasification in the reminder of the flow channel (Figures 37 and 38). However, graphite temperature in that part of the channel continues to increase slowly due to the axial conduction in the graphite sleeve from the entrance to the exit of the channel (Figure 39).

After 4.8 hrs in the gasification transient, graphite weight loss at the channel entrance reaches 10%, while that at the exit of the channel is only 1.2% (Figure 37). At that time ($t = t_{10}$), the oxygen in the bulk gas flow depletes fully all along the channel, except in the 30 cm section near the entrance (Figure 38).

Figure 39 shows the calculated axial distributions of the graphite surface temperature and that of the bulk gas flow along the channel. The solid circle symbols in this Figure indicate the axial locations in the channel at which oxygen in the bulk gas flow is consumed fully, ceasing gasification in the remainder of the channel. During the first 3.5 hrs of the transient, graphite surface temperature rises uniformly along the channel. The convective heat transfer from the graphite surface increases the temperature of the bulk gas flow with distance into the channel. At longer times, the rapid rise in the graphite surface temperature near the entrance of the channel due to the increase in the local gasification rate and the production of CO and CO_2 gases also increases that of the bulk gas flow (Figure 39).

Figure 40 compares the calculated distributions of the local gasification fluxes along the channel at various times in the simulated transient. Note the rapid rise in the local gasification near the entrance of the channel after 3.5 hrs of the transient. This is due to the increase in the graphite surface temperature by the exothermic production reactions of CO and CO_2 gases (Equations (14) and (15)). Figure 41 compares the calculated total loss of graphite in the channel at different conditions until reaching a weight loss fraction of 0.10 at the channel entrance (t_{10}). When $T_o = 900$ K and $Re_{in} = 20$, $t_{10} = 4.8$ hrs and the total graphite loss in the channel is only 14 g.

The total graphite loss by gasification increases with increasing Re_{in} (or oxygen supply), but decreases with increasing T_o. Despite the exponential effect of temperature on the gasification rate of graphite, the rapid depletion of oxygen in the bulk gas flow within a short distance into the channel decreases graphite's total weight loss.

The duration of the gasification transient until reaching a local weight loss fraction of 0.10 at the channel entrance, t_{10}, decreases with increasing Re_{in} and/or increasing T_o. The calculated results construct a gasification surface for the NBG-18 graphite in the flow channel (Figure 35).

The constructed surface in Figure 42 is a grid of constant Re_{in} curves and intersecting curves of constant initial graphite and air inlet temperature, T_o. It shows that decreasing T_o increases t_{10} as well as the total graphite loss in the channel. The gasification surface in Figure 42 also illustrates the complex interplays between the initial temperature graphite and that of air entering the channel and the corresponding oxygen supply in the inlet airflow and the rise in the graphite local surface temperature by the production of CO and CO_2 gases (Equations (14) and (15)). These parameters affect both the total and the extent of the local graphite weight loss along the channel.

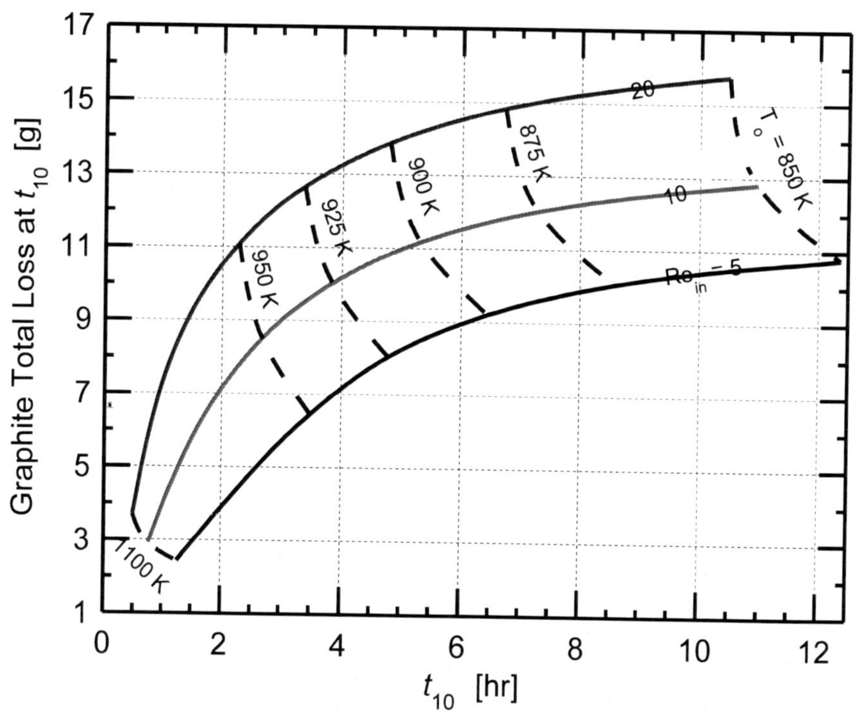

Figure 42. Effects of Inlet Air and Initial Graphite Temperature, T_o, and Inlet Air Reynolds Number, Re_{in}, on Graphite's Total Loss in channel (Figure 35) and t_{10} [89].

CONCLUSION

A massive air ingress following a break in the pipes connecting the reactor vessel to that of the closed Brayton cycle turbo machinery is a design basis accident for Very-High Temperature and High-Temperature gas-cooled Reactors. In such an accident, a partial weight loss ($\leq$10%) of graphite in the support columns in the lower plenum could compromise their mechanical strength and structure integrity, potentially resulting in a reactor core collapse. Subsequent upward flow of air under the effect of gravity through the coolant channels of the core fuel assemblies would also cause graphite gasification in the channels. Depending on the local graphite temperature, total pressure, oxygen partial pressure and velocity or Reynolds number of the atmospheric airflow or a mixture of helium and air entering the core, graphite gasification could enlarge the diameter of the coolant channels and release fission products entrapped in the graphite matrix of the fuel compacts.

In the above scenario, the prevailing mode of graphite gasification depends on its temperature. However, the transition temperatures among the three gasification modes in Figure 4 and the actual rate of gasification depend on both the graphite grade and initial mass. At low temperatures, in pores gasification is the dominate mode (Mode (a)), driven solely by the chemical reactions kinetics.

At intermediate temperatures, in pores gasification gradually becomes oxygen diffusion limited (Mode (b)). The penetration of oxygen into the open volume pores gradually decreases with increasing temperature because of the increased rate of gasification and the counter-current diffusion of the CO and CO_2 gaseous products out of the pores. The chemical-reactions kinetics also control this oxygen-limited in pores gasification. At high temperatures, graphite gasification ceases in the volume pores and shifts to the external surface, marking the transition to Mode (c).

While graphite gasification in Mode (a) and Mode (b) typically weakens the mechanical strength of the structure, gasification in Mode (c) does not affect the mechanical strength but erodes the external dimension and/or enlarge the diameter of the flow channels. Despite the high temperatures in Mode (c), which exponentially increases the chemical-reactions kinetics, the oxygen transport from the bulk gas mixture to the surface limits the actual rate of gasification.

The four chemical reactions contributing to the gasification of graphite are those of adsorption of oxygen atoms onto the surface active free sites, simultaneous formation of C-O bonds and breakup of C-C bonds to form stable surface complexes and desorption of CO and CO_2 gases. These chemical kinetics parameters for these reactions and the initial active surface area (ASA_o) of free sites depend on the grade of nuclear graphite. The gasification rate of right cylinders of graphite increases with ASA_o to a power of 1.9. And the specific surface area of active sites, S_m^o, is inversely proportional to the square root of the initial mass of graphite specimens in the experiments ($m_o \leq 50$ g). A parametric analysis investigated the effect of increasing the initial mass of right cylinders of graphite on the total gasification rate and the production rate of CO and CO_2 gasses.

Owing to the complexity of graphite gasification and its dependence on many intrinsic, chemical kinetics and operation parameters, calculating the total rate as well as the production rates of CO and CO_2 gases in the unlikely event of a massive air ingress accident in a VHTR or HTR core, is a challenging task. Many reported models for calculating the total gasification rate in a VHTR or HTR use empirical Arrhenius or Langmuir-Hinshelwood expressions with apparent activation energies and pre-exponential rate constants. Although relatively simple and easy to integrate into CFD codes for performing a reactor safety analysis, the empirical expressions are inherently limited. They are only valid for the range of temperatures and test conditions in the experiments and could not be used to calculate the individual production rates of CO and CO_2 gases, important to reactor safety and to assessing the environmental impact of their release. Furthermore, the variances in the values of the apparent activation energy and pre-exponential rate constants in the empirical expressions (Figure 3), even for the same nuclear graphite (Figure 2), could introduce large differences in the calculated gasification rates.

Reviewed are the gasification modes and the fabrication methods of nuclear graphite in VHTRs and HTRs and recommended chemical-reactions kinetics parameters for the gasification of different grades of graphite. These parameters are determined using a multi-parameter optimization algorithm from the reported measurements of the total gasification rate and transient weight loss in experiments performed by various investigators. A total of eleven chemical-reactions kinetics parameters are determined, namely:

(a) The values and Gaussian-like distributions of the average specific activation energies for adsorption of oxygen and thedesorption of CO gas (4);
(b) The average specific activation energies for desorption of CO_2 gas and the formation of stable complexes (2);
(c) The pre-exponential ratecoefficients for adsorption of oxygen, formation of stable complexes and thedesorption of CO and CO_2 gases (4); and
(d) The initial active surface area (ASA_o) of free sites (1).

The recommended chemical-reactions kinetics parameters for the gasification of nuclear graphite grades of IG-110 and NB-25, with super fine and medium size petroleum coke filler particles, are nearly the same, but different from those for nuclear graphite grades of NBG-18 and IG-430, with superfine and medium size coal tar pitch filler particles (Tables 3 and 4).

Presented are the results of validating the chemical-reactions kinetics model (Figure 21) for calculating the total gasification rate, the transient weight loss and the production rates of CO and CO_2 gases for nuclear graphite grades: NBG-18, IG-430, NBG-25 and IG-110. The good agreements of the calculations in all three modes of graphite gasification with the experimental measurements for a wide range of temperatures and weight loss fraction up to 0.40 confirm the effectiveness and accuracy of the model (Figure 22). It also confirms the applicability of the recommended chemical kinetics parameters in Table (4) and the proposed expressions for the initial specific active area of free sites (Equations (12) and (13)) for use in future safety analyses of HTRs and VHTRs.

In the diffusion-limited graphite gasification (Mode (c)), the effective diffusion velocity of oxygen through the boundary layer is calculated using a Sherwood number correlation recently developed by the authors. This correlation is consistent with reported gasification rate measurements in experiments by various investigators for different grades of nuclear graphite (Figure 22).

Results of a transient gasification analysis of NBG-18 graphite in a typical coolant channel in a VHTR prismatic fuel element are also presented. The analysis varied the temperature of the atmospheric air entering the channel and the initial graphite temperature from 800 - 1100 K and the air inlet Reynolds number from 5 to 20.

Results show that the local graphite loss by gasification is highest at the entrance and decreases rapidly with distance along the channel. In addition, the released heat in the exothermic reactions for the production of CO and CO_2 gases significantly increases the graphite temperature at channel entrance, at the expense of depleting the oxygen in the bulk gas flow. Consequently, the local graphite weight loss is highest at the entrance, but decreases rapidly with axial distance along the channel.

Calculated are the times for reaching a 10% graphite weight loss at the channel entrance (t_{10}). When the initial graphite and inlet air temperature is 900 K and the inlet air Reynolds number, $Re_{in} = 20$, the calculated $t_{10} =$ 4.8 hrs and the total graphite loss in the channel is only

14 g. The graphite's total loss increases with increasing Re_{in} (or oxygen supply), but decreases with increasing the initial graphite and air inlet temperature. Despite the exponential effect of temperature on gasification, the availability of oxygen in the bulk gas flow limits the graphite total weight loss in the channel (Figure 42).

ACKNOWLEDGMENTS

Funding for this research is provided partially by the DOE University Programs under contract No. 00044825 00002, Project No. 09-830 to the University of New Mexico and the Institute for Space and Nuclear Power Studies. The authors are grateful to Dr. C. I. Contescu of Oak Ridge National Laboratory for his valuable comments and for providing the gasification measurements for the PECA nuclear graphite.

REFERENCES

[1] MacDonald P. E., Bayless P. D., Gougar H. D., Moore R. L., Ougouag A. M., Sant R. L., Sterbentz J. W. and Terry W. K., *Proceedings Inter. Congress on Advances in Nuclear Power Plants (ICAPP '04),* Paper 3405 (2004).

[2] *Current Status and Future Development of Modular High Temperature Gas Cooled Reactor Technology,* IAEA-TECDOC-1198, 2001.

[3] Baxi C. B., Perez E., Shenoy A., Kostin V. I., Kodochigov N. G., Vasy- aev A. V., Belov S. E. and Golovko V. F., *Evolution of the Power Conversion Unit Design of the GT-MHR*, GA-A25381 (2006).

[4] *Next Generation Nuclear Plant Pre-Conceptual Design Report*, INL/ EXT-07-12967 Rev. 1, 2007.

[5] Froschauer K. G., Brähler G. and Heit W., *Proceedings Inter. Congress on Advances in Nuclear Power Plants (ICAPP '08)*, paper No. 8097 (2008).

[6] Kiryushin A. I., Kodochigov N. G., Kouzavkov N. G., Ponomarev-Stepnoi N. N., Gloushov E. S. and Grebennik V. N., "Project of the GT-MHR High-Temperature Helium Reactor with Gas Turbine," *J. Nuclear Engineering and Design* 173, 119–129 (1997).

[7] MacDonald P. E., Sterbentz, J. W., Sant, R. L., Bayless, P. D., Shultz, R. R., Gougar, H. D., Moore, R. L., Ougouag A. M., Terry, W. K., *NGNP Preliminary Point Design – Results of Initial Neutronics and Thermal-Hydraulic Assessment*, INEEL/EXT-03-00870 Rev. 1, 2003.

[8] Schultz R. R., Nigg D. W., Ougouag A. M., Terry W. K., Wolf J. R., Gougar H. D., Johnsen G. W., McEligot D. M., McCreery G. E., Johnson R. W., Sterbentz J. W., MacDonald P. E., Temitope A., Taiwo Y. C., Thomas, W. S. Yang, R. B. Vilim, W. D. Pointer and H. S. Khalil, *Next Generation Nuclear Plant-Design Methods Development and Validation Research and Development Program Plan,*. INEEL/ EXT-04-02293 Rev. 0, 2004.

[9] El-Genk M. S. and Tournier J.-M., *J. Nuclear Materials* 411, 193 (2011).

[10] El-Genk M. S. and Tournier J.-M., *J. Nuclear Materials* 420, 141 (2012).

[11] El-Genk M. S. and Tournier J.-M. (2012b), *J. Nuclear Engineering and Design* 250, 142 (2012).
[12] Kawakami H., *Tanso* 124, 26 (1986).
[13] Chi A.-H. and Kim G.-C., *J. Nuclear Materials* 381, 9 (2008).
[14] Fuller E. L. and Okoh J. M., *J. Nuclear Materials* 240, 241 (1997).
[15] Takahashi M., Kotaka M. and Sekimoto H., *J. Nuclear Science and Technology* 31(12), 1275 (1994).
[16] Kim E. S., Lee K. W. and No H. C., *J. Nuclear Materials* 348, 174 (2006).
[17] Contescu C. I., Azad S., Miller D., Lance M. J., Baker F. S. and Burchell T. D. (2008), *J. Nuclear Materials* 381, 15 (2008).
[18] Kühn K., Hinssen H.-K. and Moormann R. (2004), *Proceedings Inter. Congress on Advances in Nuclear Power Plants (ICAPP'04)*, Paper No. 4024 (2004).
[19] Moormann R., Alberici S., Hinssen H. K., Krüssenberg A. K. and Wu C. H. (1999), *Adv. Science Technology* 24, 331 (1999).
[20] El-Genk M. S. and Tournier J.-M., *International J. Nuclear Energy Science and Technology*, 7 (3), 258 (2013).
[21] Xiaowei L., Jean-Charles R. and Suyuan Y., *Nuclear Engineering and Design* 227, 273 (2004).
[22] Ogawa, M., *Int. J. Heat and Mass Transfer* 30 (5), 1017 (1987).
[23] Hinssen H.-K., Kühn K., Moorman R. B. Schlögl, M. Techter and M. Mitchell, *Nuclear Engineering and Design* 238 (11), 3018 (2008).
[24] Growcock F. B., Barry J. J., Eto M., Heiser J. H., Sastre C. A., *BNL-NUREG-51217*, 4 (1980).
[25] Wang P., Contescu C. I., Yu S. and Burchell T. D., *J. Nuclear Materials* 430, 229 (2012).
[26] Thomas J. M., *Chemistry and Physics of Carbon* 1, 121 (1965).
[27] Kelly B. T., *Irradiation Damage in Graphite due to Fast Neutrons in Fission and Fusion Systems*, IAEA-TECDOC-1154, 35 (2001).
[28] Ergun, S, *Chemistry and Physics of Carbon*, P. L. Walker ed., 3, 45, (1968).
[29] Hart P. J., Vastola F. J. and Walker P. L., Jr., *Carbon* 5, 363 (1967).
[30] Walker P. L. Jr., Taylor R. L. and Ranish J. M., *Carbon* 46, 62 (1991).
[31] Mardsen B. J., *Gas Turbine Power Conversion Systems for Modular HTGRs*, IAEA-TECDOC-1238, 177 (2001).
[32] Burchell T. D., *Carbon* 34 (3), 297 (1996).
[33] Wen K. Y., Marrow T. J. and Mardsen B. J., *Carbon* 46, 62 (2008).
[34] Ioka I., Yoda S. and Konishi T., *Carbon* 28 (6), 879 (1990).
[35] Ehrburger P., Sanseigne E. and Tahon B., *Carbon* 34 (12), 1493 (1996).
[36] Thrower P. A., *Carbon* 20, 465 (1982).
[37] Kim E. S., No H. C., Kim B. J. and Oh C. H., *Nuclear Engineering and Design* 238 (4), 837 (2008).
[38] Virgil'ev Y. S., *Soviet Atomic Energy* 84 (1), 7 (1998).
[39] Eatherly W. P. and Piper E. L., *Nuclear Graphite*, R. E. Nightingale, Ed., Academic Press, NY, 2, 22 (1962).
[40] Haag G., Delle W., Nickel H., Theymann W. and Wilhelmi G. (1987), *International Working Group on Gas-cooled Reactors*, JAERI-M 86-192, Paper III-1, 128 (1986).

[41] Haag G., Mindermann D., Wilhelmi G., Persicke H. and Ulsamer W., *J. Nuclear Materials* 171, 41 (1990).
[42] Virgil'ev Y. S., *Inorganic Materials* 37 (4), 354 (2001).
[43] Virgil'ev Y. S., *Inorganic Materials* 37 (8), 780 (2001).
[44] Pappano P. J., Burchell T. D., Hunn J. D. and Trammell M. P., *J. Nuclear Materials* 381, 25 (2008).
[45] Hrovat M. F., Grosse K. H. and Seemann R. (2008), *Proceedings 4th International Topical Meeting on High Temperature Reactor Techno- logy (HTR-2008)*, Paper No. HTR2008-58025 (2008).
[46] Engle G. B. and Kelly B. T. (1984), *J. Nuclear Materials* 122, 122 (1984).
[47] Pedraza D. F. and Koike J., *Carbon* 32 (4), 727 (1999).
[48] Snead L. L., Burchell T. D. and Katoh Y., *J. Nuclear Materials* 381, 55 (2008).
[49] Burchell T., Yahr T. and Battiste R., *Carbon* 245, 2570 (2007).
[50] Oku T., Eto M. and Ishiyama S., *J. Nuclear Materials* 172, 77 (1990).
[51] Ishiyama S., Burchell T. D., Strizak J. P. and Eto M., *J. Nuclear Mate- rials* 230, 1 (1996).
[52] Ishihara M., Sumita J., Shibata T., Iyoku T. and Oku T., *J. Nuclear Engineering and Design* 233, 251 (2004).
[53] Vreeling J. A., Wouters O. and van der Laan J. G., *J. Nuclear Materials* 381, 68 (2008).
[54] Burchell T. D. and Snead L. L., *J. Nuclear Materials* 371, 18 (2007).
[55] Albers T. L., Batty L. and Kaschak D. M., Proceedings *4th Int. Topical Meeting on High Temperature Reactor Technology (HTR-2008)*, Paper HTR2008-58284 (2008).
[56] Contescu C. I., Guldan T., Wang P. and Burchell T. D., *Carbon* 50, 3354 (2012).
[57] Takizuka T., Takada S., Yan X., Kosugiyama S., Katanishi S. and Kunitomi K., *J. Nuclear Engineering and Design* 233, 329 (2004).
[58] Essenhigh R. H., *Chemistry of Coal Utilization*, 2nd Supplementary Volume, M. A. Elliott, Ed., John-Wiley & Sons, New York, 9, 1153 (1981).
[59] Oh C. H., Kim E. S., Kang H. S., No H. C. and Cho N. Z., *Experimental Validation of Stratified Flow Phenomena, Graphite Oxidation, and Mitigation Strategies of Air Ingress Accidents*, Report No. INL/EXT-09-16465 Rev. 1, 6-1(2009).
[60] Kelly B. T., *Fusion Technology* 16, 96 (1989).
[61] Balden M., Klages K. U., Jacob W. and Roth J., *J. Nuclear Materials* 341, 31 (2005).
[62] Froberg R. W. and Essenhigh R., *Proceedings 17th International Symposium on Combustion*, 17 (1), 179 (1979).
[63] Fuller E. L. Jr., Kopp O. C. and Underwood A. D., *Proceedings Bienni- al Conference on Carbon,* 604 (1991).
[64] Moorman R., Hinssen H.-K. and Kühn K., *Nuclear Engineering and Design* 227, 281 (2004).
[65] Haynes B. S., *Combustion and Flame* 126, 1421 (2001).
[66] Moulijn J. A. and Kapteijn F., *Carbon*, 33 (3), 1155 (1995).
[67] Chen S. G., Yang R. T., Kapteijn F. and Moulijn J. A., *Industrial & Engineering Chemistry Research* 32, 2835 (1993).
[68] Fanning P. E. and Vannice M. A., *Carbon* 31, 721 (1993).
[69] Back M. H., *Canadian Journal of Chemistry* 75, 249 (1997).
[70] Zhang Z. and Yu S., *Nuclear Engineering and Design* 218, 249 (2002).
[71] Du Z., Sarofim A. F. and Longwell J. P., *Energy and Fuels* 4, 296 (1990).

[72] Ahmed S. and Back M. H., *Carbon* 23 (5), 513 (1985).
[73] Ahmed S., Back M. H. and Roscoe J. M., *Symposium on Fundamentals of Gasification* 34, 63 (1989).
[74] Laine N. R., Vastola F. J., Walker P. L. Jr., *J. Physical Chemistry* 67, 2030 (1963).
[75] Du. Z., Sarofim A. F., Longwell J. P. and Mims C. A., *Energy and Fuels* 5, 214 (1991).
[76] Brown T. C., Lear A. E. and Haynes B. S., *International Symposium on Combustion* 24, 1199 (1992).
[77] Hurt R. H. and Haynes B. S., *Proceedings Combustion Institute* 30, 2161 (2005).
[78] Sibraa A., Newburry T. and Haynes B. S., *Combustion & Flame* 120 (4), 515 (2000).
[79] Zhuang Q., Kyotani T. and Tomita A., *Energy Fuels* 9, 630 (1995).
[80] Campbell P. A. and Mitchell R. E., *Combustion and Flame* 154, 47 (2008).
[81] Blanchard A., *The Thermal Oxidation of Graphite, Irradiation Damage in Graphite due to Fast Neutrons in Fission and Fusion Systems*, IAEA-TECDOC-1154, 207 (2001).
[82] Ahmed S., Back M. H. and Roscoe J. M., *Combustion and Flame* 70, 1 (1987).
[83] Su J.-L. and Perlmutter D. D., *AIChE Journal* 31 (6), 973 (1985).
[84] Hirschfelder J. O., Curtiss C. F. and Bird R. B. (1954), *Molecular Theory of Gases and Liquids*, John Wiley & Sons, NY (1954).
[85] El-Genk M. S. and Tournier J.-M., *Progress in Nuclear Energy*, 62, 26 (2013).
[86] Xiaowei L., Jean-Charles R. and Suyuan Y., *Nuclear Science and Engineering* 151, 121 (2005).
[87] Kim E. S., Lee K. W. and No H. C., *J. Nuclear Materials* 348, 174 (2006).
[88] El-Genk M. S. and Tournier J.-M., *Personal Communications*, Unpublished Work, University of New Mexico's Institute for Space and Nuclear Power Studies, Albuquerque, NM, USA (2013).
[89] El-Genk M. S. and Tournier J.-M., *Progress in Nuclear Energy* 64, 16 (2013).
[90] Contescu C. I., *Personal Communications*, Oak Ridge National Labora tory, Oak Ridge, TN, USA (2012).

In: Graphite
Editor: Quinton C. Campbell

ISBN: 978-1-62618-576-0

Chapter 2

STRUCTURAL MIMICRY OF CARBON DRIVEN BY ULTRASHORT LASER PULSES

Sergey I. Kudryashov* and Nikolay N. Melnik
P.N. Lebedev Physical Institute,
Russian Academy of Sciences,
Moscow, Russia

ABSTRACT

The transient and permanent structural mimicry of graphite, diamond and nanotubes in response to its moderate and strong excitation by femtosecond laser pulses was experimentally studied by means of diverse complementary experimental techniques. Sub-picosecond changes of the optical probe reflection, as well as pronounced picosecond hypersonic modulation of the probe reflectivity in the photoexcited graphite were related to its transient structural transformations and melting, respectively. The final, permanent structural modification were envisioned by micro-Raman spectroscopy as atomic-scale local structural transformations (point defect generation) at moderate photoexcitation, and as well-defined graphite and diamond phases at high photoexcitation of the graphite surface, which proceeds through formation of a hot, critical electron-ion surface plasma and TPa-level plasma-driven shock-wave compressing the material and expanding it off.

Similarly, femtosecond laser-driven structural transformation toward graphite was revealed on surface and in bulk diamond. Finally, single- and multi-shot IR and visible femtosecond laser ablation of single-wall carbon nanotubes was envisioned at microscopic and mesoscopic structural levels.

Overall, the observed rich variety of carbon allotropes, their mutual transformations and the intermediate structural carbon forms are illustrated and discussed in details.

* E-mail: sikudr@sci.lebedev.ru.

1. INTRODUCTION

1.1 Allotropic Structural Phases of Carbon

Carbon is one of the most important constituent elements of novel nanoelectronic and nanophotonic materials, such as diamond, fullerenes, nanotubes and graphene. In this lieu, understanding of its behavior at elevated electronic excitation levels and temperatures, including allotropic transformations, is important to reveal their robust exploitation regimes at high electric currents, high photon or particle fluxes, resulting in extremely raised lattice temperatures. Particularly, carbon atoms exhibit unique capability to form different interatomic bonds, usually defined as in terms of sp^1, sp^2 and sp^3-hydbridization of the four outer-shell carbon electrons. The resulting two, three and four carbon-carbon bonds yield in different spatially regular carbon atomic networks – graphite, diamond, nanotubes etc. Raman scattering, being the most popular, informative and convenient method for structural characterization of various carbon allotropes, provides their highly sensitive and unambiguous identification (Figure 1.1) owing to their unique spectral "fingerprints" [1-3]. As a result, two similar Raman spectra represent two carbon materials, strongly resembling each other in their crystalline structure. In real carbon materials, because of the rich variety of the carbon hybridizations, their structure becomes rather complicated (in some sense, disordered), simultaneously including in some proportions at atomic, nano- and even micro-levels some of these known allotropic carbon forms. However, a straightforward analogy between these complex carbon structures and their simple "etalon" allotropic constituents is not adequate, since such disordered structures in many cases represent new carbon materials. Moreover, broad Raman analysis of diverse real carbon materials with disordered atomic structures, produced in different technical ways, reveals three characteristic structural groups differing in their content of sp^1, sp^2 and sp^3-hydbridized carbon atoms and corresponding grain dimensions for different allotropes (Figure 1.2).

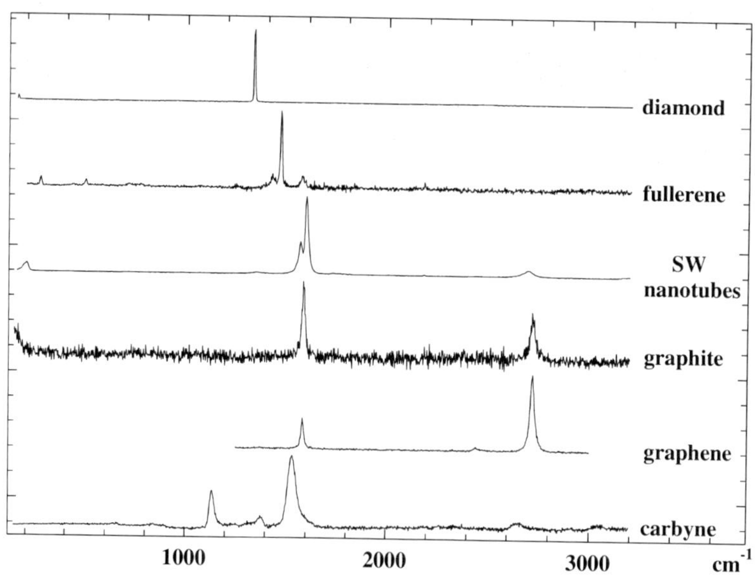

Figure 1.1. Characteristic Raman spectra of various carbon allotropes (adapted from [3]).

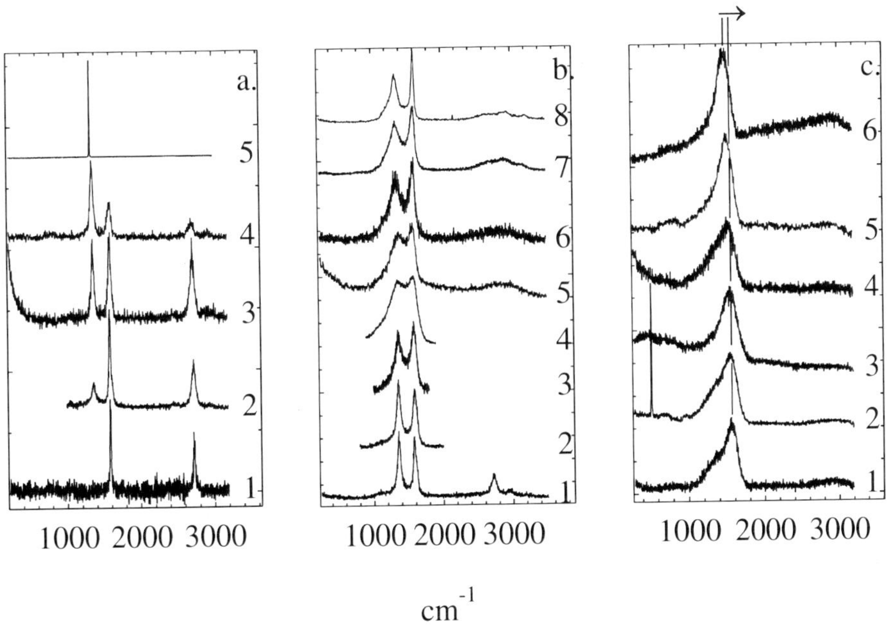

Figure 1.2. First- and second-order Raman spectra of real disordered carbon materials (see the text for details). Adapted from ref.[3].

The first group (Figure 1.2a) is represented by carbon materials with their first-order Raman spectra appearing in the range 1000см$^{-1}$÷2000см$^{-1}$ in the form of two separate sharp lines at ~1355 cm^{-1} (D, sp^3) and ~1600 cm^{-1} (G, sp^2) [1-3]. Upon disordering, the relative intensities of these D- and G-lines vary in favor of one of them, reflecting the underlying hybridization changes within the rather large graphite or diamond crystallites. For an example, two reference Raman spectra in Figure 1.2a of monocrystalline graphite (curve 1) and diamond (curve 5) can be compared to those of pyrolitic graphite (curve 2) and multi-wall nanotubes (curve 3) [3].

The second group covers other carbon materials with the first-order Raman spectra always exhibiting the same two lines, though significantly broadened in the form of characteristic D- and G-bands, which vary their half widths and band shapes upon technological treatment (Figure 1.2b) following the decreasing crystallite size. The gradual variation of the D- and G-bands in Figure 1.2b is shown for high-pressure, high-temperature treated carbon (curve 4), plasma-processed pyrolitic graphite (curve 7), carbon particle produced from soil (curve 8) etc. [3].

Finally, the third group of disordered carbon materials is represented in their Raman spectra by a broad, unresolved and asymmetric band of amorphous carbon (Figure 1.2c), containing a mixture of different – presumably, sp^2 and sp^3 – hybridizations. The typical examples in this case are the superhard polymerized fullerene (curve 3), magnetron-sputtered carbon film (curve 4), chemically-vapor deposited α-carbon (curve 6) and some others [3]. Overall, these three groups provide in terms of their structural (sp^1, sp^2 or sp^3-hydbridization) and phase (macro-, micro- or nano-crystallites) composition rather full coverage of various carbon materials, which are relevant for their different technological (especially, laser) processing regimes, and can be unambiguously characterized by Raman scattering method.

1.2. Femtosecond Laser Modification of Carbon Materials: Extreme Electronic Excitation and Temperature/Pressure Conditions

Femtosecond (fs) laser irradiation of material surfaces is known to drive the matter into extreme electronically excited, high-temperature and high-pressure thermodynamic states [4-6], being pre-requisites for ultrafast structural transformations. However, since the pioneering work of Reitze et al. [4] in early 90s, very little attention was paid to fundamental studies of carbon at high temperatures [7-9], while non-thermal pre-ablation dynamics and high-pressure structural transformations in graphite driven by ultrashort (fs) laser pulses were extensive studied using the modern time-resolved state-of-art characterization techniques [8-19]. Since 90s, the most enlightening work was made by Sokolowski-Tinten et al. [7], revealing by means of time-resolved optical reflection microscopy extended over a broad fs-ns time span, basic – spallation and fragmentation (expansion of a supercritical fluid eventually transforming in a vapor-droplet mixture [7]) – mechanisms of frontal thermal fs-laser ablation of graphite, which were intuitively assigned to sub-critical and supercritical carbon states, respectively. The later studies were mostly focused on surface profilometry and micro-Raman characterization of post-pulse fs-laser ablative craters [8, 9], since experimental techniques enabling simultaneous temporally and spatially (especially, in depth) resolved tracking of buried thermal melting, ablative spallation and fragmentation of materials heated by ultrashort laser pulses, are still missing. Moreover, historically, in previous thermophysical research high-power ultrashort- and short-pulse (femtosecond to nanosecond) laser irradiation of graphite surfaces was presumably aimed at a preparation of high-temperature sub-critical carbon states [4, 7, 20-25], and generation of strong surface shock waves transforming an underlying graphite layer into diamond [8-19, 25] or other structural phases [12]. The former laser-induced structural transformation was accurately identified by transient and stationary electron and x-ray diffraction [12-14, 16, 18-19, 24-25], or by micro-Raman scattering spectroscopy [8-10, 12, 16-17,19,25], implying that multi-GPa dynamic pressures were achieved in the near-surface graphite layer, as required for this phase transition ($\approx$ 11 GPa [20, 21] under static conditions without catalysts). However, detailed experimental characterization of such laser-driven shocks on graphite surfaces were not whenever performed yet, basing mostly on hypothetical estimates related backward to the fact of the graphite-diamond (G–D) transformation [8-10, 16, 24].

Importantly, widely used experimental techniques for characterization of initial front-side shocks at a rear sample side – mostly, interferometric VISAR [26] and its different versions [27] – are not fully applicable to graphite materials even in the form of micron-thick layers. The main reason is the highly anisotropic (layered) structure of monocrystalline graphite, not well supporting sound propagation along the normal to these atomic layers, and high dissipative losses of shock wave energy in bulk graphite on the G–D transformation, heating of the material, scattering, and related effects [28]. Moreover, similar front-side interferometric measurements would be not only inaccurate due to low optical homogeneity of high-temperature, high-pressure carbon plumes [7], but also not very informative providing only a mass velocity parameter. Likewise, rather informative side-view pump/probe shadow graphic methods, acquiring shock wave and laser plume speeds by measuring positions of their fronts regarding the sample surface at different time instants require high reproducibility of laser irradiation and sample surface preparation conditions [29, 30]. Hitherto, there is a real necessity for accurate experimental (potentially, with sub-micrometer or micrometer spatial resolution – for laser nano- and micro-scale studies in material science) characterization of basic initial (source) characteristics for such laser-driven shock waves and for evaluation of maximal shock wave pressures. Such strong shock waves are of interest for studying the high-pressure phase diagram of carbon and its potential high-pressure structural phases underexplored yet [20-21, 31].

Currently, the maximum pressure of 1.2 TPa applied so far to carbon materials using giant PALS and LULI laser facilities [31], because of its uniaxial shock wave character is just at the lower end of the pressure range (0.8-2.7 TPa), where the repeatedly predicted, but still hypothetical metallic carbon phase may appear (see Section 4.3 for details). These previous experimental structural findings were partially supported by multiple, but rather contradictory theoretical modeling works of ultrafast diamond graphitization [32-34], pressure-driven graphite-diamond transformation [35], and graphite melting [36-37] through arbitrarily chosen fs-laser generated thermodynamic states. As a result, complete and firm understanding of potential structural transitions of carbon is still missing and requires informative and reliable experimental acquisition of the underlying key transient matter characteristics (electron-hole plasma/EHP density, lattice pressure and temperature) provided by fs-laser radiation.

2. Femtosecond Laser Photoexcitation and Electronic Dynamics on HOPG Surface

Transient optical excitation and photogeneration of electron-hole plasma (EHP) through laser energy deposition, and the subsequent electronic dynamics on HOPG surface should occur in order to achieve its extreme electronic and thermodynamic states. These effects precede lattice dynamics, while controlling its amplitude.

2.1. IR Pump Self-Reflection Measurements: Transient Surface Optics, Deposited Energy Density, Pump Pulse Filamentation in Air

The important feature of an ultrashort (femtosecond/fs and short picosecond) laser pulse is assumed its interaction with matter without screening effects of transient ablative electron-ion plasma, generated by the pulse. However, previous and recent experiments have shown strong instantaneous perturbation of optical constants of the target [4-6,38] and optical vignetting effect of air plasma in single or multiple laser filaments in the focal region in front of the target [39,40]. As a result, it is important to study transient reflectance of the target regarding to the pump pulse (self-reflectance) to assure sufficient energy deposition into it and to provide its quantitative estimates. Hence, calorimetric measurements were performed at different peak fs-laser fluences F for incident and reflected 744-nm pump pulse energies for their p- and s-polarizations at the incidence angle of 45^0 on its fresh HOPG surface spots (Figure 2.1). A 10 × 10 × 1 mm optical-quality HOPG plate (mass density $\rho_0 \approx 2.18$ g/cm^3, UPV-1TMO grade) (Figure 2.1, inset) arranged on a three-dimensional motorized micro-stage under PC control by means of a USB interface and raster-moved from laser shot to shot. Its single-shot (one laser pulse per surface spot) laser irradiation at variable laser fluences F up to 20 J/cm^2 was provided by single IR (744 nm) Ti:sapphire laser pulses in TEM_{00} mode with a 110-fs full-width at half-maximum (FWHM) duration and pulse energies $E_{inc} \leq 6$ mJ varied by means of either a reflective polarization attenuator, or a combination of a quarter-wave plate and a polarizer, at a repetition rate of 10 Hz. Since p-polarization typically provides higher surface absorbance (Figure 2.2) and is more efficient for surface excitation, self-reflection measurement for the p-polarized laser pulses covered larger fluence range F = 0.02-40 J/cm^2, than for the s-polarized pulses. The resulting ablative surface craters (Figure 2.1, inset) were studied by means of a white-light optical surface profiler NewView 700s (Zygo) to provide surface fluence calibration and to extract a fluence dependence of the crater depth X.

A common bipolar trend of the self-reflectivity curves $R_{s,p}(F)$ was observed for both these polarizations at low $F < 1$ J/cm^2 (Figure 2.2). Particularly, initial dips (relative amplitudes of 1 and 5% for p- and s-polarizations, respectively) were found at the same reflectivity-corrected (effective) fluence $F_{eff} = F[1 - R_{s,p}(F)] \leq 0.05$ J/cm^2 for a photo-generated electron-hole plasma (EHP) frequency $\omega_{pl}(N_e)$ approaching the pump laser frequency ω(744 nm)[38,41], and were accompanied by a succeeding permanent increase of reflectivity at higher, supercritical EHP densities (likewise, in the case of silicon [38,41]) in this fluence range (Figure 2.2).

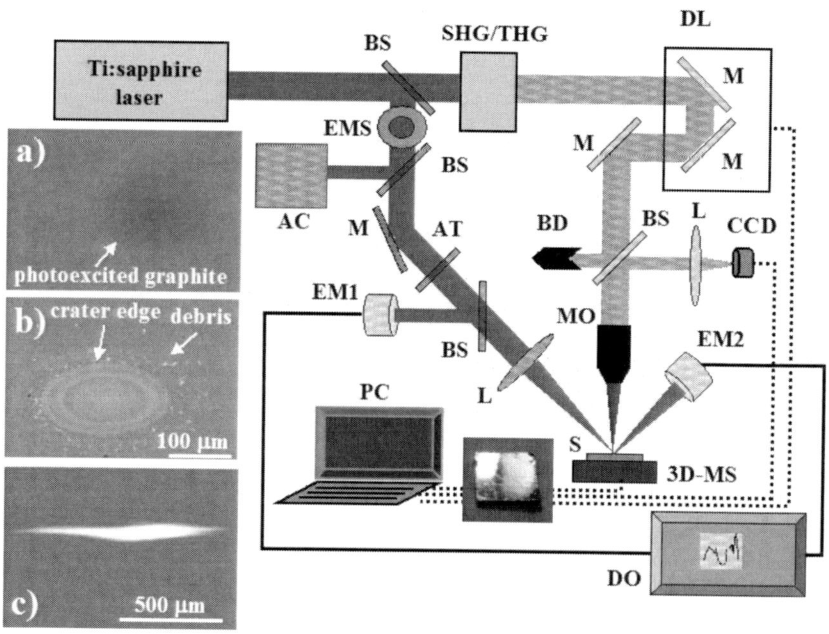

Figure 2.1. Experimental setup: BS – beam splitter, EMS – electromagnetic shutter, AC – autocorrelator, AT – attenuator (quarter-wave plate + polarizer, or polarization-sensitive reflector, or diffractive micro-optical element), EM1,2 – thermocouple energy meters, M – mirror, L – focusing silica lenses, SHG/THG – second and third harmonics generator, DL – optical probe delay line, CCD – charge-coupled device camera, BD – beam dump, DO – digital oscilloscope, S – sample with the 5 mm × 5 mm HOPG sample image nearby, 3D-MS – three-dimensional motorized micro-stage, PC – laptop for data acquisition and hardware control. Insets: the optical micrographs of the 10×10 mm HOPG piece at $\Delta t \approx +0.2$ ps (a) and the final crater image (b) at $F_0 \approx 0.8$ J/cm^2 (the image size is 200×300 µm), and of air multi-filament plasma at 2 mJ (c).

For higher values $F = 1\text{-}5$ J/cm^2 the common permanent rise of the self-reflectivity for the *s*,*p*-polarized pump pulses on a HOPG surface [42] in this work transforms into its subsequent plateau at $R_{s,p} \approx 0.35$ (Figure 2.2). For negligible EHP diffusion in HOPG between its atomic layers, the observed saturation of $R_{s,p}$ may indicate saturation of its surface electronic dynamics for several potential reasons. Namely, these are 1) transient bleaching in the photoexcited graphite because of the band filling effect [42], 2) saturation of its purely electronic EHP-driven bandgap renormalization (due to EHP screening and exchange effects [43]), 3) ultrafast bandgap collapse for the populated regions of the Brillouine zone due to the bandgap renormalization effect. More importantly, the reflectivity plateau may be alternatively related to an extrinsic effect – an onset of the target screening by absorbing and defocusing electron-ion air plasma appearing in the focal semi-volume before the target surface, where an intense electromagnetic standing wave is formed by interference of the incident and reflected laser waves [44]. To evaluate the effect of electron-ion air plasma (Figure 2.1, inset) screening on the saturation of $R_p(F)$ curve for $F = 1\text{-}5$ J/cm^2 and its final pronounced drop for $F > 20$ J/cm^2 (Figure 2.2), we experimentally verified the actual focal spot dimensions and peak laser fluences at the corresponding pulse energies E_{inc}, measuring the main radii $R_{x,y}$ of the elliptical ablative (spallation, $F_{spall} \approx 0.2$ J/cm^2, see below) damage spots on the HOPG surface arranged exactly in the geometrical (linear) focus. The resulting derivative curves $d(R_x^2)/d(\ln E_{inc}) = \sigma_x^2$ and $d(R_y^2)/d(\ln E_{inc}) = \sigma_y^2$ demonstrate the effective Gaussian parameters σ_x^2 and σ_y^2 in the pump focal region on the HOPG surface, increasing non-linearly relative to their initial, low-fluence (unperturbed) constant values (Figure 2.3), expected for the incident TEM$_{00}$ laser beam waist for at the absence of such air plasma, following the simple relationship $R_{x,y}^{\ 2} = \sigma_{x,y}^{\ 2} \ln\left(E_{inc}/E_{abl}\right)$. The resulting non-monotonous fluence calibration curve $F(E_{inc})$ for the peak fluence $F = E_{inc}/(\pi\sigma_x\sigma_y)$, according to Gaussian distribution in the convenient form $F(x,y) = \dfrac{E_{inc}}{\pi\sigma_x\sigma_y}\exp\left(-\left[\dfrac{x^2}{\sigma_x^2}\right]+\left[\dfrac{y^2}{\sigma_y^2}\right]\right)$, indicate such monotonous increase of σ_x^2 and σ_y^2 (Figure 2.3), assuming that the TEM$_{00}$ beam shape is preserved versus increasing E_{inc}. The maximal laser fluence of 20 J/cm^2 achievable on the HOPG surface is somewhat lower (by the factor of $[1 + R_p^{1/2}]^2$ [44]), than the topmost value of ≈30 J/cm^2 (fs-laser intensity ≤ 0.3 PW/cm^2) previously measured for the same laser pulses, variable focusing conditions and poor reflecting dielectric targets (polymethylmetacrylate, $R_p \approx 0.04$); the latter extreme value can be ascribed to direct air ionization (intrinsic optical breakdown) presumably by the incident laser wave. The non-linearly increasing quantities σ_x^2 and σ_y^2 in Figure 2.3 indicate air plasma defocusing and screening as major and minor effects, respectively, on the fluence reduction at the high laser pulse energies $E_{inc} > 1.5$ mJ. This fluence calibration curve in Figure 2.3 rules out air plasma defocusing as a possible reason for the saturation (plateau region) of $R_p(F)$ curve in Figure 2.2 for $F = 1\text{-}5$ J/cm^2, indicating the intrinsic character of the plateau related to the ultrafast electronic dynamics in the material. In contrast, for high incident pump energies $E_{inc} > 1.5$ mJ extrinsic air plasma defocusing and screening start predominating in the reflectivity dynamics.

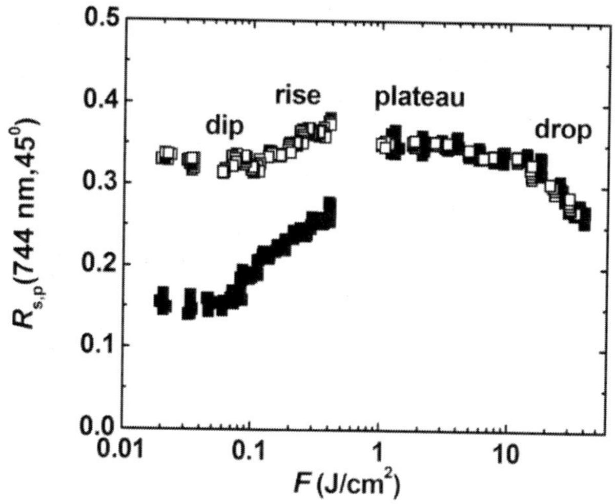

Figure 2.2. Pump self-reflectivity curves $R_{s,p}$(744 nm,45°) (s-polarization – light squares, p-polarization – dark squares) for graphite versus laser fluence F with the characteristic regions denoted as "dip", "rise", "plateau" and "drop". The horizontal arrows show the corresponding reflectivity values for unexcited graphite. The laser fluences values, exceeding 20 J/cm^2, are ill defined (see Figure 2.3 and the text).

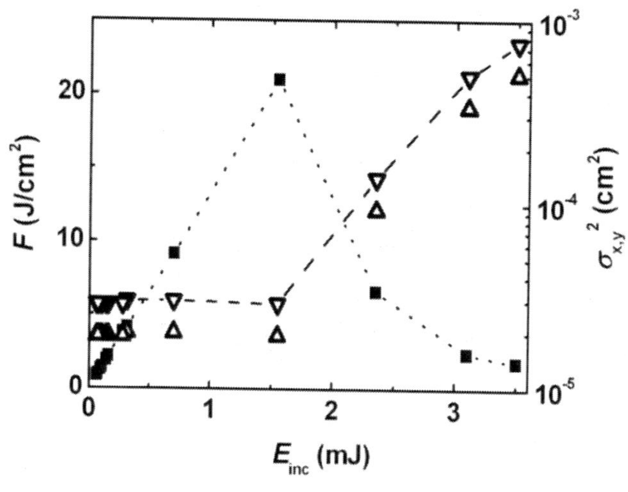

Figure 2.3. Focal Gaussian parameters σ_x^2 (up triangles) and σ_y^2 (down triangles), and the actual laser fluence F (dark squares) on the graphite surface versus pump pulse energy E_{inc}. The dashed and dotted curves are just the guides to the eye.

Altogether, $R_p(F)$ and $F(E_{inc})$ curves provide rather accurate effective fluence magnitudes $F_{eff} = F[1 - R_p(F)]$ absorbed on the HOPG surface in the broad fluence range, covering different fs-laser–matter interaction regimes considered below. It is worth also to note that the maximum intensity ≈ 0.18 PW/cm^2 ($F \approx$ 20 J/cm^2) is an order of magnitude lower, than the overestimated value of 2 PW/cm^2 previously evaluated for 800-nm fs-laser irradiation of HOPG in air [16].

2.2. Time-Resolved Pump/Probe Reflection Microscopy and Electronic Emission Measurements: Transient Optics and Electronic Dynamics

In the experiments the same 10 × 10 mm large, 1-mm thick optical-quality HOPG piece arranged on the PC-controlled three-dimensional motorized micro-stage and raster-moved from laser shot to shot (Figure 2.1), was used as a target. Its single-shot (one laser pulse per surface spot) laser irradiation was provided using a triggered electro-mechanical shutter to pick up single IR (800 nm) Ti:sapphire laser pulses in TEM_{00} mode with a 110-fs FWHM duration and maximum pulse energies of 0.2 mJ (to avoid the laser beam degradation due to its self-focusing in air and air plasma scattering/refraction), attenuated by a combination of either a quarter-wave plate, or a Glan prism/polarizer set.

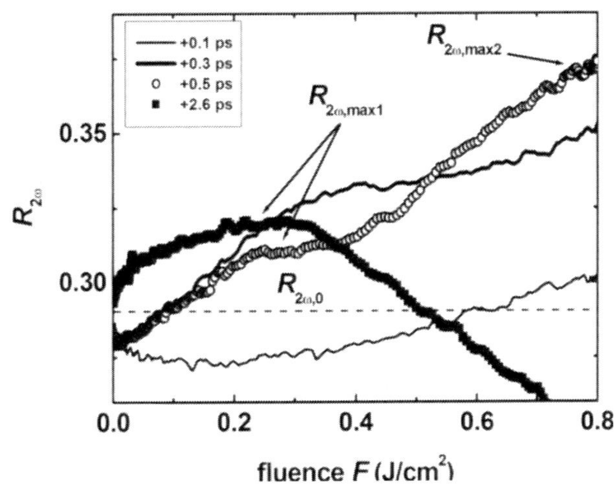

Figure 2.4. Probe reflectivity $R_{2\omega}$from HOPG surface vs. F at different delay times Δt. with the different characteristic values $R_{2\omega,0} \approx 0.29$, $R_{2\omega,max1} \approx 0.32$ and $R_{2\omega,max2} \approx 0.38$ (see the text for details).

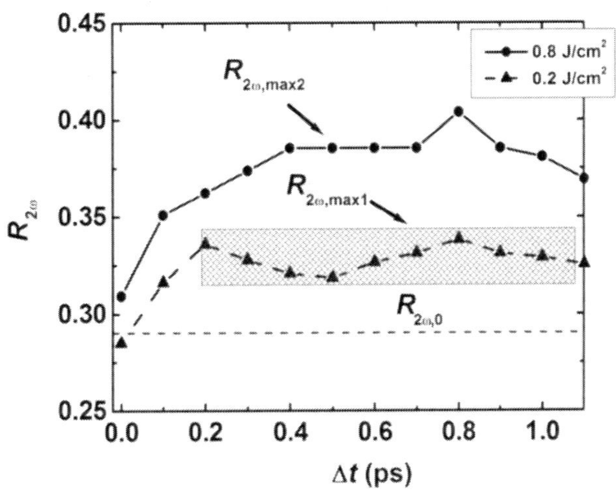

Figure 2.5. Transients $R_{2\omega}(\Delta t)$ at $F \approx 0.2$ and 0.8 J/cm^2.

The fs-laser pulses were focused in p-polarization at the angle of 45^0 by a silica lens (focal distance f = 500 mm) into a small focal spot (the Gaussian main ellipse $1/e$-radii $-\varnothing_{1/e,x} \approx 0.12$ mm and $\varnothing_{1/e,y} \approx 0.06$ mm) on the surface of the HOPG sample, arranged in front of a smaller aperture of a probe-line microscope objective (NA = 0.17) (Figure 2.1). In this time-resolved optical reflection microscopy arrangement [41,45], strong pump pulses at 800 nm and temporally delayed (delay times Δt = 0-3 ns, resolution ≈100 fs equals to the probe FWHM pulse width), much weaker second harmonic (2ω) probe pulses at 400 nm, provided through second harmonic generation in a 1.5-mm thick BBO crystal in a harmonics generator, were employed.

The incident fluence calibration, providing the peak surface fluence as a function of the incident energy and the abovementioned Gaussian beam parameters, was performed by means of an optical microscope to measure main radii of the resulting ablative surface craters (Figure 2.1, inset b). Time-resolved optical reflection microscopy images of the single-shot photo-excited HOPG surfaces demonstrate for short delay times Δt = +0.1-1 ps (Δt = 0 ps corresponds to the full overlap of the pump and probe pulses) very common reflectivity features – instantaneous monotonous small dip of the probe reflectivity $R_{2\omega}$ (–4%) and its more pronounced following increase above the initial level $R_{2\omega,0} \approx 0.29$ (Figure 2.1, inset a and Figure 2.4). Surprisingly, upon these common transient trends, two steady-state characteristic reflectivity features – a peak $R_{2\omega,max2} \approx 0.38$ for Δt> +0.4 ps and $F \approx 0.8$ J/cm^2 on a plateau $R_{2\omega,max1} \approx 0.32$ appearing for Δt> +0.3 ps and F< 0.35 J/cm^2 – were set up indicating some interesting matter states, where the $R_{2\omega}$rise from the plateau occurred for F> 0.3 J/cm^2 (Figure 2.4). These reflectivity features were also revealed in the time domain for corresponding laser fluences (Figure 2.5).

The resulting transient reflectivity modulations in Figures 2.4 and 2.5 can be related to instantaneous photo-generation of electron-hole plasma (EHP) and the simultaneous EHP-driven bandgap renormalization (BGR, δE_g) in the material, contributing through intraband and interband components of the dielectric function, respectively. Moreover, the transient reflectivity change in Figure 2.5 for the maximum $F \approx 0.8$ J/cm^2 mimics the normal incidence spectral dependence $R(0^o)$ of HOPG for photons $\hbar\omega \geq 3.1$ eV (Figure 2.6), accounting for the red spectral shift by the electronic BGR magnitude $\delta E_{g,el}$. The latter value should increase versus EHP density ρ_{eh} with a linear slope, which magnitude was measured as 2.5×10^{-22} eV×cm^3 [46], being very close to its estimate $\approx 3\times10^{-22}$ eV×cm^3 (according to previous works [43,47-48], photoexcitation of 5% valence electrons, i.e., 2×10^{22} cm^{-3} in HOPG, should provide 50% closure of the minimal direct Γ-point π-band gap ≈12 eV [49]).

More detailed simulations (for basic details, see, e.g., [50]) indicate that the instantaneous electronic BGR effect using the abovementioned slope value provides reasonably good description of the transient $R_{2\omega}$ modulation right after the pump pulse for Δt = +0.1 ps (Figure 2.7), comparing to a common Drude approach [41], combined within the expression for the complex dielectric function $\varepsilon(\Omega^*, \rho_{eh})$ for the BGR-increased probe photon frequency $\Omega^* = \Omega + \delta E_{g,el}/\hbar$ [41,51]

$$\varepsilon^*\left(\Omega^*, \rho_{eh}\right) = \varepsilon_{IB}(\Omega^*) - \frac{\Omega_{pl}^2(\rho_{eh})}{\Omega^2 + 1/\tau_e(\rho_{eh})^2}\left(1 - \frac{i}{\Omega\tau_e(\rho_{eh})}\right), \tag{2.1}$$

where the transient EHP frequency Ω_{pl} is defined as

$$\Omega^2_{pl}(\rho_{eh}) = \frac{\rho_{eh}e^2}{\varepsilon_0\varepsilon_{hf}m^*_{opt}} \tag{2.2}$$

for e and ε_0 being the electron change and vacuum permittivity, respectively. Particularly, because of the large initial interband contribution $\varepsilon_{IB}(\Omega^*)$ to the dielectric function at the 3.1-eV probe energy, the BGR and Drude effects themselves result in this figure in the only positive reflectivity changes (for the latter effect – disregarding the optical mass value), in the obvious contradiction with the distinct $R_{2\omega}$ dip at intermediate fluences. However, an introduction of band-filling (absorption saturation) factor [42], decreasing the interband contribution according to a factor $[1-\rho_{eh}/1.4\times10^{22}\ \text{cm}^{-3}]$, provides the necessary dip on the model curves. Here, the calibration value of $1.4\times10^{22}\ \text{cm}^{-3}$ denotes the EHP density, filling the π-electron bands enough to bleach the π-π interband transitions at 3.1 eV.

The monotonic decrease of the factor vs. ρ_{eh} accounts for gradual population of the high-lying UV-probed π-electronic states through, e.g., Auger-recombination. Furthermore, the Drude contribution (the high-frequency dielectric constant $\varepsilon_{hf}\approx 1$, electron-electron scattering time in the 2D electron gas $\tau_{ee}\approx 1\text{fs}\times\{10^{22}\ \text{cm}^{-3}/\rho_{eh}\}^{1/2}$ [52]) together with the band-filling (saturation) factor altogether provide positive $R_{2\omega}$ changes for the optical (e-h pair) masses $m_{opt} = \frac{m^*_e m^*_h}{m^*_e + m^*_h} < 0.08m_0$, where m_0 is the free electron mass, while the high-lying UV-probed π-electronic states correspond to $m_{opt}\geq 0.25m_0$, taking into account the e-DOS effects, as considered in [50]. As a result, it is the combination of electronic BGR, Drude and band-filling contributions that provides the most accurate fitting of the experimental $R_{2\omega}(F)$ dependence $\Delta t\approx +0.1$ ps in Figure 2.7.

Complementary electron emission studies also directed to monitoring of transient electronic dynamics, were performed in ambient air at the same fixed focusing conditions at the normal pump pulse incidence on the HOPG sample, which was raster scanned to excite its fresh spots at different values E [51]. The pump pulse focusing occurred through a 4-mm aperture in a collecting electrode mounted 1 mm in front of the grounded HOPG sample on the three-dimensional motorized micro-stage, provided this aperture of the electrode much larger than the pre-focused laser beam size.

Positive extraction potentials of 150-300 V (in the latter case, the dynamic range of the emission signal was extended by two more orders of magnitude) were applied form a high voltage source to the collecting electrode in order to extract electrons, which were almost immediately trapped on a nanosecond time scale by oxygen molecules [53]. The resulting photo-induced negative or positive currents through the sample-collector spacing were acquired in the form of image voltages U_{norm} as a function of incident laser fluence both in single- and multi-shot (N = 30-50) regimes (Figure 2.8), using a 1-MΩ low-frequency input of a digital storage oscilloscope Tektronix TDS 2014. The resulting dependences $U_{norm}(F)$ demonstrate a quasi-linear increase of the emission signal for $F< 0.2\ \text{J/cm}^2$ and non-linear $-\propto F^2$ for single-shot and $\propto F^3$ for multi-shot exposures, respectively – scaling relationships for higher fluences (Figure 2.8).

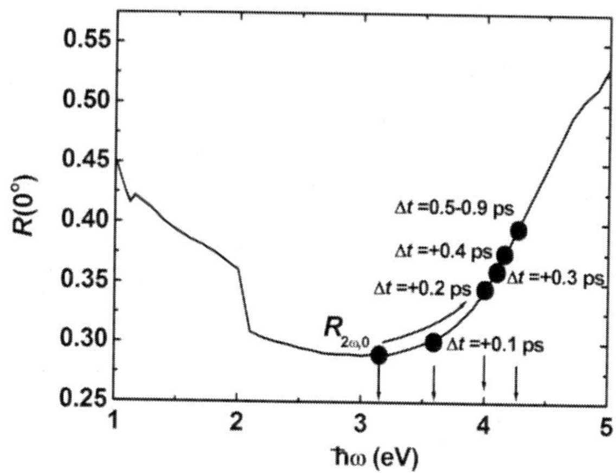

Figure 2.6. Reflectivity spectrum $R(\hbar\omega)$ of HOPG with the transient reflectivity values $R_{2\omega}(\Delta t)$ at $F\approx$ 0.8 J/cm^2 and corresponding bandgap red shifts shown by the large dots and vertical arrows, respectively.

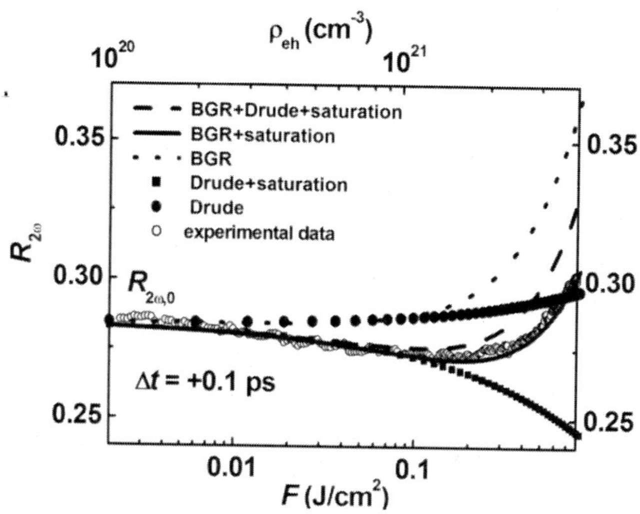

Figure 2.7. Experimental curve $R_{2\omega}(F)$ at $\Delta t\approx$ +0.1 ps (light circles, left and bottom axes) and its fitting model curves $R_{2\omega}(\rho_{eh})$ (lines and dark symbols, right and top axes) with BGR, Drude and band-filling contributions combined in different ways.

It is worth to mention specially that single-slope fitting of the entire curves provides very poor agreement at the low-fluence end. Then, such drastic break of the dependences is believed to result from the strong EHP-driven bandgap renormalization in HOPG ($\delta E_{g,el}\approx$ 0.5 eV), which is even stronger for higher-lying electronic states, including the vacuum level [51]. As a result, single-photon photoemission of high-energy non- and thermalized electrons from the conduction band or thermionic emission of electrons excited by single IR-photon to the conduction band transforms near the transition fluence $\approx$ 0.2 J/cm^2 into similar schemes of two-photon emission of electrons directly from the Fermi level due to the BGR-reduced work function, or thermionic emission of electrons excited by two IR-photon to the conduction band over the band-filled states near the Fermi level.

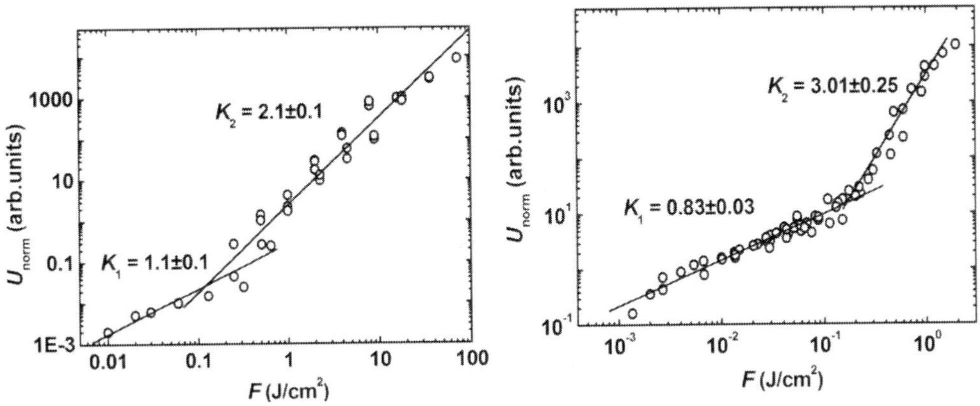

Figure 2.8. Single- (left) and multi-shot (right) image voltage U_{norm} vs. F for 744-nm photo-induced emission from HOPG at extraction voltages of +300 V (left) and +150 V (right) with the power slopes $K_{1,2}$.

Meanwhile, the rather pronounced difference in the K_2-magnitudes for single- and multi-shot irradiation regimes may result in the latter case from nanoroughness-enhanced direct 3-photon electron emission from the nanostructured HOPG (in other words, giant 3-photon emission).

3. Femtosecond Laser Driven Sub-Picosecond Structural Transformations in HOPG and Diamond

This research part was focused on experimental search for transient and permanent structurally modified states of carbon in HOPG, and permanent structurally modified states of carbon in diamond, succeeding intense fs-laser excitation of these materials.

3.1. Time-Resolved Pump/Probe Reflection Microscopy: Transient Matter States in HOPG. Micro-Raman Characterization of Laser-Driven Permanent Structural States in HOPG

Previously, in Figure 2.4 two steady-state characteristic reflectivity features – a peak $R_{2\omega,max2} \approx 0.38$ at $F \approx 0.8$ J/cm^2 for $\Delta t > +0.4$ ps on a plateau $R_{2\omega,max1} \approx 0.32$ for $\Delta t > +0.3$ ps and $F < 0.2$ J/cm^2 – were mentioned. This more detailed study of these features was performed over a broader time range (Figure 3.1). It indicates that the transient plateau $R_{2\omega,max1} \approx 0.32$ starting for $F > 0.2$ J/cm^2 at $\Delta t \geq +0.3$ ps extends until $\Delta t \sim 100$ ps, slowly decreasing in the reflectivity amplitude on the picosecond timescale. Although the sample reflectivity $R_{2\omega}$should vary very weakly in the range 0.29-0.32 (Figure 2.6) despite dramatic changes in EHP density and HOPG surface temperature over the time interval, the long-term reflectivity plateau is not a trivial spectrally dependent feature. Below, in section 4.1 the molten carbon state underlying the transient appearance of this plateau will be demonstrated.

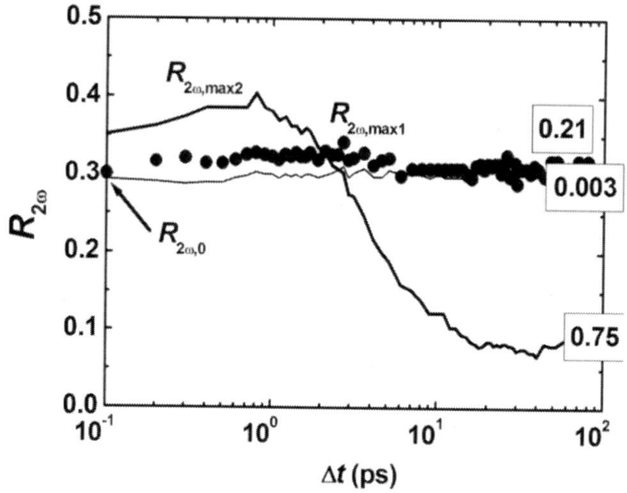

Figure 3.1. Transients $R_{2\omega}(\Delta t)$ at $F\approx$ 0.003 (reference curve), 0.21 and 0.75 J/cm^2 with the transient reflectivity maxima $R_{2\omega,max2}$ and $R_{2\omega,max1}$ shown by the corresponding marks.

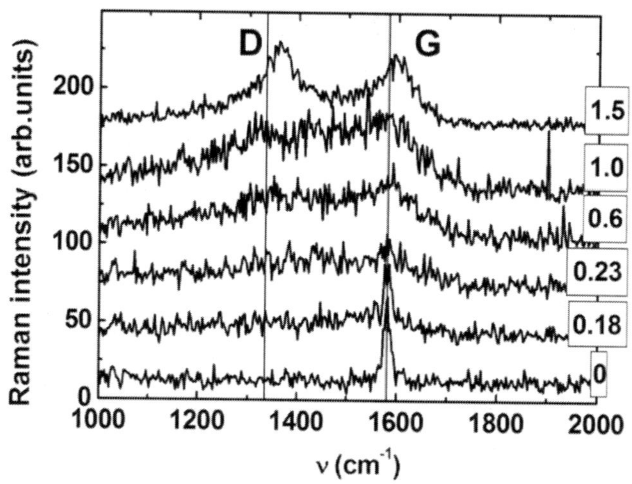

Figure 3.2. Micro-Raman spectra of single-shot excited HOPG spots at different local fluences [in J/cm^2] shown in the corresponding frames. The vertical lines and D/G-marks indicate the positions of crystalline diamond and graphite lines at 488 nm, respectively.

As shown by its post-pulse Raman micro-characterization (Figure 3.2) using a Raman spectrometer U-1000 (Jobin Yvon, 488-nm pump wavelength), during re-solidification, such molten carbon turns back to graphite almost without any changes in its surface nanotopology or electron-emission properties (Fig3.3). Comparing to the reference HOPG spectrum, the re-solidified spot ($F\approx$ 0.18 J/cm^2) preserves the position and half-width of its Raman line almost unchanged. Such ultrafast disordering/melting of HOPG during $\Delta t<$ +0.3 ps (Figures 2.5 and 3.1), which occurs for the saturated thermal motion starting at even lower fluences [14], may result from strong excitation of a "hot" low-frequency acoustic quasicontinuum through an harmonic phonon-phonon decay of highly-populated coherent in-plane Γ-E_{2g} and K-A_1' optical phonon modes [54].

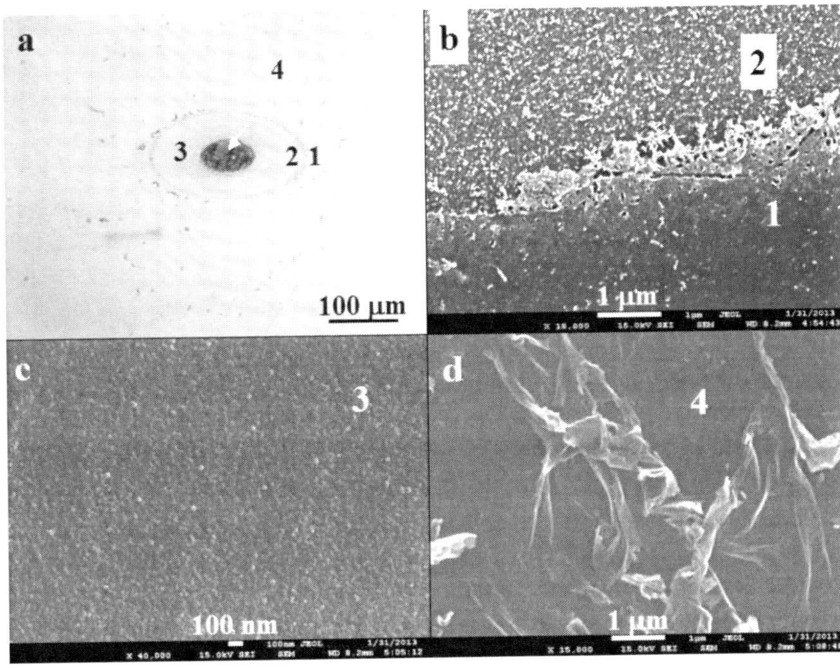

Figure 3.3. (a) Optical image of single-shot HOPG crater and SEM images of its local spots ## 1-4 at different local fluences: b – spots ## 1,2 (outside crater – 0.18 J/cm^2, inside crater – 0.23 J/cm^2), c – spot # 3 (grey internal region – 0.6 J/cm^2), and d – spot # 4 (dark central spot – ≤1.2 J/cm^2).

The permanent disordering is demonstrated by the post-pulse Raman spectra, directly showing the onset of the final G-line broadening for $F>$ 0.18 J/cm^2 through appearance of a broad and weak amorphous carbon band (Figure 3.2), representing an atomic- or nanoscale mixture of carbon atoms in various hybridization states (Figure 3.3a,b). Also, intense displacive excitation of some BGR-driving coherent optical phonon mode is illustrated by the almost continuous, purely phonon-mediated reflectivity rise right after the pump pulse in Figures 2.4 and 2.6, indicating the ongoing lattice-driven BGR, while such nearly instantaneous melting appears to be hardly feasible. Indeed, decay rates for Γ-E_{2g} and K-A_1' mode should rapidly increase during the process upon population of the "hot" low-frequency acoustic modes or even direct coupling of the electronically-softened optical modes to thermally populated low-frequency acoustic ones, which makes it a stimulated process and lowers by several times the initial E_{2g}-phonon lifetime ≈ 0.7 ps in the weakly excited HOPG [54]. The strong coupling of the optical phonon modes to the K-point electronic states, which is allowed by the quasi-momentum conservation [55], results from the structural quasi-degeneracy of the latter in the K-point owing to the initial graphite symmetry (as compared to occasional degeneracy), which will be lifted up by appearing bandgap upon strong displacive excitation of the E_{2g}-mode (the totally symmetric A_1'-mode renormalizes this gap directly without lifting up the K-point electronic quasi-degeneracy), as a pre-requisite to a symmetry-allowed structural transition (or a few of them), which changes this symmetry, most probably, in some final – cubic or hexagonal – diamond structure [8-10,12-14,16-19,24-25] without such K-valleys, i.e., with the large K-point bandgap in its e-DOS [56]. Both these modes yield in un-pairing of the in-plane carbon sub-lattices [54,55], which then can be separated along the c-axis by out-of-plane phonons to provide a structural state of close-packed carbon layers as a precursor for face-centered cubic diamond structure. Moreover, transient sub-picosecond contraction in the surface layer [14], which was assigned to surface charging, but may also unavoidably yield from EHP-lattice interaction through acoustic deformation-potential [57], should also favor transient stacking of graphitic planes. Meanwhile, such stacking with formation of sp^3-bonds becomes pronounced only on a picosecond timescale (~10 ps) [14], indicating the rather slow and, potentially, laser shockwave driven structural transformation. Indeed, at even higher fs-laser fluences $F > 1$ J/cm^2 within the ablative crater there appears a central elevated dark region with resided graphitic sheets (Figure 3.3a,d), previously observed in [11]. This region demonstrates more pronounced G- and D-bands (Figure 3.2); however, the obvious lift-off material transport during its appearance indicates the perturbation of the material by post-pulse fs-laser driven shock waves discussed below in Section 4.3. Surprisingly, the other, more pronounced reflectivity maximum $R_{2\omega,max2} \approx 0.38$ appears in Figure 3.1 for $\Delta t > +0.4$ ps and $F > 0.3$ J/cm^2, exerting a subsequent drastic drop on a picosecond timescale due to the following thermal expansion in the form of a micro-heterogeneous vapor-droplet mixture. Such additional sub-picosecond reflectivity rise may indicate a next vibronic structural transformation step, which brings the excited matter closer to the diamond structure, as shown by the more pronounced, broad amorphous carbon band in the Raman spectra in Figure 3.2, though in the nanocrystalline state (Figure 3.3c).

3.2. Permanent Structural States of Buried Laser-Written Tracks Inside CVD Diamond and Their Micro-Raman Characterization

Femtosecond lasers provide unique opportunities for precise bulk nano- and micro-structuring of transparent materials for optoelectronic and photonics applications, enabling fabrication of micro-devices integrating for different purposes, e.g., optical waveguides, Bragg gratings, birefringent regions, memory bits and micro-fluidic channels [58-62]. Owing to high peak powers of their pulses at low pulse energies, upon strong focusing at pre-determined focal depths such lasers produce in transparent materials sub- or micro-scale sub- or critical laser plasmas via their multi-photon, impact or tunnel ionization, with plasma energy densities sufficient for bulk structural transformation of the solids or their buried ablation [63-65] without macroscopic (micrometer-sized) thermal or shock-affected zones [66], or any other damage along the laser beam path excepting the focal volume. Recently these unprecedented fs laser micro-machining opportunities have been applied in fabrication of buried linear optical micro-structures (graphitic wires) in natural and chemical vapor deposition (CVD) grown diamonds [67-69], representing due to their high refractive index {n_{diam}(589 nm) ≈ 2.42 [70,71]} a perspective class of optical materials for optoelectronic and photonics applications such as, e.g., bulk buried THz photonic crystals and similar microstructures. Such high-contrast lines of various lengths were written at different sub-μJ pulse energies (*sub-critical* peak powers P) and longitudinal scanning rates for the natural and CVD diamond samples, arranged on a three-dimensional translation stage [67,68].

Since such lines were formed by fs-laser pulses with low pulse energies and *sub-criticalP* magnitudes (regarding the critical power for self-focusing in diamond, $P_{crit,diam} \approx$ 0.4-1 MW [69,72]) through *non-filamentary* cumulative damage and growth of the lines due to propagation of a graphitization front, rather low (sub-μm/pulse) writing rates were achieved [68]. In contrast, another – transversal – fs laser writing geometry have been developed sometime ago demonstrating in this geometry precise, rapid and efficient micro-structuring of silica glasses [58-62,73] by reducing – at much shallower writing (focusing/self-focusing) depths – influence of relevant spherical aberrations and filamentation effects. Previously, this geometry was employed in study of Ashcom et al. [69], enabling fast raster scanning of diamond samples and writing of linear tracks of separate *non-filamentary*, micron-sized point-like damage features at different depths within the samples also at low (*sub-critical*) fs laser *P* magnitudes, in order to determine their optical damage (writing) thresholds.

Microfabrication was performed in the transverse geometry by 744-nm, 100-fs Ti: Sa laser pulses with energies in the range 0.5-13 μJ at the repetition rate of 10 Hz. The diamond sample arranged on a motorized 3D-microstage was scanned at the velocity of 2 μm/s. The fs-pulses were focused by a microscope objective with the numerical aperture NA = 0.65, providing the waist radius (without filamentation) $w_{calc} = \dfrac{\lambda\sqrt{n_{diam}^2 - NA^2}}{\pi NA} \approx 1$ μm and the waist length $l_{R,calc} = \dfrac{\lambda\left(n_{diam}^2 - NA^2\right)}{\pi NA^2} \approx 3.6$ μm for the 744-nm refractive index of diamond $n_{diam} \approx$ 2.5. The buried microtracks were written at the depths of 300 and 500 μm. Monitoring and inspection of the microfabrication process were carried out at magnification of x6, x15 and x60, using an optical microscope Levenhook Bio View 630 equipped with a PC-connected CCD-camera DCM 310. Samples of CVD diamond Monodie-12 (Element Six Ltd.) were used as deposited without preliminary pre-polishing of their surfaces.

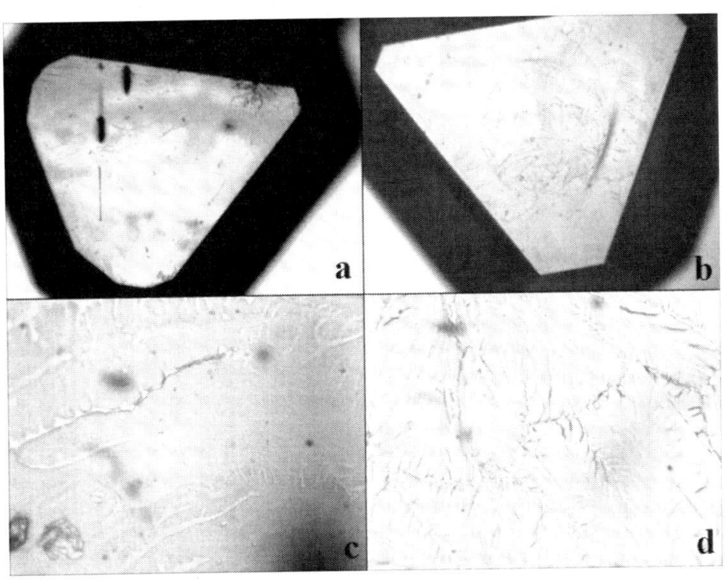

Figure 3.4. General (×6, frame size – 2.3×2.9 mm) and closer (×60, frame size – 0.23×0.29 mm) views of the moderate- (a,c) and high-roughness (b,d) CVD diamond surfaces.

However, before microfabrication the samples were divided into groups with moderate and high surface roughness (Figure 3.4), using the optical microscope. This is an important step, since due to the high refractive index of diamond (n ~ 2.5), even negligible surface micro-roughness may significantly distort a wave front of a strongly focused fs-pulse and result in its short-scale degradation through self-focusing.

Then, the pulse waist increases and the writing threshold energy increases, accordingly. Two characteristic samples with moderate- and high-roughness surfaces are shown in Figure 3.4, while the former (left) sample was chosen for microfabrication. Surface microfabrication was performed for the moderate-roughness sample in order to study its basic features, including the modification threshold energy and induced stresses, as well as the spatial focusing limits in air. The surface track written at the variable pulse energy of 0.5-8 μJ, decreasing from the bottom to the top, with the simultaneous decrease in the track diameter from 10 to 2 μm (Figure 3.5a,b), while the single-shot modified spots producing the track are clearly visible on the image b. The minimal, 0.5-μJ track width demonstrates the spatial focusing into 2-3 μm and the single-shot surface modification threshold of ≈5 J/cm^2, which is in agreement with some previous single-shot fs-laser damage threshold measurements ranging from 4 to 45 J/cm^2 [67,74,75]. Raman characterization of the surface track with a 5-μm broad 488-nm pump beam indicates that, because of enhanced compressive stresses, the Raman D-band at $\nu_D \approx$ 1332 cm^{-1} (at the absent G-band) near the surface track shifts up in its peak position (Figure 3.6b), as well as the band half width (FWHM) (Figure 3.5c,5c), but significantly decreases in amplitude (Figure 3.6a). The near-edge compressive (–0.03 GPa) and internal tensile (+0.15 GPa) stresses σ related to the surface track, resulting in the positive ($\Delta\nu_D \approx$+0.06 cm^{-1}) and negative ($\Delta\nu_D \approx$–0.3 cm^{-1}) shifts, respectively, were evaluated as follows [67]

$$\Delta\nu[\mathrm{cm}^{-1}] = -2\sigma[\mathrm{GPa}] \quad (3.1)$$

Likewise, inside the track the Raman amplitude decreases even more (Figure 3.6a), while the band position and half width decrease owing to the stress relaxation via structural modification (Figure 3.6b,c).

During fs-laser microfabrication in bulk diamond the laser focal point was buried 300 and 500 μm in depth, although even 1-mm depth is not a principal limitation for such fs-laser writing procedure. Comparing to the diamond surface, upon such tight (NA = 0.65) focusing bulk tracks were absent for pulse energies up to a few tens μJ, possibly, due to laser wave front degrading on the rough surface and succeeding filamentation inside the sample. In contrast, such microfabrication proceeds well if was started on the already existing bulk track or defect (Figures 3.7 and 3.8).

In such a manner, rather thick bulk track started at 15 μJ can be extended during the focal point motion within the sample at the 300-μm depth and even upon dramatic pulse energy decrease down to 1-2 μJ or less and simultaneous decrease in its diameter to 2-3 μm (Figures 3.7, 3.8 and 3.9a). Likewise, much thinner tracks – even well below the spatial resolution of the optical microscope within the diamond (≤ 1.5 μm) – were written at this depth at lower pulse energies till 0.5 μJ (Figure 3.8c). Raman characterization of the thick and thin buried tracks in Figures 3.7 and 3.8a,b demonstrates again the pronounced decrease of the D-band amplitude (though, more slower for the thick track) (Figure 3.6a), while the D-band position shifts up at the track edge (shift $\Delta\nu_D \approx +0.4$ cm^{-1}, i.e., compressive stress $\sigma \approx -0.2$ GPa) and then slightly relaxing inside the thin track, similarly to the surface one, while remains almost the same inside the thick bulk track (Figure 3.6b).

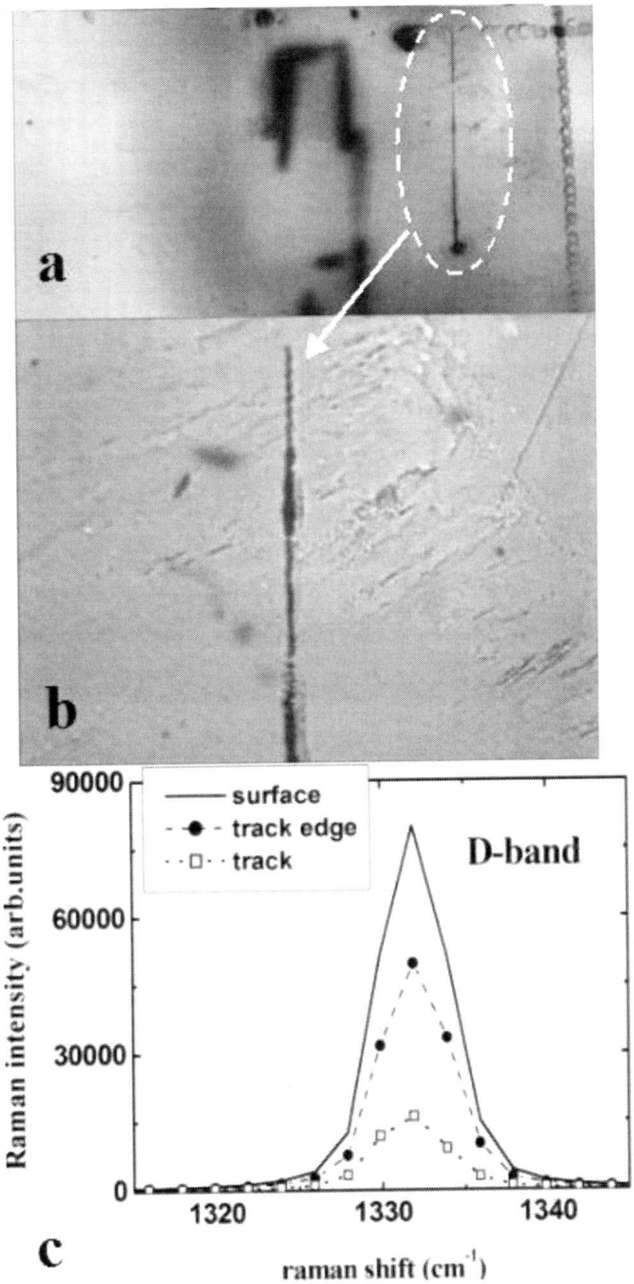

Figure 3.5. Surface track (a – general, b – closer view) written in the energy range of 0.5-8 μJ and highlighted by the dashed ellipse, and D-band Raman spectra related to it (c). The frame sizes are 0.9×1.2 mm (a) and 0.23×0.29 mm (b), respectively.

These trends indicate more relaxed states of the thin surface and bulk tracks, comparing to the reference diamond and, especially, to their edges, while, in contrast, no relaxation occurs upon the modification in the thick track.

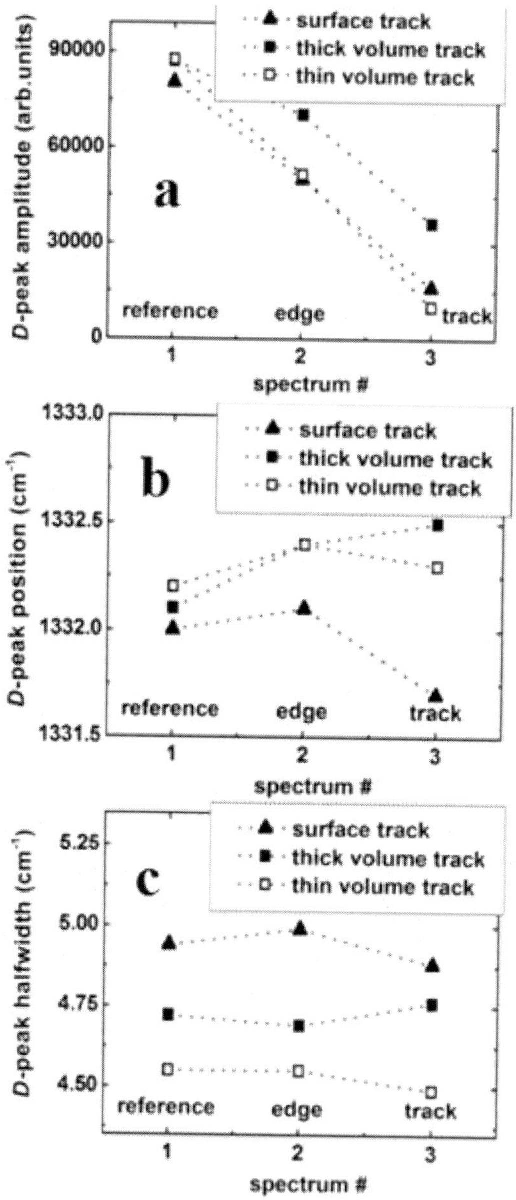

Figure 3.6. Summary of D-band Raman amplitudes (a), peak positions (b) and half widths (FWHM) (c) for the different positions regarding the surface track, bulk thick and thin tracks.

Finally, at larger, 500-μm depth inside the sample the track width can be also strongly varied versus fs-laser pulse energy (Figure 3.9b,c), e.g., at the 13-μJ pulse energy such width increase to ~50 μm, but is reduced to somewhat larger (~5 μm) magnitudes, as expected for deeper focusing, following the pulse energy decrease till 0.8 μJ (Figure 3.9c). In this case, at the maximum writing depths the real-time visual process control enables to avoid the track overexposure owing to sufficient (smaller) exposure times, thus, yielding in thinner surface and bulk tracks comparing to our previous experiments (see Figure 3.10) [76].

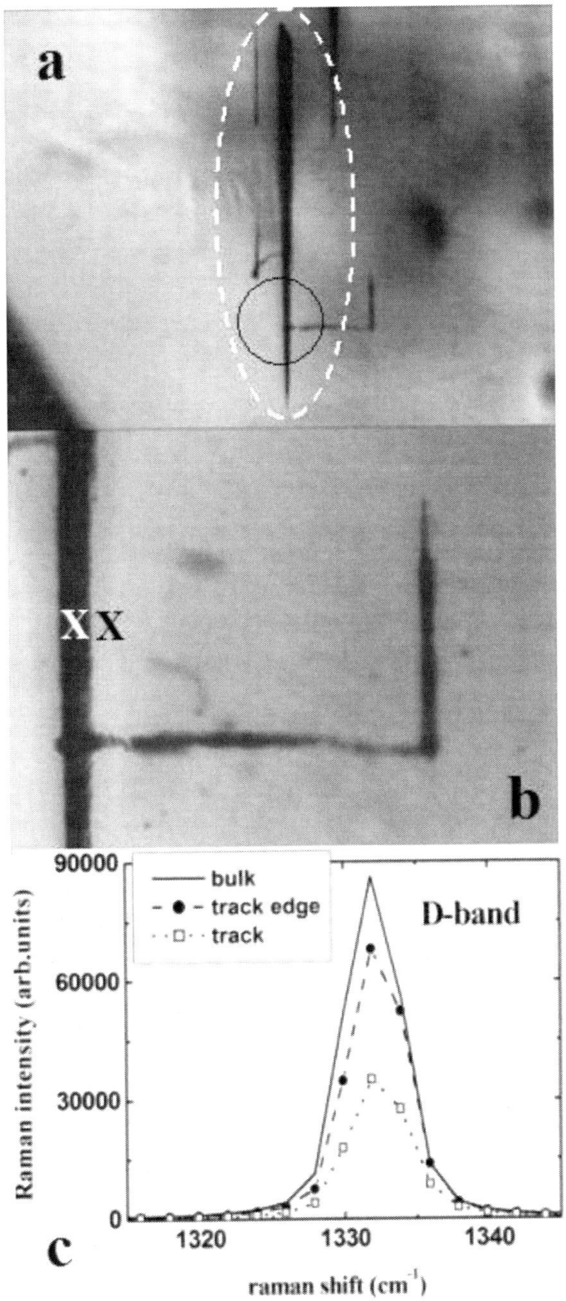

Figure 3.7. Bulk track (a – general, b – closer view) written at the 300-μm depth in the energy range of 0.5-8 μJ and highlighted by the dashed ellipse, and D-band Raman spectra related to it (c). The frame sizes are 0.9×1.2 mm (a) and 0.23×0.29 mm (b), respectively.

In comparison with other similar studies [67,74-80], the fabricated minimal surface and bulk tracks were written at similar sub-μJ pulse energies, providing track widths ~1-2 μm comparable with the best results achieved so far. However, in contrast, significant graphitization of the irradiated diamond was prevented in the high optical contrast tracks preserving its initial diamond structure (Figure 3.11).

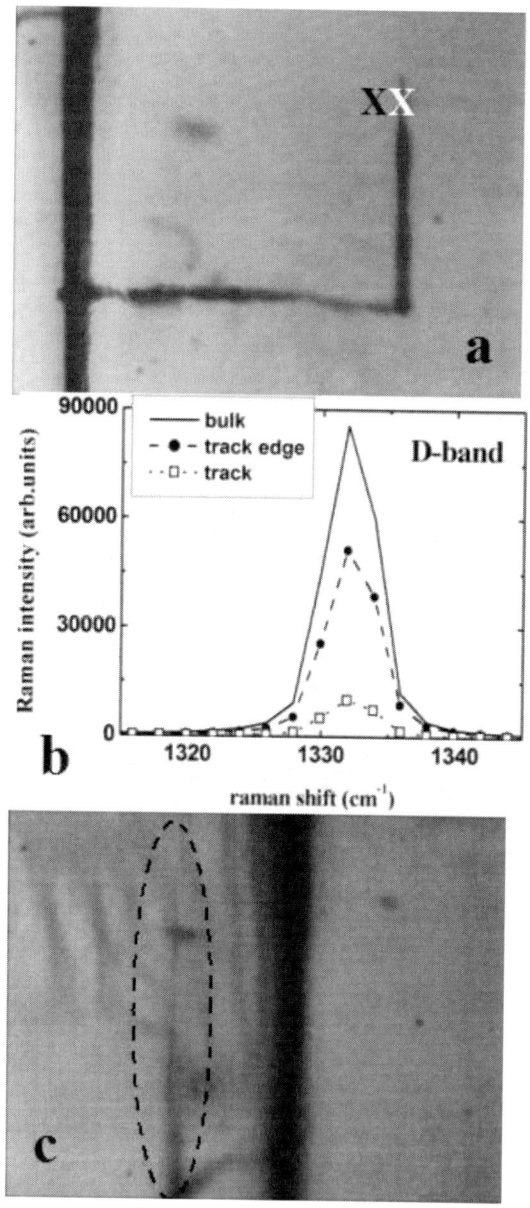

Figure 3.8. a – thin bulk track in Figure 3.6 written at the 300-μm depth in the energy range of 1-2 μJ and the different positions for Raman probing marked by the crosses; b – D-band Raman spectra related to it; c – another, ultrathin bulk track written at the 300-μm depth in the energy range of 0.5-2 μJ and highlighted by the dashed ellipse. The frame sizes are 0.23×0.3 mm.

4. Femtosecond Laser Melting, Ablation and Plasma Formation on HOPG Surface

Thermal melting, ablation and plasma formation on HOPG surface occur through a sequence of ultrafast solid-liquid and liquid-vapor transitions dynamically envisioned by means of time-resolved optical microscopy in its different versions, but result also in simultaneous generation of ultra-strong shock waves both in ambient air and the ablated graphite transforming the latter in diamond under appropriate irradiation conditions.

4.1. Melting

In these material modification experiments the same experimental arrangement and experimental parameters, as well as the same HOPG sample (Figure 2.1), previously described in Section 2.2, were used.

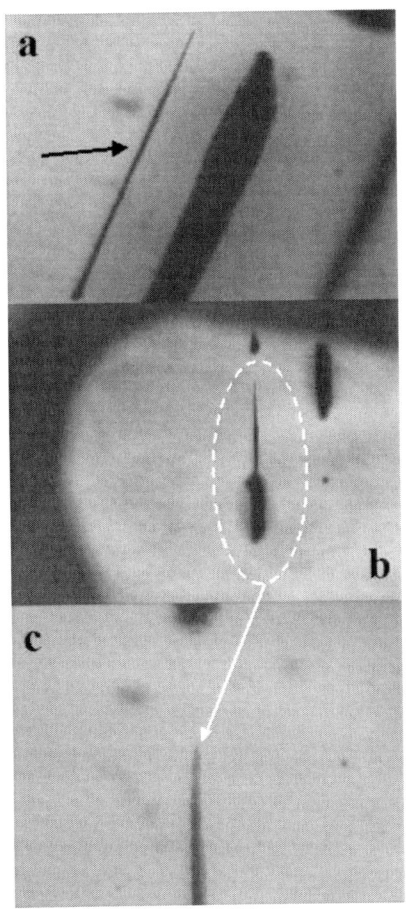

Figure 3.9. a – thin bulk track written at the 300-μm depth in the energy range of 1-4 μJ and shown by the arrow; another bulk track (b – general view, frame size – 0.9×1.2 mm; c – closer view, frame size – 0.23×0.3 mm) written at the 500-μm depth in the energy range of 0.8-13 μJ and highlighted by the dashed ellipse.

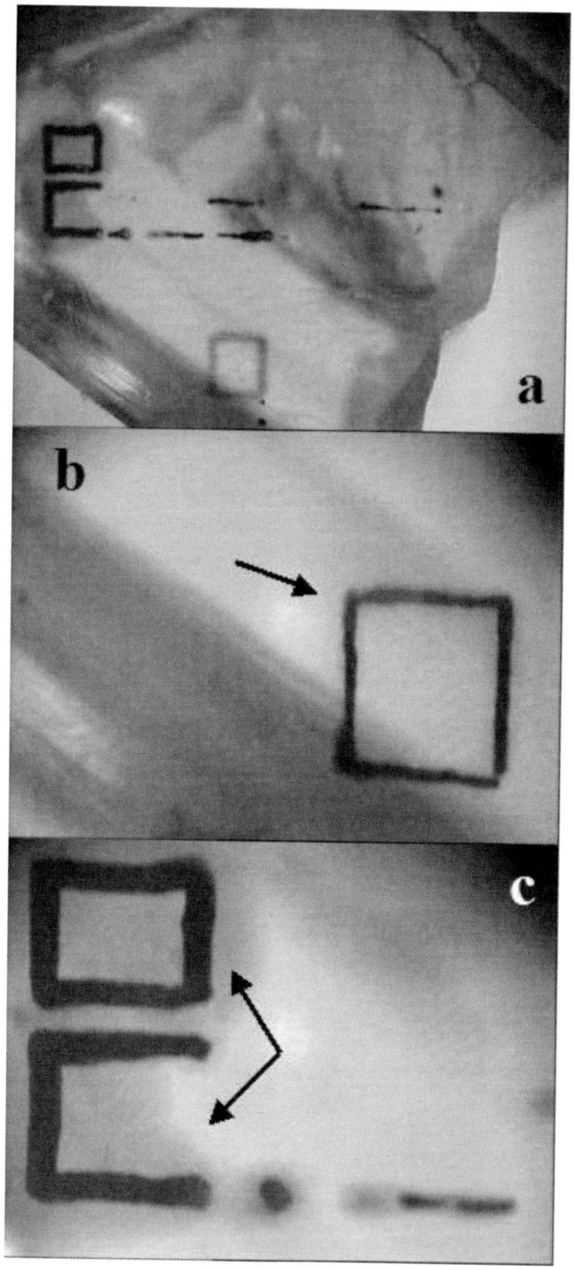

Figure 3.10. Different bulk tracks and symbols written at the depths of 0.36-1 mm inside a natural diamond piece at different energies in the range of 7.5-30 μJ. The space mark is the buried "□" symbol with the 240-μm long side.

The incident fluence calibration, providing the peak surface fluence as a function of the incident energy and the abovementioned Gaussian beam parameters, was performed by means of an optical microscope to measure main radii of the resulting ablative surface craters (Figure 4.1), while their maximal depths X at variable peak laser fluences F up to 0.8 J/cm^2 were measured using a white-light optical surface profiler New View 700s (Zygo).

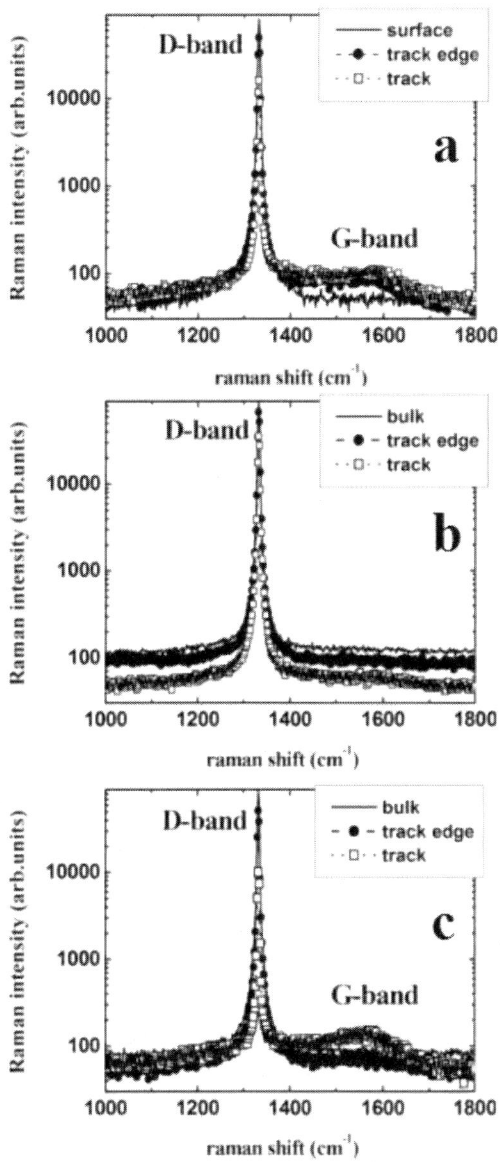

Figure 3.11. Full Raman spectra of the surface (a) and bulk – thick (b) and thin (c) – tracks, showing the dominating D-band and the very minute (see the logarithmic ordinate scale) emerging G-band.

Time-resolved optical reflection microscopy images of the single-shot photo-excited HOPG surfaces in Figure 4.1 demonstrate very common reflectivity features – monotonous short-term (multi-picosecond) drop of the probe reflectivity $R_{2\omega}$, upon its initial picosecond rise to $R_{2\omega,melt}\approx 0.35 > R_{2\omega,0}$ due to HOPG melting ($F_{melt}\approx 0.15$ J/cm^2, cfg., $F_{melt}\approx 0.13$ J/cm^2 for 624-nm fs-laser pulses [4]) and thermal melt expansion [4]. Particularly, HOPG melting was identified in the form of the increased surface reflectivity $R_{2\omega,high}>R_{2\omega,0}\approx 0.30$ up to 0.35 (Figure 4.2), appearing in this work at $\Delta t\approx 1$ ps for $F>F_{melt}\approx 0.15$ J/cm^2 (compare, e.g., to the previously measured HOPG melting threshold ≈ 0.13 J/cm^2 for 632-nm fs-laser pulses [4]) and persisting from picoseconds till nanoseconds.

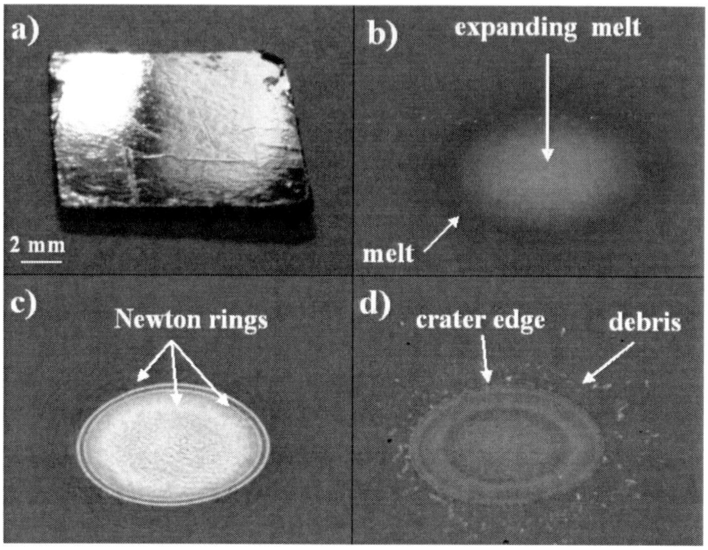

Figure 4.1. Images of HOPG sample (a), its fs-laser excited spot at Δt = 10 ps (b), 300 ps (c) and the final crater (d). The image size in (b-d) is 500 μm × 300 μm.

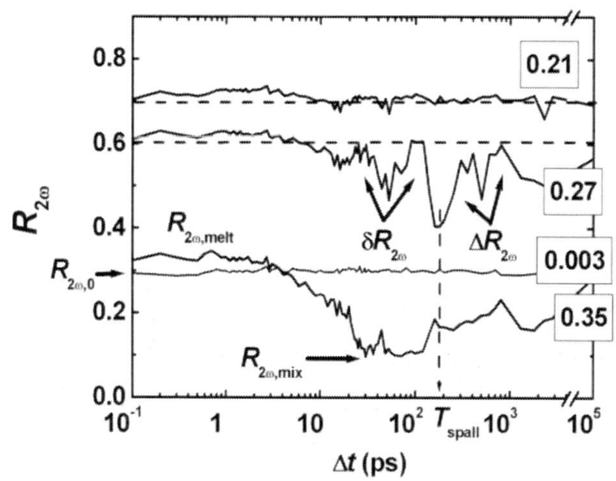

Figure 4.2. Transients $R_{2\omega}(\Delta t)$ for F = 0.0.003 (non-offset thin solid reference curve), 0.21 (solid curve offset up by 0.4), 0.27 (solid curve offset up by 0.3), and 0.35 (non-offset curve) J/cm^2. $R_{2\omega,0}$, $R_{2\omega,\mathrm{melt}}$ and $R_{2\omega,\mathrm{mix}}$ indicate the initial HOPG reflectivity, as well as transient reflectivity of molten HOPG and its vapor/droplet mixture, while T_{spall}, $\delta R_{2\omega}$ and $\Delta R_{2\omega}$ denote the spallation onset (F = 0.27 J/cm^2), the amplitudes of acoustically-driven and interference reflectivity oscillations, respectively.

The corresponding energy density $\varepsilon(F_{\mathrm{melt}}) = (1 - R_{\mathrm{p}}(F_{\mathrm{melt}}))F_{\mathrm{melt}}V_{\mathrm{m}} / \delta_{800} \approx 2.0\times10^5$ J/mole for the fluence-dependent p-polarization pump reflectivity $R_{\mathrm{p}} \approx 0.24$ [81], the HOPG molar volume $V_{\mathrm{m}} \approx 5.5$ cm^3/mole and the fluence-independent penetration depth for the 800-nm laser pulses $\delta_{800} \approx 32$ nm [71] {the vertical (in-depth) heat transport is negligible for HOPG on a sub-nanosecond timescale [42]}, is in good agreement with the carbon melt enthalpy ≈ (2.0±0.2)×10^5 J/mole [82].

More detailed inspection of the obtained $R_{2\omega}(\Delta t)$ transients reveals a new type of oscillations (Figure 4.2), besides the common sub-nanosecond reflectivity oscillations due to the optical interference [83], which appear on a picosecond timescale in the narrow range $0.16 < F < 0.38$ J/cm^2(i.e., $F_{melt}<F<F_{frag}$), i.e., upon surface melt formation ($F> F_{melt}$) during its thermal expansion. These picosecond reflectivity oscillations exhibited eventually increasing periods T_{rev} and amplitudes $\delta R_{2\omega}$ in a time domain with thermal expansion of the surface HOPG melt.

As a result, these short-term oscillations can be associated with acoustic reverberations, modulating the probe reflectivity via a piezoreflection effect [57]. Such reverberations are initiated in the melt, apparently, by the very fast mass density change associated with the ultrafast melting process [84] and supported by the acoustic impedance mismatch between HOPG and the molten carbon phase. Starting from the melting threshold $F_{melt} \approx 0.15$ J/cm^2, the increasing magnitudes T_{rev} (Figure 4.3) represent the melt thickness, eventually increasing and saturating versus Δt until the spallative lift-off of the melt layer, as follows from propagation of the melting front in the fs-laser pre-heated HOPG (at the almost absent inter-layer heat conduction [42]). Indeed, the maximal magnitudes $T_{rev,max}$ taken during the saturation stage or just prior spallation (Figure 4.3), monotonously increase in the range F = 0.16–0.3 J/cm^2 (Figure 4.4), showing, to some numerical factor, the same rising functional dependence as a fluence dependence of the maximal melt thickness Y_{max} calculated using a simple Beer-like relationship $Y_{max}(F) = \delta_{800}\ln(F/F_{melt})$.

In the calculations, transient HOPG reflectance and its absorption depth $\delta_{800} \approx 32$ nm [71] were approximately considered as fluence-independent quantities during the 800-nm pump pulse. An additional indication of the single-shot HOPG melting for $F \geq F_{melt} \approx 0.15$ J/cm^2 was obtained by optical microscopy inspection of the final fs-laser irradiated spots, exhibiting pronounced higher-reflection surface regions with well-defined edges, starting from $F \geq F_{melt}$ (Figure 4.5).

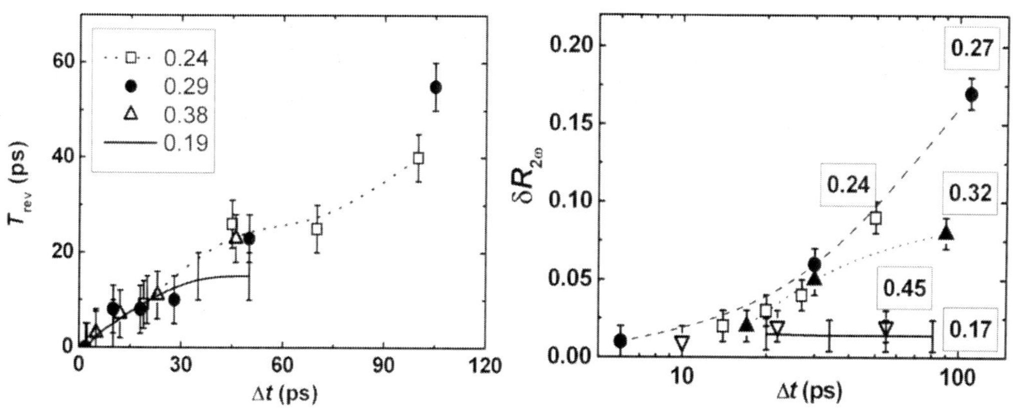

Figure 4.3. (left) Temporal variation of acoustic reverberation periods T_{rev} for F = 0.19 (solid curve), 0.24 (light squares), 0.29 (dark circles) and 0.38 (light triangles) J/cm^2; (right) temporal variation of acoustic reverberation amplitudes $\delta R_{2\omega}$ for F = 0.17 (bottom solid curve), 0.24 (light squares), 0.27 (dark circles), 0.32 (dark triangles) and 0.45 (light triangles) J/cm^2.

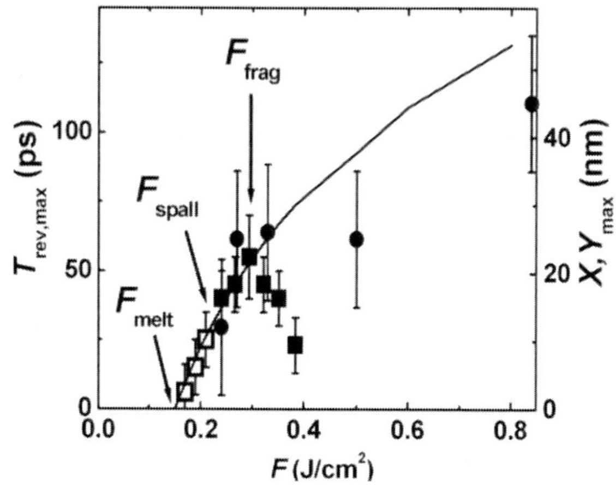

Figure 4.4. Fluence dependences of the maximal reverberation period $T_{rev,max}$ (left axis, light and dark squares), final crater depth X (right axis, dark circles) and calculated maximal melt thickness Y_{max} (right axis, solid curve), with the arrows showing the melting (F_{melt}), spallation (F_{spall}) and fragmentation (F_{frag}) thresholds for HOPG.

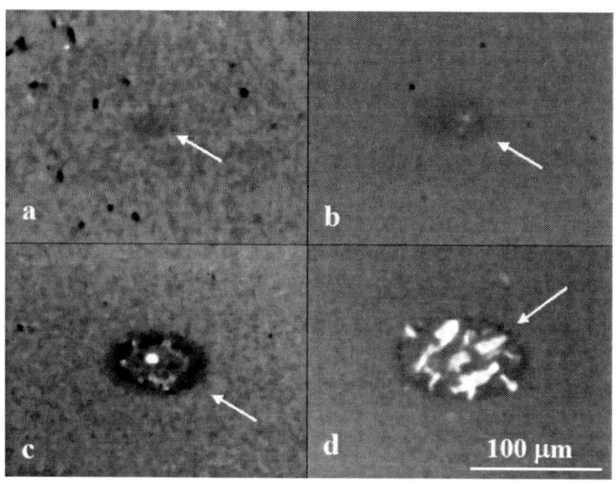

Figure 4.5. Optical microscopy images of HOPG surface irradiated by single fs-laser pulses at $F = 0.16$ (a), 0.18 (b), 0.19 (c) and 0.23 (d) J/cm^2, with the darker re-crystallized regions and brighter spalled fragments (shown by the arrows) being more and less reflective, respectively.

In the range $F_{melt} < F < F_{spall}$ these modified surface regions have elliptical shapes (Figure 4.5a,b) because of the oblique laser pulse incidence, while $F > F_{spall}$ they appear as rings around spallative craters (Figure 4.5a,b). Their higher reflectivity may result from thermal annealing of the HOPG surface upon its melting and re-crystallization, as implied by the threshold-like appearance of the effect in Figure 4.5. Simultaneously, time- and spectrally-integrated fs-laser induced emission from the HOPG surface (Figure 4.6) was acquired through the same time-resolved optical reflection microscopy scheme (Figure 2.1) with the probe pulse blocked and corrected to the CCD spectral response function $S_{CCD}(\lambda)$ in the range of 300-700 nm and the silica micro-objective transmission spectrum $T_{MO}(\lambda)$.

The acquired thermal emission images were graphically processed to obtain a series of emission intensity, I_{exp}, dependences on F taking spatial – vertical or horizontal – cross-sections of the glowing spots on the graphite surface (Figure 4.6). The peak I_{exp} values plotted versus corresponding peak F values exhibit rather smooth increase (Figure 4.7) and can be well fitted by the thermal emission (Planck) formula for temperature-dependent, spectrally-integrated energy flux $I_{calc}(T)$ in the temperature range T = (4-20)×10^3 K, which was calculated as follows

$$I_{calc} = \int_{0}^{\infty} \int_{300nm}^{700nm} T_{MO}(\lambda) S_{CCD}(\lambda) \varepsilon(\lambda,T) \frac{C_1 \lambda^{-5}}{\exp(C_2/\lambda T) - 1} d\lambda dt \quad , \qquad (4.1)$$

where the common factors C_1 = 37418 W·μm^4/cm^2 and C_2 = 14388 μm·K for the emission wavelength λ in μm, and the carbon emissivity $\varepsilon(\lambda,T) \approx 0.3$ was taken constant, as it is the only $\varepsilon(\lambda)$ which is known only for sub-melting temperatures [71].

Interestingly, in the intermediate range F = 0.13-0.2 J/cm^2 the curve $I_{exp}(F)$ demonstrates a distinct rise and plateau, which is accompanied by appearance of luminous fragments (sheets) covering the corresponding region of the surface laser spot (Figure 4.6); these ultrathin spalled carbon sheets become semitransparent afterwards (Figures 4.5 and 4.6). The observed plateau of the $I_{exp}(F)$ curve can be considered as a classical indication of a phase transformation, while the corresponding plateau temperature $T_{pl} \approx (4.8 \pm 0.4) \times 10^3$ K indicates melting of graphite within the temperature range 3700-5000 K, covering all representative measurements of the carbon melting temperature T_m (see Refs. 85,86 for broad bibliography).

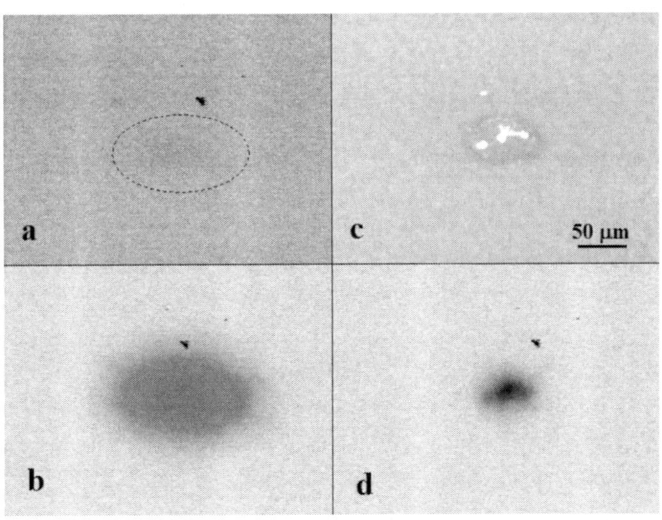

Figure 4.6. Transient emission images from HOPG surface irradiated by single fs-laser pulses at F = 0.18 (a), 0.4 (b) J/cm^2. The images c and d provide the comparison of the final crater and the transient emission region at F = 0.21 J/cm^2, showing the position correlation between the bright spalled fragments (c) and the strong emission spots in (d).

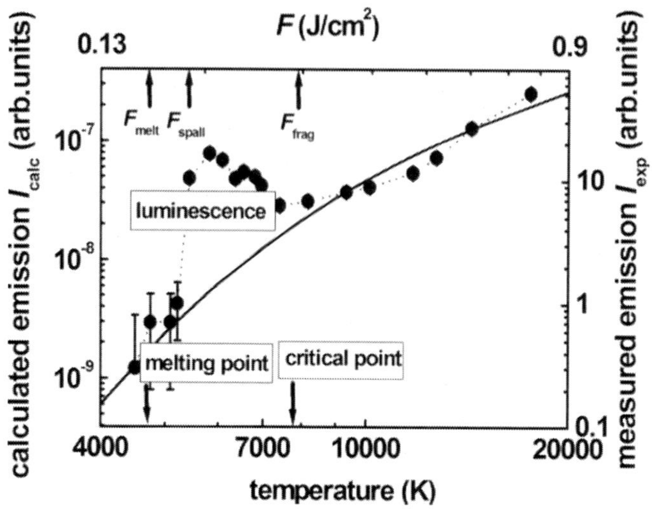

Figure 4.7. Dependences of experimental emission intensity I_{exp} on F (dark circles, right and upper axes) and of calculated emission intensity I_{calc} on surface temperature T (continuous curve, left and bottom axes) for HOPG, with the melting (F_{melt}), spallation (F_{spall}) and fragmentation (F_{frag}) thresholds, and the expected melting and critical point positions. The luminescence mark shows the abnormal emission trend on the curve $I_{exp}(F)$ in the spallation regime.

The fact of graphite melting is also confirmed by the observation of the spalled carbon layers in transient and final images of the irradiated surface in Figures 4.5-4.6, with such spallation occurring at slightly higher fluences F = 0.2-0.3 J/cm^2 (see the next Section). This monotonous rising trend for $I_{exp}(F)$ is interrupted in the range $F_{spall} \approx 0.18$ J/cm$^2 < F < F_{frag} \approx 0.3$ J/cm^2, where the smooth emission spots, characteristic for $F \leq F_{spall}$ (Figure 4.6a,b), are disturbed by much stronger inhomogeneous emission (Figure 4.6c), spatially correlating with the final permanent spalled fragments residing on the surface (Figure 4.6d). The intense transient emission from the fragments may result from fractoluminescence during solidification of the spalled (apparently, molten) carbon sheets [87], or, simply, recombination luminescence of thermal carriers in the molten carbon during its structural transformation back to solid graphite with annealing of sp^3-coordinated or just broken carbon atoms, which were present in the transient molten phase or appear during re-solidification, respectively.

Hence, the using of the time-integrated thermal emission for transient temperature characterization of the fs-laser excited HOPG surface provides consistent (potentially effective) results both at the low- and high-temperature ends, excluding the only spallation region, where the non-thermal luminescence predominates. Moreover, a few features of the $I_{exp}(F)$ curve – a plateau for $F_{melt} \leq F \leq F_{spall}$ (no temperature change at the increasing F and the deposited energy density) and the well-fitted monotonous change $F > F_{frag}$ (no luminescence of the spalled fragments) – provide reasonable estimates for the melting ($T_{melt} \approx 4.8\times10^3$ K, for the reference see [85,86]) and the critical ($T_{crit} = 7\times10^3$ K, for the reference see [88]) temperatures (Figure 4.7), which are in reasonable agreement with estimates for molten and critical carbon enthalpies above and below in the text, respectively. The general monotonous rising trend in the entire laser fluence range provides temperature estimates for sub- and supercritical carbon states.

4.2. Sub-Critical Spallation

Time-resolved optical reflection microscopy images of the photo-excited HOPG surfaces demonstrate within the molten region, as the most remarkable visual features, strong picosecond decrease of the probe reflectivity $R_{2\omega}$ and succeeding appearance of elliptical interference fringes (Newton rings [7, 83]), surrounding a central low-reflectivity region (Figure 4.1c). Specifically, for laser fluences slightly above the HOPG melting threshold ($F>$ 0.21 J/cm^2) $R_{2\omega}$ magnitude drops on a fluence-dependent timescale Δt = 40-400 ps and then slowly restores back during $\Delta t<$ 10 ns (Figure 4.2). In the latter time window, starting from the fluence-dependent instant T_{spall}, the $R_{2\omega}(\Delta t)$ transients are modulated by pronounced long-term (sub-nanosecond) oscillations with fluence-dependent amplitudes $\Delta R_{2\omega}$ and periods, disappearing for $F>$ 0.3 J/cm^2. These temporal oscillations represent in the time domain the abovementioned interference Newton rings (Figure 4.1c), produced through interference between the probe pulse replica reflected from the lifting-off spalled melt layer and the underlying molten or hot HOPG surface, which are separated just by highly transparent sub-critical carbon vapor (the expected refraction index $n^* \approx 1$ [89]). As a result, these oscillations can be related to spallative ablation of the surface carbon melt occurring in the range F = 0.21-0.3 J/cm^2$>F_{spall}\approx$ 0.20 J/cm^2 (see the crater in Figure 4.1d and the spalled fragments in Figures 4.5 and 4.6), in good agreement with the spallation threshold $\approx$ 0.18 J/cm^2 previously reported for 800-nm fs-laser pulses [7]. In contrast, at higher fluences $F\geq F_{frag}$ the lower reflectivity within the micro-heterogeneous central region (Figure 4.1c) appear in the center of the Newton rings in the form of a scattering and absorbing ablative plume. This plume represents picosecond hydrodynamic expansion and fragmentation of a supercritical carbon fluid (SCF) [7] during propagation of a rarefaction wave in the strongly heated surface carbon layer with a thickness of tens of nanometers. The corresponding fragmentation threshold, $F_{frag}\approx$ 0.3 J/cm^2, is in good agreement with its previously measured magnitude of 0.27 J/cm^2 [7]. The time instants Δt_{NR} when the $R_{2\omega}(\Delta t)$ transients exhibit their extrema (see, e.g., Figure 4.2) – maxima and minima, are readily transformed to the corresponding lift-off distances $n^* L_{spall}$ (Figure 4.8), using the simple conditions $n^* L_{spall} = 0.5m\lambda_{2\omega}, n^* L_{spall} = 0.25[2m+1]\lambda_{2\omega}$ – for the m-th order interference maxima and minima, respectively. As a result, slopes of the resulting curves $n^* L_{spall}(\Delta t_{NR})$ for different fluences in Figure 4.8 provide the corresponding lift-off velocities of the spalled melt layer $\approx$1-2 km/s.

Then, the good coincidence of the $Y_{max}(F)$ and $X(F)$ curves in Figure 4.4, depicting the almost complete melt removal until the melt/solid HOPG boundary, provides a hint that it is a tensile rupture at the former rigid boundary that may be responsible for its spallation with the residual rather smooth crater (Figure 4.1d), rather than nanofoaming within the HOPG melt, providing nanorough craters [90]. Indeed, simple analysis of $T_{rev,max}(F)$, $Y_{max}(F)$ and $X(F)$ curves in the range $F\leq F_{frag}$ makes possible quantitative evaluation of the longitudinal sound velocity in the carbon melt and general characterization of its acoustic impedance, comparing to that of solid graphite $\rho_s C_{l,s}$ = 2.18×10^3 kg/m^3 × 2.3 km/s ≈ 5×10^6 kg/m^{2}s (where the longitudinal sound velocity in graphite $C_{l,s}\approx$ 2.3 km/s [91]). Particularly, if an acoustic impedance of the molten carbon, $\rho_m C_{l,m}$, is lower than $\rho_S C_{l,S}$ (the rigid melt/HOPG boundary [92]), then the short-term oscillations appear in $R_{2\omega}(\Delta t)$ transients in the same phase only after two round-trips, i.e., for $C_{l,m}T_{rev,max}\approx 4X$ and $C_{l,m}\approx$ 1.6 km/s $<C_{l,s}\approx$ 2.3 km/s.

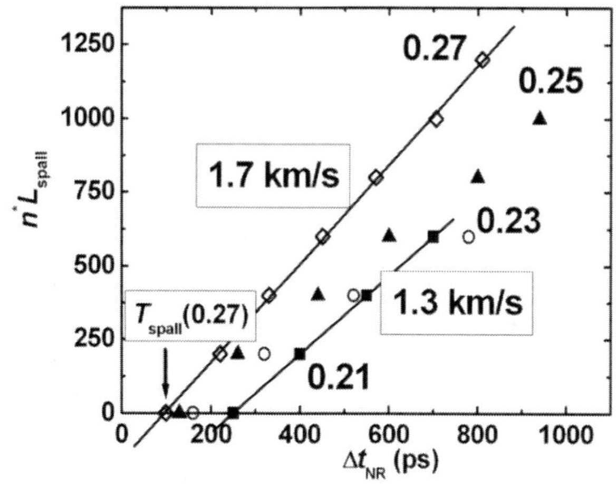

Figure 4.8. Lift-off history $n^*L_{spall}(\Delta t_{NR})$ of the spalled carbon melt layer for different magnitudes F (shown at the corresponding symbols) and the linear fitting curves for the two extreme spallation values F ($F \geq F_{spall}$ and $F \leq F_{frag}$) with the slopes indicating the lift-off velocities V_{spall} in the frames.

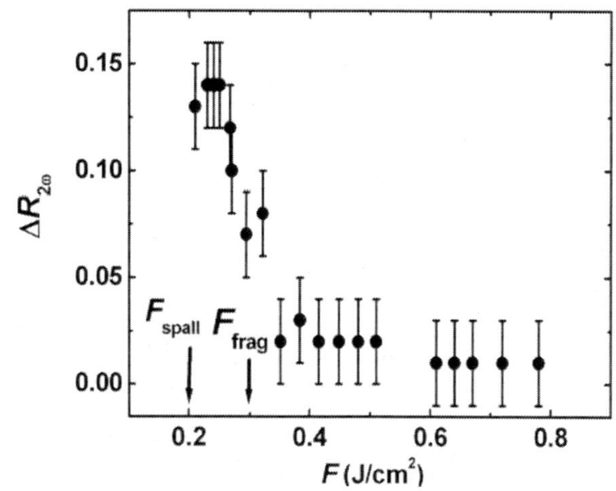

Figure 4.9. The maximum contrast (reflectivity modulation amplitude) factor $\Delta R_{2\omega}$ for Newton rings versus F in the range between F_{spall} and F_{frag}.

Alternatively, another inequality $\rho_m C_{l,m} > \rho_S C_{l,S}$ (the free melt/HOPG boundary [92]) requires only one round-trip for the oscillation to emerge in the same phase, and yields in $C_{l,m}T_{rev,max} \approx 2X$ and $C_{l,m} \approx 0.8$ km/s. The latter case with $\rho_m C_{l,m} > \rho_S C_{l,S}$ and $C_{l,m} < C_{l,s}$ appears to be hardly possible for the reason of the extraordinary melt density $\rho_m > 6\times10^3$ kg/m^3 required for the melt, especially, accounting for the most probable decrease of its mass density during by the simultaneous decrease of its reflectivity till $R_{2\omega,mix} \approx 0.07$ on the picosecond thermal expansion timescale (Figure 4.2), which is more typical for rarefication-driven Mott-like metal-insulator transitions {in contrast, potentail transformation to a diamond-like (sp^3-coordinated) melt would result increase $R_{2\omega}$ magnitudes owing to the high refractive index of diamond [70,71]}.

In a proof, the derived lower impedance of the carbon melt implies that the rarefaction wave, when reflected at the melt/HOPG interface, gets its front part superimposed with its tail part in the same phase, thus increasing (maximum, doubling) its effective instantaneous amplitude at the very interface and facilitating the tensile rupture. Likewise, the abovementioned gradual increase of the amplitude $\delta R_{2\omega}$versus Δt and $F_{spall} \leq F \leq F_{frag}$ during the short-term acoustic reverberations (Figure 4.3, right) occurs exactly during the thermal expansion stage of the HOPG melt with the increasing thickness $Y(\Delta t) = \frac{1}{4}C_{l,m}T_{rev}(\Delta t)$. This implies that the acoustic impedance of such expanding melt eventually decreases and each double round-trip of the pressure wave in the melt layer can be considered, to the squared product of the reflection coefficients – $R_{air/melt} \approx 1$ and $R_{melt/HOPG} < 1$ – for the air/melt and melt/HOPG interfaces, as its straightforward propagation in a virtual medium with a spatially-distributed – particularly, monotonously decreasing – acoustic impedance. Then, according to the continuity principles, the pressure wave propagating along the acoustic impedance gradient will increase its vibration velocity, appearing in the optically probed absorption layer in the melt in the form of its enlarged strain, modulating the probe reflectivity through the piezo reflection effect [57]. Though the previously unknown value of a longitudinal sound velocity in molten carbon, $C_{l,m} \approx 1.6$ km/s, was derived above from our experimental data, its transient mass density is not known yet, making more detailed quantitative description of this effect impossible.

Finally, it is instructive to discuss for HOPG and other common semiconductors – GaAs [90] and Si [93] – the difference in relationships between their rupture position inside surface melt, as well as spallation timescales (100-250 ps vs. < 800 ps and < 500 ps) and lift-off velocities (>1 km/s vs. <0.8 km/s and >0.7 km/s) in the fluence range $F_{spall} \leq F \leq F_{frag}$, apparently associated with somee difference in their spallation mechanisms. Above, we have shown that the carbon melt with its acoustic impedance lower than that of HOPG, most probably, spalls when the transient rarefaction wave increases its instantaneous amplitude at the very melt/HOPG interface during reflection because of the constructive (in-phase) interference of the incident and reflected waves, which is also the case for molten GaAs and Si. However, in the case of GaAs and Si spallation is more delayed and occurs at much lower or comparable lift-off velocities, respectively; moreover, short-term acoustic reverberations are also present in molten GaAs and Si, though exhibiting not such high relative modulation, possibly, because of considerably stronger acoustic damping in the almost twice thicker cavitating top melt layer [90]. Together with the finding of the rupture position within the GaAs melt closer to its air/melt interface, rather than at the melt/solid interface (the HOPG and Si case), these facts indicate that in the GaAs case its spallation is more strongly driven by transient cavitation of more volatile As atoms, rather than by transient rarefaction, thus more tending to the corresponding critical temperature. Meanwhile, basic reasons of the pronounced difference are yet to be understood.

4.3. Supercritical Fragmentation

The disappearance of the interference Newton rings in the central low-reflectivity region of the fs-laser spot on the HOPG surface for higher $F \geq 0.3$ J/cm^2 (Figure 4.1c), accompanied by a drastic drop of the intereference amplitude $\Delta R_{2\omega}$ (Figures 4.2 and 4.9), can be associated with highly-scattering nano- or microheterogeneous vapor-droplet mixture, which is clearly visible in Figure 4.1c. Such mixture is known to result from fragmentation of supercritical fluid (SCF) of carbon during its adiabatic expansion [7,93] ("fragmentation" mechanism [94,95]) for fluences, exceeding the corresponding fragmentation threshold F_{frag} (Figure 4.10). Indeed, our estimate of the deposited energy density $\varepsilon(F_{frag}) = (1 - R_p(F_{frag}))F_{frag}V_m / \delta_{800} \approx$ 3.8×10^5 J/mole for the *p*-polarization pump reflectivity $R_p(F_{frag}) \approx 0.28$ [81], considering the penetration depth $\delta_{800} \approx 32$ nm as a fluence-independent quantity, reasonably (by 20%) exceeds the previously evaluated critical carbon enthalpy $\approx (3.1\pm0.3)\times10^5$ J/mole [82], allowing the supercritical fluid to expand adiabatically enough prior its transformation to the irreversibly lifting-off vapor-droplet mixture.

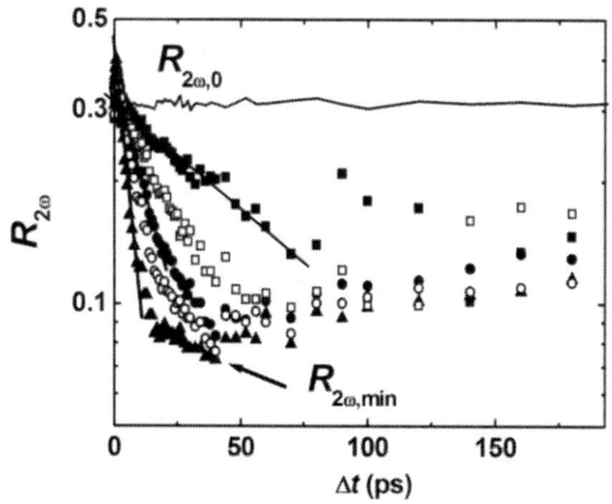

Figure 4.10. Dynamics of probe reflectivity $R_{2\omega}(\Delta t)$ for $\Delta t \leq 200$ ps and different laser fluences F [J/cm^2]: 0.16 (top solid curve), 0.35 (dark squares), 0.50 (light squares), 0.69 (dark circles), 0.92 (light circles), 2.0 (dark triangles). The straight lines show the linear fitting curves in the coordinates $\lg R_{2\omega}-\Delta t$ for some of these transients in the initial region of $R_{2\omega}$ drop. The arrow shows the minimum value $R_{2\omega,min} \approx 0.07$.

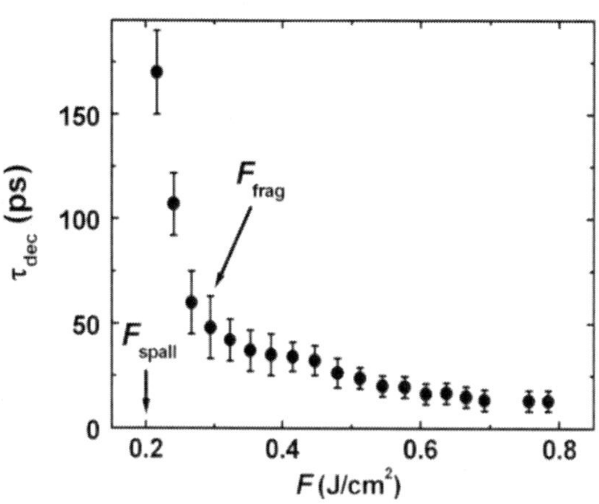

Figure 4.11. Dependence of the reflectivity decay time τ_{dec} (dark circles) on $F>F_{frag}$ fitted with the exponential decay function (solid curve).

Moreover, the derived characteristic fragmentation threshold fluence $F_{frag} \approx 0.3$ J/cm^2 for the disappearance of Newton rings in Figure 4.1c is consistent with its previous evaluation $\approx$ 0.27 J/cm^2 [7]. In a time domain, this fragmentation ablation path appears in the form of a picosecond monotonous decrease of $R_{2\omega}$ magnitude, which proceeds at a decay time τ_{dec} monotonically decreasing versus increasing F (Figures 4.10-4.11) and yields in the minimum value $R_{2\omega,min} \approx 0.07$, persisting over tens of picoseconds until the following gradual recovery of $R_{2\omega}$ (Figure 4.2).

The minute value $R_{2\omega,min}$ indicates, according to basic principles of density-mediated Mott transitions [96], considerable adiabatic expansion of supercritical carbon fluid, apparently resulting in its simultaneous fragmentation (see below). Moreover, approaching the maximum value about 50 ps at 0.3 J/cm^2, i.e., near the fragmentation threshold, the dependence $T_{rev,max}(F)$ in Figure 4.3 then starts to diminish, declining from the $Y_{max}(F)$ curve and rapidly going almost to zero near 0.4 J/cm^2. In the context of the pronounced correspondence between the maximal reverberation times $T_{rev,max}$ and Y_{max}, this falling trend of the $T_{rev,max}(F)$ curve may indicate transient partial contraction of the melt layer already during the melt front propagation ($\Delta t < 60$ ps) and the simultaneous thermal expansion of its hotter top layer, apparently, in the supercritical state for $F > F_{frag}$ (see, e.g., the corresponding reflectivity curve for $F \approx 0.38$ J/cm^2 in Figure 4.2). This interpretation is supported by direct comparison between the $T_{rev,max}(F)$ dependence and the corresponding final crater depth dependence $X(F)$ (Figure 4.4). Particularly, it demonstrates for X gradually increasing above F_{spall}, that the non-monotonous fluence-dependent variation of $T_{rev,max}$ reflects the maximal melt thickness Y_{max}, remaining just prior the spallation onset after the preceding explosive removal of the top melt layer in the supercritical state, i.e., both fragmentation and spallation ablation mechanisms proceed in the surface melt in its different regions (hotter top or cooler bulk layers, respectively) and at different (picosecond vs. sub-nanosecond, respectively) timescales [90,93].

This contrasts with previous molecular dynamics (MD) simulations [97] showing that the spalled melt layer permanently becomes thinner for fs-laser fluences increasing above the corresponding spallation threshold even without preceding supercritical fluid expansion ("fragmentation"), since shallower, rather than deeper position of a rupture interface in melt is assumed for increasing fs-laser fluences [97]. Such interplay of fragmentation and spallation processes was not observed in the previous theoretical (mostly, MD simulations) studies [94-95,97-100], where pure spallation and fragmentation (sometimes, also hypothetical intermediate sub-critical phase explosion and high-fluence direct liquid-vapor transformation [94]) paths were mostly modeled, neglecting the typically inhomogeneous spatial energy density distribution and the dynamic stress-strain picture in a surface melt.

The reflectivity dynamics $R_{2\omega}(\Delta t)$ for different $F > F_{frag} \approx 0.3$ J/cm^2 (Figure 4.10) enable us to evaluate parameters of SCF hydrodynamic expansion (rarefaction) at delay times $\Delta t < 100$ ps at the low-fluence end $F \sim 1$ J/cm^2. Particularly, the high-fluence reflectivity transients exhibit a strong decrease of $R_{2\omega}$tending at higher F to a minimal constant reflectivity level, $R_{2\omega,mix} \approx 0.07$ (Figure 4.10), with different fluence-dependent temporal exponents $\tau_{dec} \approx$ 15-110 ps (Figure 4.11). Since in-depth (vertical) diffusion is almost absent in highly oriented pyrolitic graphite [42], fs-laser energy deposition can be considered to occur presumably in its surface absorbing layer $\delta_{744nm} \approx 32$ nm [71]. Then, these exponents τ_{dec} represent, apparently, different initial linear expansion (rarefaction) rates V_{hydro} of the SCF plume emerging for $\Delta t \leq 3$

ps as a surface SCF layer with a certain thickness X, in the form $V_{hydro} \approx X/\tau_{dec}$. Particularly, magnitudes V_{hydro} can be evaluated for the normal Beer-like energy deposition in the form $V_{hydro} \approx \frac{\delta_{744nm}}{\tau_{dec}} \ln\left(\frac{F}{F_{frag}}\right)$, evaluating the effective thickness of the SCF layer $\approx \delta_{744nm} \ln\left(\frac{F}{F_{frag}}\right)$. As shown below, the calculated curve $V_{hydro}(F) < 5$ km/s upon its extrapolation to higher fluences reasonably tends to the SCF-driven shock wave speeds C_S, as well as the initial SCF plume pressures $P_{hydro} \sim \rho_0 V_{hydro}^2 < 30$ GPa tending at higher fluences to the shock wave pressure dependence $P_S(F)$ (see Figure 4.14 below).

Additionally, tabletop generation of super-strong shock waves on a graphite surface during SCF expansion was studied by its intense femtosecond (fs) laser ablation at variable laser fluences and their characterization, using the well-established non-contact ultrasonic technique [101], in order to evaluate not only SCF temperatures, but also SCF pressures. In this experimental geometry (Figure 4.12), fs-laser pulses were focused in p-polarization at the angle of 45^0 by a silica lens (focal distance $f = 100$ mm) into a small focal spot on the surface of the HOPG sample, arranged upstream the air plasma filament and in front of an active surface of a piezoelectric ultrasonic transducer MiniWAT-2 (PVDF film, bandwidth < 30 MHz, sensitivity of 10 V/atm) [102].

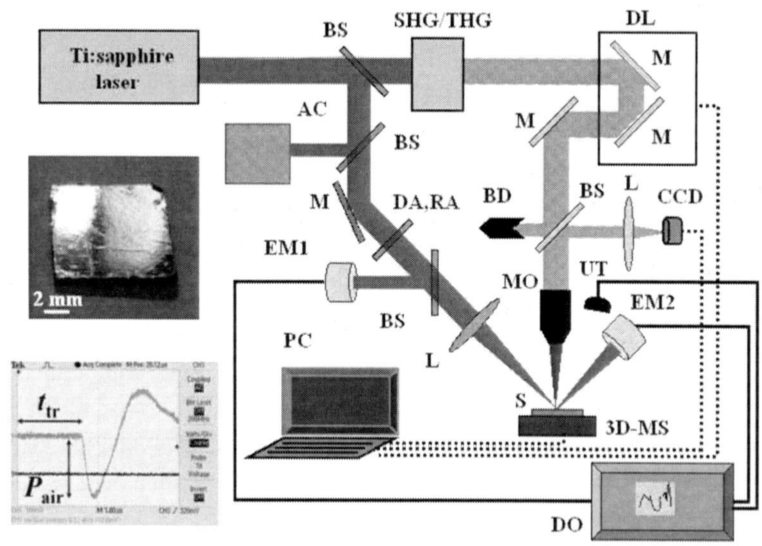

Figure 4.12. Experimental setup: BS – beam splitter, AC – autocorrelator, DA, RA – variable diffractive and reflective energy attenuators, EM1,2 – thermocouple energy meters, M – mirror, L – focusing silica lenses, SHG/THG – second and third harmonics generator, DL – optical probe delay line, CCD – charge-coupled device camera, BD – beam dump, DO – digital oscilloscope, UT – ultrasonic transducer, 3D-MS – three-dimensional motorized micro-stage, PC – laptop for data acquisition and hardware control. Insets: the optical micrograph of the HOPG piece with dimensions of 10 mm × 10 mm × 1 mm (S), and oscilloscope image of the typical ultrasonic transient with the arrows schematically indicating the transit time t_{tr} and pressure amplitude P_{air}.

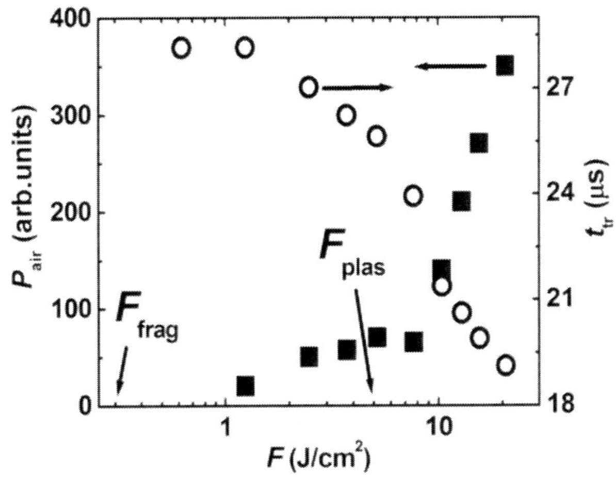

Figure 4.13. Ultrasonic pressure amplitude P_{air} (left axis, dark squares) and air transit time t_{tr} (light circles) versus F. The vertical arrows show the SCF and plasma formation thresholds, F_{frag} and F_{plas}, respectively.

During our non-contact far-field ultrasonic studies, laser-generated transient ablative pressure waves propagated a 9.7-mm distance l_{tr} between the HOPG surface and the transducer, while then the transducer fed a 50-Ω input of a digital storage oscilloscope TDS-2024, which was triggered by an electric pulse from a fast photodiode DET-210 illuminated by a weak split laser pulse. The far-field (time-differentiated) ultrasonic transients acquired only for $F>F_{frag}$ (Figure 4.13), exhibited presumably bipolar waveforms with compressive pressure amplitudes P_{air} and transit times t_{tr} (Figure 4.12, inset), representing, upon restoring their actual waveforms through temporal integration [102], unipolar compressive pulses with a sharp front and weak subsequent tail, which are characteristic of propagating air shock waves [28] driven, apparently, by ablative plume/plasma contact fronts. These compressive pressure pulses demonstrated a monotonously increasing dependence of P_{air} on F with two logarithmic-like regions linked at $F\approx$ 7 J/cm^2 (Figure 4.13). Simultaneously, the experimentally measured air transit times t_{tr} started to decrease for $F>$ 1 J/cm^2, corresponding to supersonic average propagation velocities (Figure 4.13) (in air, the sound velocity equals to $C_0\approx$ 0.346 km/s at the normal conditions [70]), with a drastic decrease for $F>$ 5-7 J/cm^2. Comparing to long-pulse laser-driven shocks [101], it appears more accurate to consider the observed ultrafast explosive fs-laser ablation as a point-like, three-dimensional explosion [103-104] characterized by other expansion rules [105,106]. This approach was successfully used earlier in the case of fs-laser ablation of aluminum [107]. For such point-like three-dimensional explosion of ideal gas with the initial energy E deposited within the sphere of the mass density ρ_S with the initial radius $R_S \approx \sigma_{x,y}\sqrt{\ln\left(F/F^*\right)}$, the radial variation relationship for the shock wave pressure P and velocity D are much stronger (even neglecting the related dissipative losses for shock wave excitation of air molecules) [106]

$$P(r) \propto \frac{E}{r^3} \approx P_S\left(\frac{R_S}{r}\right)^3, \; D \propto r^{-3/2} \tag{4.2}$$

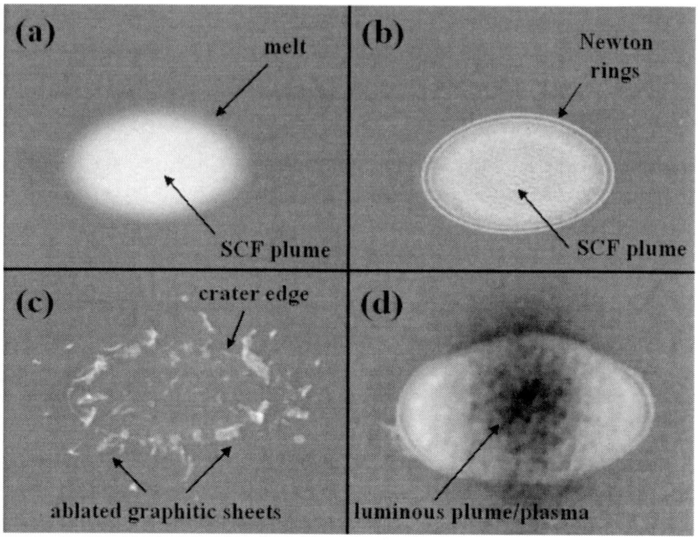

Figure 4.14. Images $R_{2\omega}(x,y)$ of low-reflectivity (bright) transient SCF (a,b) and luminous carbon plasma (d) acquired at different fluences and delay times: a $-\Delta t \approx 20$ ps, b $-\Delta t \approx 350$ ps, c $-\Delta t \approx \infty$ (all – $F = 2.5$ J/cm^2), d $-\Delta t \approx 500$ ps, $F = 6.3$ J/cm^2. The frame sizes are 300×250 μm.

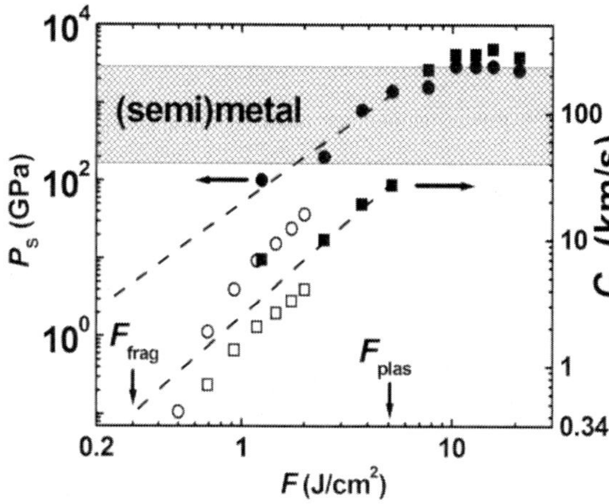

Figure 4.15. (left axis) Derived source pressure P_S (dark circles) and pressure P_{hydro} evaluated assuming hydrodynamic expansion of the heated skin-layer (light circles) vs. F, with the dashed line showing the extrapolated GPa-level source pressures near the threshold F_{frag}. The shadowed area indicates the pressure region of hypothetical high-pressure diamond-(semi)metal structural transformations of carbon (after [116-121]). (right axis) Derived source velocity C_S (dark squares) and velocity V_{hydro} evaluated assuming hydrodynamic expansion of the heated skin-layer (light squares) vs. F, with the dashed line showing their extrapolation at F_{frag} to the normal sound velocity ≈ 0.34 km/s in air.

The initial (source) values of the shock wave pressure P_S and velocity $C_S = D(R_S)$ were approximately related, following [106], as [31]

$$P_S \approx \frac{\gamma_{1,2}+1}{2}\rho_S C_S{}^2 \quad (4.3)$$

In the first, low-fluence ablation regime through hydrodynamic SCF expansion, with its threshold $F_1^* = F_{frag} \approx 0.3$ J/cm^2 and $0.3 < F < F_{plas} \approx 5$ J/cm^2 (Figure 4.14), the initial (source) density $\rho_{S,1}$ can be chosen equal to the HOPG density $\rho_0 \approx 2.18$ g/cm^3, since the fs-laser energy deposition occurs presumably in the solid target, with the adiabatic ratio $\gamma_1 \approx 5/3$ for carbon atoms or 1.38 for diatomic carbon molecules in plume. In contrast, in the second, high-fluence regime of ablation through plasma expansion with $F > F_2^* = F_{plas} \approx 5$ J/cm^2, fs-laser energy deposition is known to occur at the critical plasma front in the near-surface double electron-ion plasma layer. This layer appears and expands on the fs-laser pulse scale [108], being characterized by $\rho_{S,2}$ equal to the critical plasma density $\rho_{cr} \approx 0.034$ g/cm^3 [the critical plasma density $N_{cr}(744\text{ nm}) \approx 1.7\times10^{21}$ cm^{-3} and the carbon atom mass $M \approx 2\times10^{-23}$ g] and $\gamma_2 \approx 5/3$ for carbon atomic ions. This expansion regime (the first propagation stage) for the resulting shock wave is valid until the transition pressure [106]

$$P_{trans} \approx \frac{\gamma_0 + 1}{\gamma_0 - 1} P_0 , \qquad (4.4)$$

where $P_0 = 1$ atm is the atmospheric air pressure and $\gamma_0 \approx 1.38$ for diatomic air molecules [106], yielding in $P_{trans} \approx 6$ atm.

During our calculations by use of the computer code PEFS-2 [107], the magnitude D decreased down to the normal sound speed $C_0 \approx 0.346$ km/s already for $P/P_{trans} \sim 10$, so the succeeding pressure wave propagation toward the ultrasonic detector (the second propagation stage) was considered as purely sonic and spherical. Thus, the overall transit time, t_{tr}, was taken as a sum of the latter, sonic transit time, t_{sonic}, over the radial distance l_{sonic} where $D \approx C_0$, and the former, supersonic transit time, t_{super}, over the radial distance ($l_{tr} - l_{sonic}$), where D drops down to C_0 following the relationship $D \propto R^{-3/2}$. The initial source pressures P_S were chosen to satisfy the equality $t_{tr} \approx t_{super} + t_{sonic}$, with the corresponding source shock wave velocities $C_S = D(R_S)$ calculated using Equation 4.3.

The proposed consideration of the fs-laser HOPG surface ablation in the form of the point-like explosion is justified by the pressure magnitudes P_S evaluated by means of this two-stage expansion model for different fs-laser fluences (Figure 4.15). At the low-fluence end ($F \sim 1$ J/cm^2), the derived pressures $P_S \sim 10$-10^2 GPa are reasonably higher than the previously estimated thermoelastic stresses in graphite (~ 10 GPa [109,110]). Then, at the high-fluence end ($F \sim 10$ J/cm^2) the derived pressures at $F \approx 21$ J/cm^2 approach magnitudes on the order of 3 TPa, in reasonable agreement with the previous theoretical estimates [111,112] and a few previously measured values [107,113]. The intermediate pressure values $P_S \sim 10^2$-10^3 GPa in Figure 4.15 are also consistent with experimental ones reported in literature for comparable fs-laser fluences [107,111,113-115], though for other materials (particularly, for aluminum). Furthermore, these derived pressures P_S are in good correspondence with the deposited surface energy density $\varepsilon_{dep}(F) = F[1 - R_p(F)]/\delta_{744nm} \sim 10^5$-$10^7$ J/cm^3 in the fluence range 0.3-20 J/cm^2, where the optical penetration depth in the unexcited graphite $\delta_{744nm} \approx 32$ nm [71] and the $R_p(F)$ dependence is given in Figure 2.2, since typically the relation $P_S \sim \varepsilon_{dep}$ holds. Likewise, at the intermediate fs-laser fluences $F \leq F_{plas}$, the evaluated C_S values fall in the range 10-30 km/s being in reasonable agreement with other previously experimentally measured initial plume/plasma velocities [107, 111, 113-115]. However, for $F>F_{plas}$ the evaluated plasma expansion velocities C_S are much higher for light carbon ions, than the present literature data [107, 111, 113-115] despite of comparable plasma shock pressures $P_S\sim$ 1 TPa. Surprisingly, pressures $P_S\sim$ 1 TPa evaluated in the high-fluence critical plasma absorption regime (Figure 4.15), are comparable to pressures, P_{cr}, in *sub-nanosecond* laser-driven isothermal critical plasmas (with the atomic mass M and average ion charge $\langle Z\rangle$), estimated using the known scaling relationships [108,122-124], e.g.,

$$P_{cr} = 2\rho_{cr}C_{plas}^{\ 2} = 12\left(\frac{I_{las}}{10^2\,\mathrm{TW/cm^2}}\right)^{2/3}\left(\frac{1\,\mu\mathrm{m}}{\lambda_{las}}\right)^{2/3}\left(\frac{M}{2\langle Z\rangle}\right)^{1/3}\ [Mbar] \tag{4.5}$$

where the initial plasma speed C_{plas} is defined at the critical plasma surface in the form

$$I_{las} = 4\rho_{cr}C_{plas}^{\ 3}, \tag{4.6}$$

approaching, for the used fs-laser intensities $I_{las} = (0.5\text{-}2)\times10^2$ TW/cm^2 (F = 5-21 J/cm^2) at the same laser wavelength λ_{las} = 744 nm, the magnitudes $C_{plas}\approx$ (1.3-2.3)$\times10^2$ km/s comparable to the high-fluence magnitudes $C_S\sim$ (2-3)$\times10^2$ km/s in Figure 4.15. Simultaneously, the plasma pressure values $P_{cr}\approx$ 1.5-4 TPa evaluated for this laser intensity range, using Equation 4.5, are reasonably consistent with the shock wave pressures $P_S\sim$ 2-3 TPa generated in the critical plasma regime.

Importantly, the maximal plasma pressures $P_S\sim$ 3 TPa achieved on the HOPG surface upon the *femtosecond* laser irradiation in the tabletop arrangement are higher than the previous pressure loads ($\leq$ 1.2 TPa) produced on carbon targets using multi-kJ (PALS, Czech Republic) or even MJ-level (LULI, France) *sub-nanosecond* laser macro-facilities [31]. Moreover, these maximal P_S magnitudes cover the entire pressure range 0.8-2.7 TPa (Figure 4.14), where a number of hypothetical high-pressure carbon phases – solid metal at 0.17 [116], 0.84 [117], 1.1 [118], 1.2 (B-8) [119] or 2.7 TPa [120], and BC-8 semi-metallic phase at 1.1 TPa [121] – may exist. Hence, one can potentially use such and similar tabletop femtosecond laser facilities and, e.g., non-contact ultrasonic pressure acquisition diagnostics, to directly verify the existence of these hypothetical high-pressure carbon phases in their transient or quenched states. Under such laser irradiation circumstances, the well-known advantage of femtosecond pulses to deposit into targets comparable surface energy densities at the same laser intensities, as in the case of sub- and nanosecond laser pulses, but at much lower laser fluences (and pulse energies), is employed. Furthermore, these measurements provide a useful reference point, in terms of relationship between surface ablative pressures and underlying fs-laser intensities, for future tabletop high-pressure structural and other studies of carbon and diverse materials at even higher pressures, which can be achieved during fs-laser ablation of solids in rarefied gases or vacuum.

As an example, a new interesting topological feature – the elevated dark spot inside the single-shot HOPG crater for $F>$ 1 J/cm^2 [11] (Figure 3.3a), with resided graphitic sheets and almost equal, well-defined D- and G-bands in its Raman spectrum (Figure 3.2) was observed and interpreted in terms of strong fs-laser driven shock-wave effects. Particularly, the initial source pressure in the fs-laser excited surface layer for $F > 1$ J/cm^2 corresponds to the range of 10-100 GPa (Figure 4.15), being sufficient for high-pressure G-D transformation. On the other hand, strong shock compression of the lower-density HOPG may result in its transient compressive heating and the final ablative expansion observed in Figure 3.3a, which is well-known effect for shockwave compressed low-density materials [28] (see also Section 5 and Ref.146). Apparently, this is the specific case at the high-fluence irradiation of rather low-density HOPG, thus, yielding not in a quenched high-density metallic carbon, but in the well-defined lower-density graphite and diamond nanocrystalline phases. Moreover, as transient sp^3-bonding appears only at ~10 ps [14], the laser shockwave driven process appears to be very possible way for the graphite-diamond transformation. Likewise, multi-shot ($N \sim 10^2$) moderate-fluence fs-laser exposure also yields in sp^3-coordinated carbon generated through high-temperature, high-pressure near-critical thermodynamic conditions on the HOPG surface. In the first, spallative ablation regime with $F_{spall} < F < F_{frag}$ characterized by weak shock wave generation (Figure 4.15), only very small, background intensity of the corresponding D-line (i.e., slightly modified graphite surface with the minor presence of sp^3-coordinated carbon atoms) was measured by micro-Raman analysis (Figure 4.16, left).

This fact may be related either to ultrafast disordering of HOPG during its potential vibronic graphite-diamond transformation, resulting in an atomic-scale mixture of sp^2- and sp^3-coordinated carbon atoms after one shot (Fig.3.2, $F \approx 0.23$ J/cm^2) and, finally, in multi-shot annealing or multi-cycle shock-wave compression of the intermediate carbon phase, which leads eventually to diamond and graphite nanocrystallites in uneven proportions (Figure 4.16, left). In contrast, in the second, intense fragmentation ablation regime characterized by sub-TPa laser-generated shock wave generation (Figure 4.15), even for $F \geq F_{frag}$ sp^2- and sp^3-coordinated carbon atoms appear already in comparable fractions, according to their comparable D- and G-band Raman intensities (Figure 4.16, right).

5. Femtosecond Laser Ablation of Single-Wall Carbon Nanotubes: Microscopic versus Mesoscopic Mechanisms Envisioned by SEM and Micro-Raman Characterization

Carbon materials, such as pyrolytic graphite, carbon nanotubes, and graphene are of permanently increasing importance for nanoscience and nanotechnology [125, 126]. Specifically, thin films of single-wall carbon nanotubes (SWCNTs) are a promising material for efficient field emitters [127] (especially, upon their laser ablation pre-patterning [128-130]), key elements of passive light modulators for femtosecond lasers [131,132]. Individual SWCNTs or their bundles may work as seeding "ablation centers" in laser lithography schemes with improved spatial resolution [133], or as specifically adhesive, highly optically absorbing species for nano-surgery, capable to trigger ablation in nanoscale intracellular regions [134]. Nevertheless, ablation mechanisms of SWCNT-based materials by short (nanosecond) and ultrashort (sub- and picosecond) laser pulses are not well understood yet, accounting for both individual (microscopic) nanotube features – semiconducting or metallic character, variable diameter and chirality, resulting in the feature-dependent optical absorption spectra [135], and mesoscopic CNT packaging in the materials, e.g., in the form of bundles.

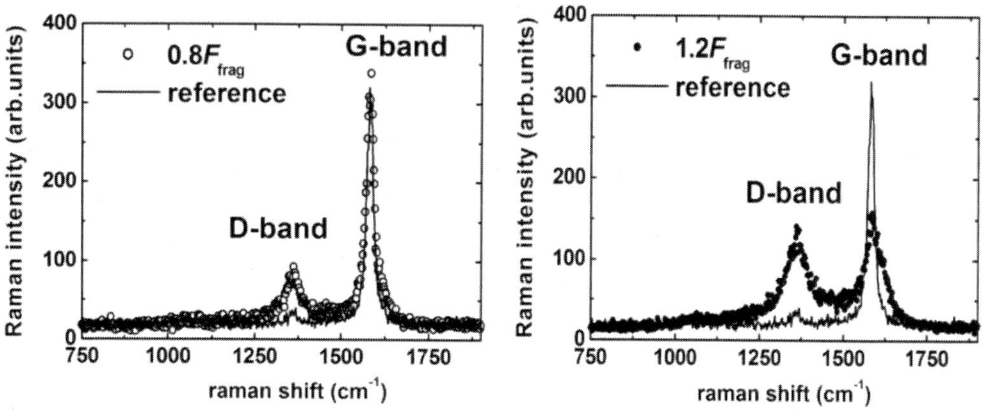

Figure 4.16. Micro-Raman spectra of reference (un-irradiated) and laser-irradiated (left – sub-threshold $F = 0.8F_{frag}$, right – above-threshold $F = 1.2F_{frag}$) HOPG spots with the G,D-bands shown by the corresponding mark.

As a result, ablation thresholds, differing by an order of magnitude – either exceeding 0.6 J/cm^2 [128,130], or falling down to ten times lower magnitudes ≈ 50 mJ/cm^2 [133,136-137] were reported. Apparently, within the hierarchical organization of bulk CNT-based materials, microscopic laser ablation of individual CNT species will require high laser fluences owing to high optical and thermal SWCNT strength, in comparison to lower-fluence mesoscopic laser ablation of their bundles.

Potential indications of such different ablation paths can be provided by Raman spectroscopy of laser ablated CNT materials (see Figure 5.1), exhibiting, e.g., the permanently increasing structural damage to individual CNTs with femtosecond laser fluence increasing from 50 mJ/cm^2 to 2 J/cm^2 [136]. Furthermore, even microscopic ablation itself may proceed via a non-thermal (selective cap opening) [138] ablation path, or through thermal one [139]. As a result, in this particular study our goal was to clarify the basic microscopic and mesoscopic laser ablation mechanisms for SWCNT materials. A 0.5-mm thick, 10 mm × 10 mm wide SWCNT sample was prepared using nanotubes synthesized in He atmosphere of an arc discharge between two graphite electrodes with the Ni catalyst [140].

The obtained residue in a toluene was enriched in respect to SWCNT content from 20% to 80% through ultrasonic treatment and ultracentrifugation [141]. The SWCNT suspension was deposited on a filter and dried. Structural characterization of the SWCNT substrate was performed at 488 nm using a micro-Raman spectrometer (U-1000, Jobin Yvon), indicating a characteristic SWCNT radial breathing mode (RBM) line at $\nu_{RBM} \approx 180$ cm^{-1} and a G^- and G^+ doublet at 1565 and 1590 cm^{-1} (Figure 5.1). Altogether, these spectral signatures defined the SWCNT semiconducting character and the average diameter $d_{RBM} = 1.36 \pm 0.04$ nm, following the relationship for ν_{RBM} [142]

$$\nu_{RBM} = Ad_{RBM} + B \quad , \tag{5.1}$$

and for the frequencies ν^+_G and ν^-_G [143]

$$\left(\nu_G^+ - \nu_G^-\right) d_{RBM}{}^2 = C_G, \tag{5.2}$$

where the calibration factors A = 116 nm×cm^{-1} and B = 6.5 cm^{-1}, while the experimental factor C_G = 46.8 ± 0.9. Its direct comparison with the same standard factors for semiconducting (C_G = 47.7) and metallic (C_G = 79.5) SWCNT species [143] unambiguously demonstrates the semiconducting character of out SWCNTs.

The SWCNT substrate arranged on a PC-controlled, three-dimensional motorized micro-stage and raster-moved from laser shot to shot to provide its single- or multi-shot exposure, was ablated, upon focusing by a 50-mm focal length silica lens, at different peak femtosecond laser fluences F_0 in TEM_{00} mode.

Laser irradiation was provided by linearly-polarized 320-fs fundamental IR (1030 nm) or 230-fs second harmonic visible (515 nm) pulses of a Yb:doped fiber oscillator/amplifier laser system, running at repetition rate of 0-500 kHz and pulse energies ≤ 10 μJ varied by means of an output acousto-optical modulator [144]. The SWCNT substrate was inspected for the ablated spot diameters d_{abl} and topologies using a scanning electron microscope JEOL 7001F. The single-shot IR/vis fs-laser ablated spots both demonstrate the significant mesoscopic perturbation of the initial surface nanotexture with appearance of chaotically oriented SWCNT bundles with the well-defined ablation edges (Figures 5.2 and 5.3). Nevertheless, the corresponding single-shot mesoscopic ablation threshold magnitudes $F_{mes}(1)$ are wavelength-dependent – 25±3 and 94±17 mJ/cm^2 for 1030-nm and 515-nm pulses, respectively.

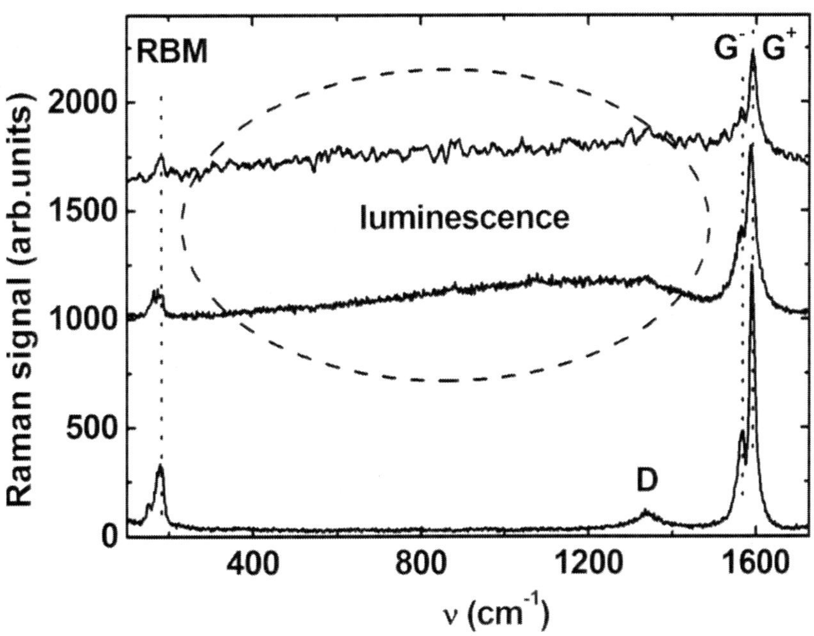

Figure 5.1. Raman spectra taken from the non-ablated region (bottom curve), from the center of the ablative crater after 30 shots at 1030 nm and (F≈F_0≈ 1.25 J/cm^2, top curve) and from the standing SWCNTs at the crater edge (F≈ 0.07 J/cm^2, medium curve). The upper curves are offset for clarity.

Such spectral selectivity may represent the sharp spectral optical absorption features (van Hove singularities of the joint density of states) of the band spectrum for the 1.36-nm wide semiconducting SWCNTs [143], with the IR radiation in the resonance and the visible radiation slightly off-resonance, thus, resulting in the observed considerable difference in the single-shot mesoscopic ablation thresholds. The measured fs-laser ablation thresholds, as well as the observed mesoscopic ablation topologies are reasonably consistent with the previously obtained ablation thresholds about 50 mJ/cm^2 at 800 nm for the slightly distrurbed surface nanotexture [133,136]), accounting for the radius- and chirality-dependent optical absorption features for SWCNTs [143]. Moreover, such shallow single-shot ablative surface disintegration of the SWCNT substrate in the form of the chaotically oriented nanotubes or their bundles results in the very negligible damage to the remaining or resided nanotubes, as shown by the low-intensity D-band of defects in Figure 5.1. This becomes obvious in comparison to the distinct characteristic RBM- and G^-,G^+-lines and provides another indication of the presumably mesoscopic single-shot ablation mechanism, involved into such disintegration of the nanotube packaging texture within the material.In striking contrast to the single-shot trend, much more distinct ablation craters and accompanying lower-fluence ablative lateral features – pores and extensive debris – emerge on the substrate surface at both these laser wavelengths upon higher laser exposures N = 30-10^3 pulses per spot (Figures 5.2-5.5). The most pronounced, drill-through craters appear unexpectedly in the very central parts of the laser focal spots at 1030 nm for $N \geq$ 30 pulses and the local fluence $F > F_{crat}(30) \approx$ 0.8 J/cm^2 (Figure 5.4a,b and Figure 5.6), while at 515 nm and $F_0 \approx$ 0.3 J/cm^2 much shallower craters were produced even for $N \geq$ 10^3 pulses (Figure 5.4c,d and Figure 5.7). In part, this yields from the much higher – almost by four times (cfg., the 1030-nm and 515-nm single-shot mesoscopic ablation thresholds $F_{mes}(1)$ = 25±3 mJ/cm^2 and $F_{mes}(1)$ = 94±17 mJ/cm^2, respectively) – absorption coefficient at 1030 nm.

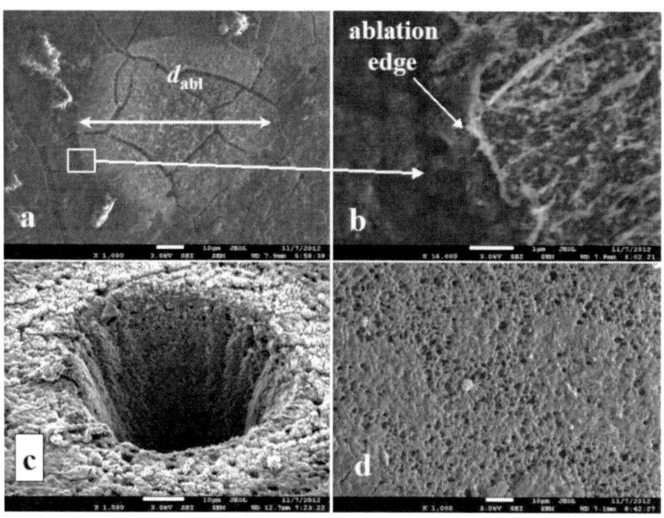

Figure 5.2. SEM images of the SWCNT substrate: (a) the surface spot ablated by a single shot at 1030 nm and $F_0 \approx$ 1 J/cm^2 (the scale bar is 10 μm); (b) its edge (the scale bar is 1 μm); (c) multi-shot (N = 1000) crater ablated at 1030 nm and $F_0 \approx$ 1 J/cm^2 and viewed at the 45°-angle of incidence (the scale bar is 10 μm); (d) the reference non-ablated surface spot (the scale bar is 10 μm).

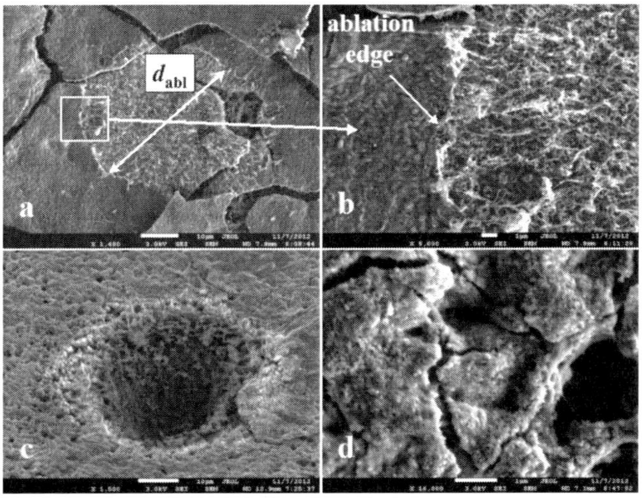

Figure 5.3. SEM images of the SWCNT substrate: (a) the surface spot ablated by a single shot at 515 nm and $F_0 \approx 0.3$ J/cm^2 (the scale bar is 10 μm); (b) its edge (the scale bar is 1 μm); (c) multi-shot (N = 1000) crater ablated at 515 nm and $F_0 \approx 0.3$ J/cm^2 and viewed at the 45°-angle of incidence (the scale bar is 10 μm); (d) the reference non-ablated surface spot (the scale bar is 1 μm).

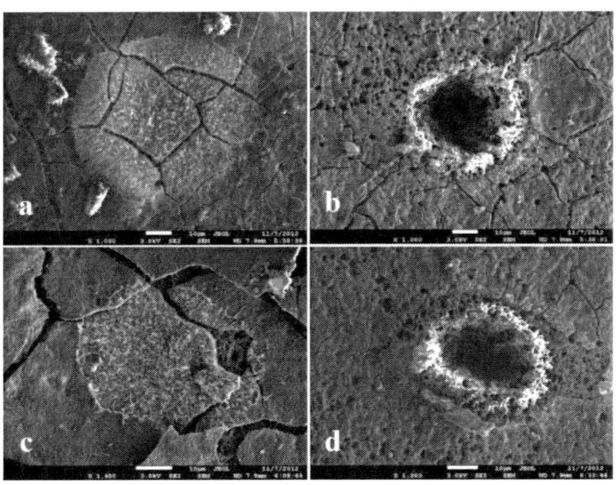

Figure 5.4. SEM images of the SWCNT substrate: (a,b) the surface spots ablated by the single and multiple (N = 30) shots at 1030 nm and $F_0 \approx 1$ J/cm^2; (c,d) the surface spots ablated by the single and multiple (N = 1000) shots at 515 nm and $F_0 \approx 0.3$ J/cm^2. The focusing conditions during the single and multi-shot exposures are the same. All scale bars are 10 μm.

However, it is still obvious that, for comparable IR and visible laser fluences – $F_{crat}(100) \approx 0.45$ J/cm^2 and $F_0 \approx 0.3$ J/cm^2 at $N \geq 10^3$ pulses, respectively, the other, much stronger ablation mechanism with multi-micron ablation rates per pulse works in the high peak fluence IR-laser irradiation regime. The accompanying micro-pores were emerged laterally around the craters (Figure 5.5) at much lower, almost N-independent fluences – F_{por} = (0.03-0.04)±0.01 J/cm^2 at 1030 nm (Figure 5.6) and F_{por} = (0.1-0.2)±0.05 J/cm^2 at 515 nm (Figure 5.7), apparently, through inhomogeneous ablation of the substrate.

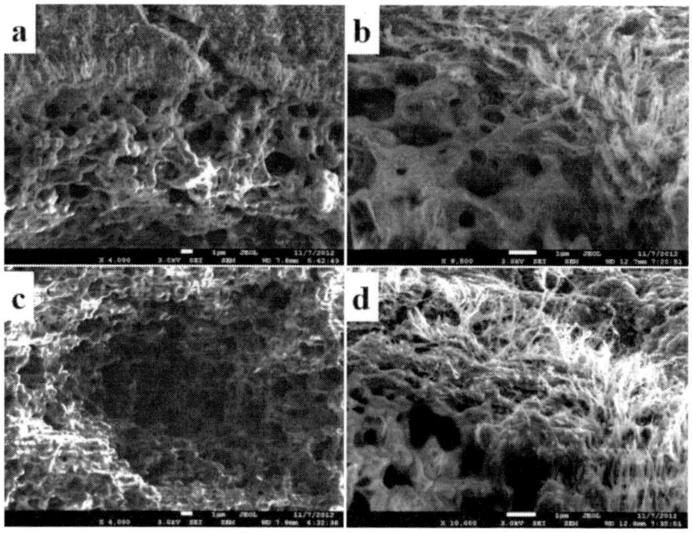

Figure 5.5. High-magnification SEM images of the multi-shot craters on the SWCNT substrate: the surface spot ablated by multiple (N = 100) shots at 1030 nm and $F_0 \approx$ 1 J/cm^2 (a,b) or at 515 nm and $F_0 \approx$ 0.3 J/cm^2 (c,d), respectively, and viewed at the normal (a,c) and 45°-angle (b,d) of incidence. All scale bars are 1 μm.

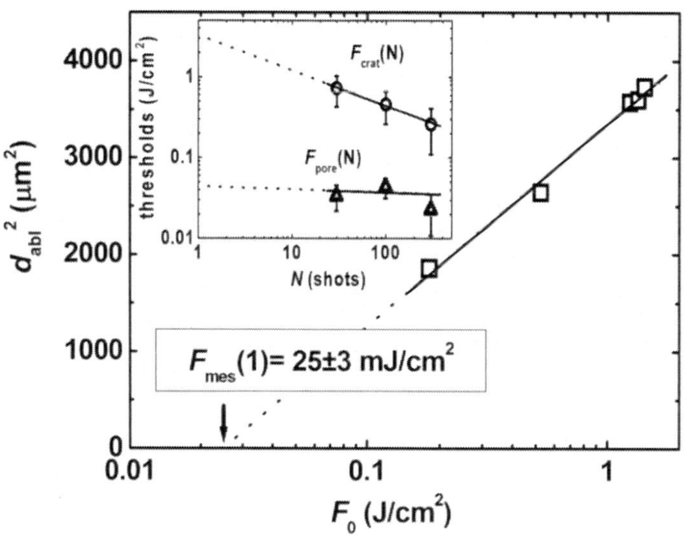

Figure 5.6. Fluence dependence of squared single-shot ablation spot diameter, $d_{abl}^2 - F_0$, at 1030 nm, and the corresponding wavelength-dependent ablation threshold value F_{mes}(l). Inset: dependences F_{crat} (light circles) and F_{pore} (light triangles) versus N.

Again, the relationship between the multi-shot thresholds at 1030 and 515 nm is close to ≈4, as before for the single-shot mesoscopic ablation thresholds $F_{mes}(1)$, supporting this our previous observation. Both these highly exposed ablated regions exhibit in their Raman spectra very severe structural damage to the remaining individual surface SWCNTs, featured in the form of their visible luminescence near 488 nm (Figure 5.1) because of the defective C-C bonds and strongly damped characteristic RBM- and G^-,G^+-bands.

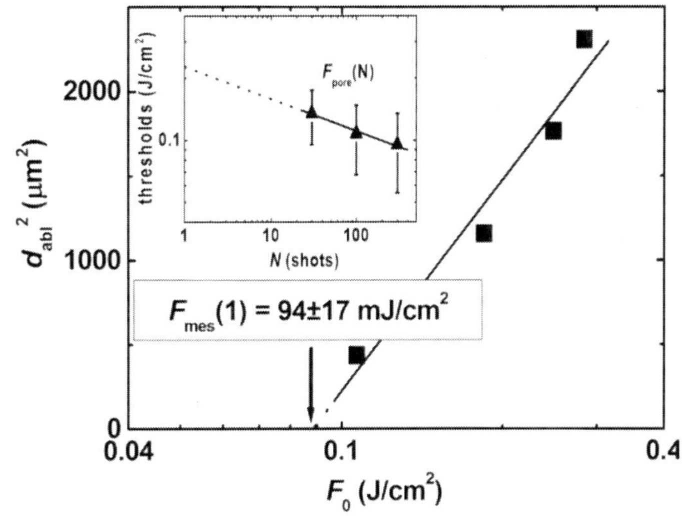

Figure 5.7. Fluence dependence of squared single-shot ablation spot diameter, dabl2 – F0, at 515 nm, and the corresponding wavelength-dependent ablation threshold value Fmes(l). Inset: dependence Fpore (dark triangles) versus N.

Since the multi-micron deep intense IR fs-laser ablation of the SWCNT substrate becomes visible at the same local fluences only upon multi-shot exposures, it is obvious that the ablation is triggered by some cumulative effects, according to the cumulative dependence of $F_{crat}(N)$ in Figure 5.6. Moreover, such multi-micron deep intense IR fs-laser ablation of the SWCNT substrate contradicts the much shallower (a few tens of nanometer) IR optical penetration depth in it. This contradiction can be resolved assuming some surface "seeding" ablation mechanism, which then triggers the bulk removal of the material. In our opinion, one potential triggering process could be intense internal ablative disintegration of SWCNT species, launching into the bulk substrate a strong (~ 10^{11} Pa) shock wave [145] (see also Section 4.3 and Figure 4.15), which strongly compressively heats the low-density material, yielding in its expulsion [28], as previously demonstrated during fs-laser ablation of low-density graphite foils [146] (see Figure 5.8). The high anticipated single-shot threshold for such ablation, $F_{crat}(N=1) = 3\pm1$ J/cm^2 (Figure 5.6), is consistent with the previous estimates of the microscopic CNT ablation threshold > 0.6 J/cm^2 [130] and does indicate the underlying severe fs-laser ablation of SWCNTs. The evaluated threshold $F_{crat}(N=1)$ is comparable to typical thresholds for ablation of highly-reflective metals (e.g, aluminum [107]) through fragmentation of their supercritical fluids, but far exceeds the same fragmentation ablation threshold for crystalline graphite, ≈ 0.3 J/cm^2 [7] (see also Section 4.3), possibly, because of very low van der Waals cohesion between atomic sheets of graphite.

In the framework of this two-step ablation mechanism, qualitatively similar ablation character may be implied for the transition micro-porous surface regions in Figure 5.5, where the ablation mechanism operates at the lower fluences not totally, but locally – in a fluctuation manner, and yielding in multiple micro-pores instead of a single multi-micron crater. Such fluctuation-driven initiation of the multi-shot deep craters at 515-nm lower laser fluences, starting from a few micro-pores, is illustrated in Figure 5.9. Again, the almost N-independent thresholds F_{pore} for the appearance of the micro-porous surface regions, surrounding the deep central craters, indicate that the micro-pore opening could be the initial, potentially – triggering – step during the intense cumulative ablation. Surprisingly, at both these laser wavelengths for lower, near-threshold laser fluences the lateral crater-edge debris appear rather organized, being represented by irregular arrays of standing SWCNTs (Figures 5.5b,d). According to their micro-Raman characterization, such almost undamaged standing nanotubes resemble the original ones, with the characteristic structural parameters – the average diameter d_{RBM} = 1.40±0.02 nm and the semiconducting factor C_G = 47.5± 0.8 versus 1.36±0.04 nm and 46.8± 0.9 for the initial species, respectively. Such partial cumulative alignment of the initial chaotic collection of the SWCNT species, as compared to the single-shot near-edge chaotic arrangement of nanotube bundles in Figures 5.2 and 5.3, may result from their polarization-sensitive ablation [139], which is much more efficient for the nanotubes with the higher projection to the polarization vector.

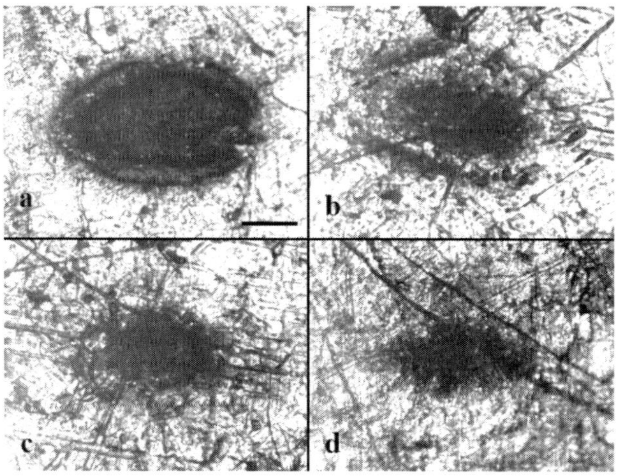

Figure 5.8. Optical images of low-density graphite foil ablated by single 800-nm, 100-fs laser shots at F_0≈ 0.7 (a), 0.5 (b), 0.3 (c) and 0.2 (d) J/cm^2. The scale bar in the image (a) is 50 μm.

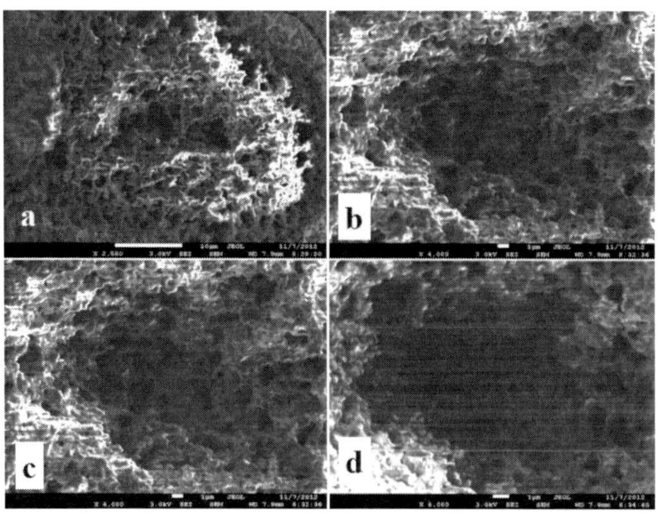

Figure 5.9. SEM images of the SWCNT substrate ablated by 30 (a), 100 (b), 300 (c) and 1000 (d) successive shots at 515 nm and F0 ≈ 0.3 J/cm2. The scale bars are 10 (a) and 1 μm (b-d).

Moreover, the standing orientation of the individual SWCNT species and of their bunches on the substrate surface may be repetitatively produced by the outgoing gaseous ablative fluxes. The hair-like texture and semiconducting character of the produced SWCNT surface patterns makes their large-scale surface arrays very promising in photocatalitic and photonic sensing applications, owing to their enhanced optical absorbance and abnormously enlarged effective surface. Comparing to other patterning methods of carbon nanotube based materials for production of carbon nanotube based devices, e.g. high-density oxygen plasma etching, the low-fluence mesoscopic ablative laser patterning provides large-scale, high-volume rates with good pattern quality and doesn't damage these species microscopically, while oxygen plasma can generate defects in the carbon nanotubes and can further react with the defects [147,148].Therefore, the femtosecond laser patterning process for carbon nanotubes, which is compatible with CMOS technology, high-speed, and damage-free on a flexible substrate is very promising.

CONCLUSION

Femtosecond laser interaction with basic carbon materials – graphite, diamond and nanotubes – under extreme electronic excitation, pressures and temperatures results in their structural transformations into the intermediate structurally different excited state, representing an atomic- or nano-scale mixture of different carbon hybridizations (Figure 6.1), as envisioned by micro-Raman spectroscopy. Though such interaction starts from their ultrafast electronic excitation, main lattice dynamics occurs on sub- and picosecond timescales, e.g., as sub-picosecond disordering/melting of graphite and its picosecond transformation to diamond through intermediate structural states with mixed hybridization, while in the latter case a shockwave impact may predominate over an electronic lattice softening effect during the transformation. Meanwhile, the intermediate excited state can be quenched into different final structural carbon forms with mixed hybridization states (Figure 6.1), depending on the post-pulse evolution history, which may include melting, ablation, shock-wave impact and all conjugate relaxation processes, such as re-solidification, re-deposition, acoustic relaxation etc.

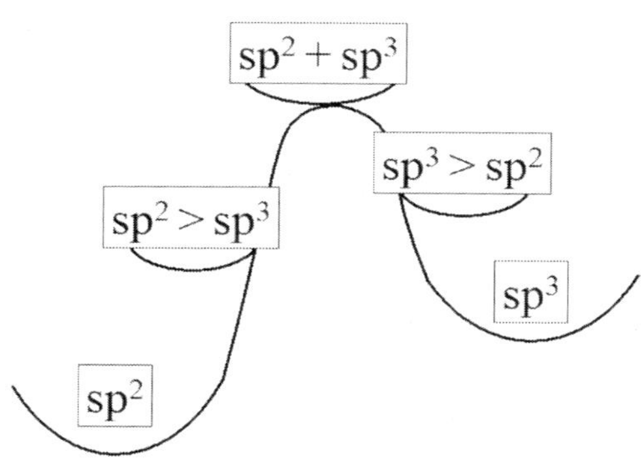

Figure 6.1. Schematic of basic structural states of carbon and the excited transition state.

In contrast, due to their ultimate thermal stability of carbon nanotubes, their microscopic ablative disordering in occurs in more complex way only at much higher fs-laser fluences upon multi-shot exposures, potentially driving strong solid-state shock waves, which destroy in bulk the mesoscopic nanotube package without any damage to individual nanotubes.

ACKNOWLEDGMENTS

The authors acknowledge the incredible contributions and technical assistance in their experimental work, invaluable discussions and friendly support from their numerous past and present co-workers – Prof. Alexander Karabutov (Moscow State University), Prof. Dietrich von der Linde and Prof. Klaus Sokolowski-Tinten (both – Duisburg University), Prof. Bërbel Rethfeld (University of Kaiserslautern), Prof. Andrey Ionin, Dr. Roman Khmelnitskiy, Dr. Leonid Seleznev, Dr. Dmitriy Sinitsyn and many our students from P.N. Lebedev Physical Institute of Russian Academy of sciences. The authors are also grateful to Russian Foundation for Basic Research (grant nos. 08-08-00756 and 10-08-00941) and PC "Kristall" (Smolensk) for their financial support.

REFERENCES

[1] Reich, S.; Thomsen, C.; *Phis. Trans. R. Soc. Lond.* 2004, 362, 2271-2288.
[2] Ferrari, A.C.; Robertson, J.; *Phis. Trans. R. Soc. Lond.* 2004, 362, 2477-2512.
[3] Karavanskiy, V.A.; Mel'nik, N.N.; Zavaritskaya, T.N.; *JETP Lett.* 2001, 74, 186; see also Melnik, N.N. *Raman scattering – 80 years of investigations*; FIAN: Moscow, 2008, 147-161 (in Russian).
[4] Reitze, D.H.; Ahn, H.; Downer, M.C.; *Phys. Rev.* B 1992, 45, 2677.
[5] Saeta, P.; Wang, J.-K.; Siegal, Y.; Bloembergen, N.; *Phys. Rev. Lett.* 1991, 67, 1023.
[6] Sokolowski-Tinten, K.; Schulz, H.; Bialkowski, J.; von der Linde, D. *Appl. Phys.* A 1991, 53, 227.
[7] Sokolowski-Tinten, K.; Kudryashov, S.I.; Temnov, V.; Bialkowski J.; *Springer Series in Chemical Physics* 2000, 66, 425.
[8] Shirk, M.D.; Molian, P.A.; *Carbon* 2001, 39, 1183.
[9] Lenner, M.; Kaplan, A.; Huchon, C.; Palmer, R.E.; *Phys. Rev.* B 2009, 79, 184105.
[10] Dallas, T.; Holtz, M.; Ahn, H.; Downer, M.C.; *Phys. Rev.B* 1994, 49, 796.
[11] Kudryashov, S.I.; *J. Appl. Phys.* 2006, 100, 036103.
[12] Hu, A.; Rybachuk, M.; Lu, Q.-B.; Duley, W.W.; *Appl. Phys. Lett.* 2007, 91, 131906.
[13] Carbone, F.; Baum, P.; Rudolf, P.; Zewail, A.H.; *Phys. Rev. Lett.* 2008, 100, 035501.
[14] Raman, R.K.; Murooka, Y.; Ruan, C.; Yang T.; *Phys. Rev. Lett.* 2008, 101, 077401.
[15] Nishioka, K.; Nasu, K.; *Phys. Rev. B* 2009, 80, 235420.
[16] Sano, T.; Takahashi, K.; Sakata O.; *J. Phys. D: Conf. Ser.* 2009, 165, 012019.
[17] Miyamoto, Y.; Zhang, H.; Tomanek, D.; *Phys. Rev. Lett.* 2010, 104, 208302.
[18] Sciani, G.; Miller, R.J.D.; *Rep. Prog. Phys.* 2011, 74, 096101.
[19] Nüske, R.; Jurgilaitis, A.; Enquist, H.; Harb, M.; Hakanson, U.; Larsson, J.; *Appl. Phys. Lett.* 2012, 100, 043102.

[20] Bundy, F.P.; *Physica A* 1989, 156, 169.

[21] Bundy, F.P.; Basset, W.A.; Weathers, M.S.; Hemley, R.J.; Mao H.K.; *Carbon* 1996, 34, 141.

[22] Malvezzi, A.M.; Bloembergen, N.; Huang, C.Y.; *Phys. Rev. Lett.* 1986, 57, 146.

[23] Baitin, A.V.; Lebedev, A.A.; Romanenko, S.V.; Senchenko, V.N.; *High Temp.-Press.* 1990, 21,157.

[24] Bonelli, M.; Miotello, A.; Ossi, P.M.; Pessi, A.; Gialanella, S.; *Phys. Rev. B* 1999, 59, 13513.

[25] Basharin, A.Yu.; Dozhdikov, V.S.; Dubinchuk, V.T.; Kirillin, A.V.; Lysenko, I.Yu.; Turchaninov, M. A.; *Tech. Phys. Lett.* 2009, 35, 428.

[26] Eliezer, S.; Moshe, E.; Eliezer, D.; *Laser Part. Beams* 2002, 20, 87.

[27] Celliers, P.M.; Bradley, D.K.; Collins, G.W.; Hicks, D.G.; *Rev. Sci. Instr.* 2004, 75, 4916.

[28] Graham, R.A. *Solids under high-pressure shock compression*; Springer: New York, 1993.

[29] Mingareev, I.; Horn, A.; *Appl. Phys. A* 2008, 92, 917.

[30] Wu, Z.; Zhu, X.; Zhang, N.; *J. Appl. Phys.* 2011, 109, 053113.

[31] Batani, D.; Strati, F.; Stabile, H.; Tomasini, M.; *Phys. Rev. Lett.* 2004, 92, 065503.

[32] Kern, G.; Hafner, J.; *Phys. Rev. B* 1998, 58, 13167.

[33] Jeschke H.O., Garcia M.E., Bennemann K.H., *Phys. Rev. B* 1999, 60, R3701.

[34] Wang C.Z., Ho K.M., Shirk M.D., Molian P.A., *Phys. Rev. Lett.* 2000, 85, 4092.

[35] Scandolo S., Bernasconi M., Chiarotti G.L., Focher P., Tosatti E., *Phys. Rev. Lett.* 1995, 74, 4015-4018.

[36] Silvestrelli P.L., Parinello M., *J. Appl. Phys.* 1998, 83, 2478.

[37] Garcia M.E., Jeschke H.O., *Appl. Surf. Sci.* 2003, 208-209, 61.

[38] Hulin, D.; Combescot, M.; Bok, J.; Migus A.; *Phys. Rev. Lett.* 1984, 52, 1998.

[39] Ionin, A. A.; Kudryashov, S.I.; Makarov, S.V.; Seleznev, L.V.; Sinitsyn, D.V.; *JETP Lett.* 2009, 90, 423.

[40] Geints, Yu.E.; Zemlyanov, A.A.; Ionin, A.A.; Kudryashov, S.I.; Seleznev, L.V.; Sinitsyn, D.V.; Sunchugasheva, E.S.; *JETP* 2010, 111, 724.

[41] Sokolowski-Tinten, K.; von der Linde, D.; *Phys. Rev. B* 2000, 61, 2643.

[42] Siebert, K.; Cho, G.C.; Kütt, W.; Kurz, H.; Reitze, D.H.; Dadap, J.I.; *Phys. Rev. B* 1990, 42, 2842.

[43] Kudryashov, S.I.; Kandyla, M.; Roeser, C.; Mazur, E.; *Phys. Rev. B* 2007, 75, 085207.

[44] Bulgakova, N.M.; Zhukov, V.P.; Vorobyev, A.Y.; Guo, C.; *Appl. Phys.A* 2008, 92, 883.

[45] Kudryashov, S.I.; Emel'yanov, V.I.; *JETP* 2002, 94, 94.

[46] Pagliara, S.; Galimberti, G.; Mor, S.; Montagnese, M.; Ferrini, G.; Grandi, M.S.; Galinetto, P.; Parmigiani, F.; *J. Am. Chem. Soc.* 2011, 133, 6318.

[47] Spataru, C.D.; Benedict, L.X.; Louie, S.G.; *Phys. Rev. B* 2004, 69, 205204.

[48] Apostolova,T.; Ionin, A.A.; Kudryashov, S.I.; Seleznev, L.V.; Sinitsyn, D.V.; *Opt. Eng.* 2012, 51, 121808.

[49] Tatar, R.C.; Rabii, S.; *Phys. Rev. B* 1982, 25, 4126.

[50] Golosov, E.V.; Ionin, A.A.; Kolobov, Yu.R.; Kudryashov, S.I.; Ligachev, A.E.; Makarov, S.V.; Novoselov, Yu.N.; Seleznev, L.V.; Sinitsyn, D.V.; Sharipov, A.R.; *Phys. Rev. B* 2011, 83, 115426.

[51] Ionin, A.A.; Kudryashov, S.I.; Makarov, S.V.; Saltuganov, P.N.; Seleznev, L.V.; Sinitsyn, D.V.; Sharipov, A.R; *JETP Lett.* 2012, 96, 375.

[52] Bigot, J.-Y.; Portella, M.T.; Schoenlein, R.W.; Cunningham, J.E.; *Phys. Rev. Lett.* 1991, 67, 636.

[53] Bruschi, L.; Santini, M.; Torzo, G.; *J. Phys. B: At. Mol. Phys.* 1984, 17, 1137.

[54] Schafer, S.; Liang, W.; Zewail, A.H.; *New J. Phys.* 2011, 13, 063030.

[55] Samsonidze, G.G.; Barros, E.B.; Saito, R.; Jiang, J.; Dresselhaus, G.; Dresselhaus, M.S.; *Phys. Rev. B* 2007, 75, 155420.

[56] Madelung, O. *Physics of Group IV Elements and III-V Compounds*; Landolt-Bernstein New series 1982, 17a, 36.

[57] Yu, P.; Cardona, M. *Fundamentals of Semiconductor Physics*; Springer: New York, 1996.

[58] Davis, K.M.; Miura, K.; Sugimoto, N.; Hirao, K.; *Opt. Lett.* 1996, 21, 1729.

[59] Marshall, G.D.; Ams, M.; Withford, M.J.; *Opt. Lett.* 2006, 31, 2690.

[60] Bricchi, E.; Klappauf, B.G.; Kazansky, P.G.; *Opt. Lett.* 2004, 29, 119.

[61] Glezer, E.N.; Milosavljevic, M.; Huang, L.; Finlay, R.J.; Her, T.-H.; Callan, J.P.; Mazur, E.; *Opt. Lett.* 1996, 21, 2023.

[62] Osellame, R.; Maselli, V.; Vazquez, R.M.; Ramponi, R.; Cerullo, G.; *Appl. Phys. Lett.* 2007, 90, 231118.

[63] Kudryashov, S.I.; Joglekar, A.; Mourou, G.; Herbstman, J.F.; Hunt, A.J.; *Appl. Phys. Lett.* 2007, 91, 141111.

[64] Kudryashov, S.I.; Zvorykin, V.D.; Ionin, A.A.; Mizeikis, V.; Juodkazis, S.; Misawa, H.; *Appl. Phys. Lett.* 2008, 92, 101916.

[65] Hu, X.; Qian, B.; Zhang, P.; Wang, X.; Su, L.; Qiu, J.; Zhu, C.; *Laser Phys. Lett.* 2008, 5, 394.

[66] Gamaly, E.G.; Joudkazis, S.; Nishimura, K.; Misawa, H.; Luther-Davis, B.; Hallo, L.; Nicolai, P.; Tikhonchuk, V.T.; *Phys. Rev.* B 2006, 73, 214101.

[67] Kononenko, T.V.; Meier, M.; Komlenok, M.S.; Pimenov, S.M.; Romano, V.; Pashinin, V.P.; Konov, V.I.; *Appl. Phys. A* 2008, 90, 645.

[68] Neff, M.; Kononenko, T.V.; Pimenov, S.M.; Romano, V.; Luthy, W.; Konov, V.I.; *Appl. Phys. A* 2009, 97, 543.

[69] Ashcom, J.B. PhD thesis, Harvard University, 2003; http://mazur-www.harvard.edu/publications.php?function=displayandrowid=169.

[70] Grigor'ev, I.S.; Meylikhova, E.Z. (eds.), *Physical Quantities*; Energoatomizdat: Moscow, 1991.

[71] Palik, E.D. (Ed.) *Handbook of Optical Constants of Solids*; Academic Press: Orlando, 1998.

[72] Boyd, R.W. *Nonlinear Optics*; Academic Press: San Diego, 1992; Kaminskii, A.A.; Ralchenko, V.G.; Konov, V. I.; *Laser Phys. Lett.* 2006, 3, 171.

[73] Cerullo, G.; Osellame, R.; Taccheo, S.; Marangoni, M.; Polli, D.; Ramponi, R.; Laporta, P.; De Silvestri, S.; *Opt. Lett.* 2002, 27, 1938.

[74] Hsu, E.M.; *Applied Physics A* 2010, doi: 10.1007/s00339-010-5986-4.

[75] Zalloum, O.H.Y.; *Opt. Exp.* 2010, 18, 13122.

[76] Ionin, A.A.; Kudryashov, S.I.; Mikhin, K.E.; Seleznev, L.V.; Sinitsyn, D.V.; *Laser Physics* 2010, 20, 1778.

[77] Simmonds, R.D.; *Opt. Exp.* 2011, 19, 24122.

[78] Kononenko, T.V.; *Diam. Relat. Mater.* 2009, 18, 196.
[79] Kononenko, T.V.; Diam. Relat. Mater. 2010, doi: 10.1016/j.diamond.2010.12.013.
[80] Lee, G.C.B.; *J. Exp. Nanoscience* 2012, 7, 662.
[81] Kudryashov, S.I.; Emel'yanov, V.I.; *JETP Lett.* 2001, 73, 487.
[82] Kudryashov, S.I. *Thermodynamic characteristics of the metastable liquid-vapor equilibrium during pulsed laser vaporization of polycrystalline graphite*; PhD thesis: Moscow State University, 1999.
[83] von der Linde, D.; Sokolowski-Tinten, K.; Bialkowski, J.; *Appl. Surf. Sci.* 1997, 109-110, 1.
[84] Sokolowski-Tinten, K.; Blome, C.; Dietrich C.; *Phys. Rev. Lett.* 2001, 87, 225701; Enquist, H.; Navirian, H.; Hansen, T.N.; *Phys. Rev. Lett.* 2007, 98, 225502; Nicoul, M.; Shymanovich, V.; Tarasevich, A.; von der Linde, D.; *Appl. Phys. Lett.* 2011, 98, 191902.
[85] Colonna, F.; Los, J.H.; Fasolino, A.; Meijer, E.J.; *Phys. Rev. B* 2009, 80, 134103.
[86] Basharin, A.Yu.; Lysenko, I.Yu.; Turchaninov, M.A.; *High Temp.* 2012, 50, 464.
[87] Sokolowski-Tinten, K.; Ziegler, W.; von der Linde, D.; Siegal, M.P.; Overmyer, D.L.; *Appl. Phys. Lett.* 2005, 86, 121911.
[88] Leider, H.R.; Krikorian, O.H.; Young, D.A.; *Carbon* 1973, 11, 555.
[89] Temnov, V.V.; Sokolowski-Tinten, K.; Zhou, P.; *J. Opt. Soc. Am. B* 2006, 23, 1954.
[90] Ionin, A.A.; Kudryashov, S.I.; Seleznev, L.V.; Sinitsyn, D.V.; *JETP Lett.* 2011, 94, 753.
[91] Bosak, A.; Krisch, M.; Mohr, M.; Maultzsch, J.; Thomsen, C.; *Phys. Rev. B* 2007, 75, 153408.
[92] Gusev, V.E.; Karabutov, A.A. *Laser Optoacoustics*; AIP: New York, 1993.
[93] Ionin, A.A.; Kudryashov, S.I.; Seleznev, L.V.; Sinitsyn, D.V.; Bunkin A.F.; Lednev, V.N.; Pershin, S.M.; *JETP* 2013, 143, 347.
[94] Perez, D.; Lewis, L.J.; *Phys. Rev. B* 2003, 67, 184102.
[95] Upadhyay, A.K.; Inogamov, N.A.; Rethfeld, B.; *Phys. Rev. B* 2008, 78, 045437.
[96] Mott, N.F. *Metal-Insulator Transitions*; Taylor-Francis: Boston, 1990; Rodolakis, F.; Hansmann, P.; Rueff, J.-P.; Toschi, A.; *Phys. Rev. Lett.*2010, 104, 047401.
[97] Zhakhovskii, V.V.; Inogamov, N.A.; Nishihara, K.; *JETP Lett.* 2008, 87, 423.
[98] Leveugle, E.; Ivanov, D.S.; Zhigilei, L.V.; *Appl. Phys. A* 2004, 79, 1643.
[99] Inogamov, N.A.; Zhakhovskii, V.V.; Ashitkov, S.I.; *JETP* 2008, 107, 1.
[100] Demaske, B.J.; Zhakhovsky, V.V.; Inogamov, N.A.; *Phys. Rev. B* 2010, 82, 064113.
[101] Ionin, A.A.; Kudryashov, S.I.; Seleznev, L.V.; *Phys. Rev. E* 2010, 82, 016404.
[102] Mizeikis, V.; Juodkazis, S.; Balciunas, T.; Misawa, H.; *J. Appl. Phys.* 2009, 105, 123106.
[103] Zeng, X.; Mao, X.; Wen, S.B.; Greif, R.; Russo, R.E.; J. Phys. D: *Appl. Phys.* 2004, 37, 1132.
[104] Henley, S.J.; Crey, J.D.; Sylva, S.R.P.; Fuge, G.M.; *Phys. Rev. B* 2005, 72, 205413.
[105] Sedov, L.I. *Similarity and dimensional methods in mechanics*; Cleaver Hume: London, 1959.
[106] Zel'dovich, Y.B.; Raizer, Y.P. *Physics of shock waves and high-temperature hydrodynamic phenomena*; Dover: New York, 2002.
[107] Ionin, A.A.; Kudryashov, S.I.; Makarov, S.V.; Seleznev, L.V.; Sinitsyn,D.V.; *JETP Lett.* 2011, 94, 34.

[108] Dachraoui, H.; Husinsky, W.; *Phys. Rev. Lett.* 2006, 97, 107601.
[109] Bundy, F.P.; Basset, W.A.; Weathers, M.S.; Hemley, R.J.; Mao, H.K.; *Carbon* 1996, 34, 141.
[110] Gustafson, P.; *Carbon* 1986, 24, 169.
[111] Yang, J.; Zhao, Y.; Zhang, N.; Liang, Y.; Wang, M.; *Phys. Rev. B* 2007, 76, 165430.
[112] Eidmann, K.; Meyer-ter-Vehn, J.; Schlegel, T.; Hüeller, S.; *Phys. Rev. E* 2000, 62, 1202.
[113] Zhang, N.; Zhu, X.; Yang, J.; Wang, X.; Wang, M.; *Phys. Rev. Lett.* 2007, 99, 167602.
[114] Evans, R.; Badger, A.D.; Fallies, F.; Mahdieh, M.; Hall, T.A.; *Phys. Rev. Lett.* 1996, 77, 3359.
[115] Huang, L.; Yang, Y.; Wang, Y.; Zheng, Z.; Su, W.; J. Phys. D: *Appl. Phys.* 2009, 42, 045502.
[116] van Vechten, J.A.; *Phys. Rev. B* 1973, 7, 1479.
[117] Yin, M.T.; *Phys. Rev. B* 1984, 30, 1773.
[118] Biswas, R.; Martin, R.M.; Needs, R.J.; Nielsen, O.H.; *Phys. Rev. B* 1984, 30, 3210.
[119] Fahy, S.; Louie, S.G.; *Phys. Rev. B* 1987, 36, 3373.
[120] Ruoff, A.L.; Luo, H.; *J. Appl. Phys.* 1991, 70, 2066.
[121] Mao, H.K.; Bell, P.M.; *Science* 1978, 200, 1145.
[122] Mora, P.; *Phys. Fluids* 1982, 25, 1051.
[123] Lindl, J.; *Phys. Plasmas* 1995, 2, 3933.
[124] Atzeni, S.; Plasma Phys. *Control. Fusion* 2000, 42, B143.
[125] Singh, S.C.; Zeng, H.B.; Guo, C.; Weiping, C. (Eds.) *Nanomaterials: processing and characterization with lasers*; Wiley: Weinheim, 2012.
[126] Lichtfouse, E.; Schwarzbauer, J.; Robert, D.; *Environmental chemistry in a sustainable world*; Springer: New York, 2012.
[127] Rinzler, A.G.; Hafner, J.H.; Nikolaev, P.; Nordlander, P.; Colbert, D.T.; Smalley, R.E.; Lou, L.; Kim, S.G.; Tománek, D.; *Science* 1995, 269, 1550.
[128] Chae, J.; Jin, H.; Jain, K.; *Mater. Lett.* 2009, 63, 1823.
[129] Elmer, J.W.; Yaglioglu, O.; Schaeffer, R.D.; Kardos, G.; Derkach, O.; *Carbon* 2012, 50, 4114.
[130] Chae, J.; Ho, X.; Rogers, J.A.; Jain, K.; *Appl. Phys. Lett.* 2008, 92, 173115.
[131] Solodyankin, M.A.; Obraztsova, E.D.; Lobach, A.S.; Chernov, A.I.; Tausenev, A.V.; Konov, V.I.; Dianov, E.M.; *Opt. Lett.* 2008, 33, 1336.
[132] Ouyang, C.; Shum, P.; Wang, H.; Wong, J.H.; Wu, K.; Fu, S.; Li, R.; Kelleher, E.J.R.; Chernov, A.I.; Obraztsova, E.D.; *Opt. Lett.* 2010, 35, 2320.
[133] Guo, S.X.; Ben-Yakar, A.; J. Phys. D: *Appl. Phys.*2008, 41, 185306.
[134] Huang, N.; Wang, H.; Zhao, J.; Lui, H.; Korbelik, M.; Zeng, H.; *Lasers in Surgery and Medicine* 2010, 42, 638.
[135] Saito, R.; Dresselhaus, G.; Dresselhaus, M.S. *Physical Properties of Carbon Nanotubes*; Imperial College Press: London, 1998.
[136] Hashida, M.; Shimizu, S.; Sakabe, S.; *J. Phys.: Conf. Ser.* 2007, 59, 487.
[137] Lednev, V.N.; Pershin, S.M.; Obraztsova, E.D.; Kudryashov, S.I.; Bunkin, A.F.; J. Phys. D: *Appl. Phys.* 2013, 46, 052002.
[138] Dumitrică, T.; Garcia, M.E.; Jeschke, H.O.; Yakobson, B.I.; *Phys. Rev. B* 2006, 74, 193406.
[139] Kocabas, C.; Meitl, M.A.; Gaur, A.; Shim, M.; Rogers J.A.; *NanoLett.*2004, 4,2421.

[140] Obraztsova, E.D.; Bonard, J.-M.; Kuznetsov, V.L.; Zaikovskii, V.I.; Pimenov, S.M.; Pozharov, A.S.; Terekhov, S.V.; Konov, V.I.; Obraztsov, A.N.; Volkov, A.P.; *Nanostruct. Mater.* 1999, 12, 567.
[141] Il'ichev, N.N.; Obraztsova, E.D.; Garnov, S.V.; Mosaleva, S.E.; *Quant. Electron.* 2004, 34, 572.
[142] Alvarez, L.; Righi, A.; Guillard, T.; Rols, S.; Anglaret, E.; Laplaze, D.; Sauvajol, J.-L.; *Chem. Phys. Lett.* 2000, 316, 186.
[143] Dresselhaus, M.S.; Dresselhaus, G.; Jorio, A.; Souza, A.G.; Saito, F.R.; *Carbon* 2002, 40, 2043.
[144] http://www.amplitude-systemes.com/satsuma-fiber-laser.html
[145] Kudryashov, S.I.; Ionin, A.A.; Makarov, S.V.; Mel'nik, N.N.; Seleznev, L.V.; Sinitsyn, D.V.; *AIP Conf. Proc.* 2012, 1464, 244.
[146] Kudryashov, S.I.; Ionov, S.G.; *Physics and chemistry of materials processing* 2001, 4, 66 (in Russian).
[147] Lu, S.; Panchapakesan, B.; *Appl. Phys. Lett.* 2006, 88, 253107.
[148] Lu, S.; Liu, Y.; Shao, N.; Panchapakesan, B.; *Nanotechnology* 2007, 18, 065501.

In: Graphite
Editor: Quinton C. Campbell
ISBN: 978-1-62618-576-0

Chapter 3

NANOSCALE $SP^2 \rightarrow SP^3$ CONVERSION BY VISIBLE LIGHT IRRADIATION IN GRAPHITE

L. Radosinski[1,*], K. Nishioka[2], K. Nasu[3] and D. Wojt[4]
[1]Wroclaw University of Technology, The Department of Chemistry, Group of Chemical and Biochemical Processes, Wroclaw, Poland
[2]Chuo University, Department of Physics, Higashinakano, Hachioji-shi, Tokyo, Japan
[3]High Energy Accelerator Research Organization (KEK), Oho, Tsukuba, Japan
[4]Imagination Technologies Limited, Pl. Wolnosci 7/301, Wroclaw, Poland

ABSTRACT

Various recent experiments indicate us, that a graphite photo-stimulated with femtosecond laser pulse results in the formation of a quite new, but stable domain with sp^3 interlayer bonds. Here, we will theoretically review this mysterious non-equilibrium phase transition phenomenon, in terms of the lattice relaxation and the proliferation of photo-generated inter-layer charge transfer excitations. The excitation takes place in the visible region, and at the Franck-Condon state, the resulting electron-hole pair spanning neighboring two layers, is quite unstable, being easily dissipated into the 2d-semimetallic continuum electronic states, as positive and negative free carriers. However, by a small but finite probability, the electron and the hole become stable, as the lattice relaxation proceeds, being bound with each other, just like an exciton, through the inter-layer Coulomb attraction. This exciton state will be shown to self-localize, contracting the inter-layer distance and buckling the six membered ring of graphite, only around it. Thus, local inter-layer σ bonds are formed, or a tiny sp^3 nano-domain appears in the semi-metallic continuum. The Local Density Approximation (LDA) electronic structure analysis reveals that the electronic state of the new phase is an insulator domain immersed in a semimetal. In the next, in terms of molecular dynamics calculations we will explain proliferation of this domain and will show that the overall process of the transition becomes a multi-photon one. Furthermore using LCBOP potential and molecular dynamics, we will explain a temperature-induced ablation of the two bottom

* E-mail: lukasz.radosinski@pwr.wroc.pl.

layers with the diaphite inside. Finally a possible application of diaphite as hydrogen adsorption surface will be described.

1. INTRODUCTION

Carbon based materials have been recently studied intensively due to their interesting electronic and structural properties [1-5]. Various applications have been found in nano-electronics, medicine and material engineering triggered by the discovery of carbon structures such as fullerens, nanotubes and single graphite sheets, "graphenes" [1].

In nature however, the carbon element crystal manifests itself in two stable forms - graphite and diamond, both of them totally different structurally and electronically.

Because of the deep technological interest, this graphite-diamond transition has been studied for a long time [6, 7]. However, most approaches were focused on treating graphite pills with high pressure followed by strong irradiation or high temperature [7]. These methods may be energetically expensive and the results will be unsatisfactory due to number of impurities. Recently a new approach has been proposed by the so called Photo Induced Phase Transition (PIPT) [8], where graphite-diamond transition supposed to be induced by careful visible light irradiation. The key concept in this idea is multi-stability where the material possesses, besides its true ground state, also a false one separated by an energy barrier. This barrier is high enough to rule out possible transition due to ordinary thermal fluctuations. It is possible however to perform such a transformation via photo-excitation and lattice relaxation. This process followed by proliferation may end in a macroscopic, stable domain resulting in a new material. One may relate this concept with the mythical idea of a philosopher's stone driving generations of early scientists into modern chemistry and physics. The idea of transforming one element into another via mystical method – the philosopher's stone - may be found in photoinduced phase transition phenomena where subtle electronic excitation triggers massive structural-electronic response, so called "domino effect" [8]. Henceforth photoinduced phase transitions are sometimes called a modern alchemy.

The principal intend of the paper is to theoretically review this sp^2 to sp^3 conversion in graphite, and also to partly show recent experimental discoveries of photoinduced transformation of hybridization in graphite resulting in new stable phase called "diaphite" [2, 3, 5].

In the first two paragraphs of the paper a short presentation of graphite and diamond structure will be given followed by brief description of experimental and theoretical concepts of graphite-diamond transformation. The graphite-diamond transformation will be described in the context of PIPT [8].

In the third paragraph the key experimental finding will be introduced showing a photoinduced graphite-diaphite transition.

In the fourth paragraph a detailed description of adiabatic energy surface calculation of graphite will be shown. The large scale calculations were performed using Brenner's empirical potential. The geometrical structure of diaphite will be given as well as estimation of energy barrier between graphite and diaphite and LDA calculations for diaphite.

In the fifth paragraph a detailed description of initial mechanism of transformation will be presented. The model developed by us i.e. a photoinduced electron-hole excitation followed by its localization and leading via Coulomb interactions to diaphite domain formation will be given. In the sixth paragraph a proliferation of diaphite phase will be described. The large scale Molecular Dynamics (MD) [9] calculation prove that this transformation is a multiphoton one and only by femtosecond excitation. In the seventh paragraph the influence of diaphite domain formation on the remaininng graphite will be investigated using Long Range Carbon Bond Order Potential (LCBOP) [10, 11]. Furthermore temperature behavior of graphite-diaphite system will be studied. In the eight paragraph a possible application of diaphite will be discussed namely as hydrogen adsorbent. The last paragraph will be dedicated for discussion and future perspectives.

2. Graphite-Diamond Transformation

The carbon element manifests itself in nature in two stable crystal forms: graphite and diamond. A somewhat surprising fact is that, although diamond is the hardest natural material known on Earth, graphite is a little more stable than diamond having its formation energy ≈ 0.02eV lower than diamond. Although both materials are nearly energetically degenerated, transition between one another is energetically expensive and difficult. Graphite is classified according to the way how the subsequent layers are stacked.

Following this ordering graphite structure is put into three groups:

- hexagonal or Bernel structure with an AB-type stacking, where half of the atoms are located above the atoms from adjacent plane and half of them are located at the center of an above hexagonal ring in the adjacent plane. The space group symmetry of this structure is D^4_{6h}.
- rhombohedral structure with ABC-type stacking, where half of the atoms are located above the atom from adjacent plane however, below the center of hexagonal ring from above plane. The space group symmetry of this structure is D^5_{d3}.
- simple hexagonal structure with an AA-type stacking, where all of the atoms from adjacent planes are located directly above each other. The space group symmetry of this structure is D^1_{6h}.

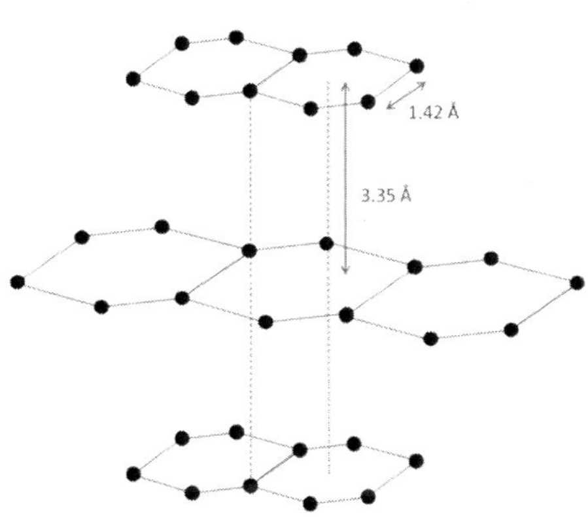

Figure 1. Graphite structure. The interlayer distance is 3.35 Å whereas the distance between carbon atoms of the same layer is 1.42 Å.

Each graphite layer is formed by a hexagonal net of carbon atoms, with equal nearest neighbor distance a= 1.42 Å. Rather strong and short sp^2-type covalent bonds between intra layer atoms are confronted with weak and long (3.35 Å) interaction's distances between neighboring layers. The huge difference between interlayer and intra layer bonds lengths and types is manifested in graphite's properties. Since electrons within the layers are strongly delocalized graphite is a conductor and it finds various applications in electronics. However due to the weak interlayer Van der Waals interactions graphite is no resistive to stress. Recently graphite attracted a great interest due to discovery of possibility of forming nano-structures out of single graphite sheets. Structures like fullerenes C_{60} [13] found variety of applications in medicine. Another example is a nanotube discovered as a mid-step material during fullerenes production. They have found extensive uses in electronics and novel materials. Finally single graphite sheets known as graphene exhibit many interesting electronic properties.

In comparison to graphite, diamond exhibits full tetrahedral symmetry and its structure can be divided into two groups:

- Cubic diamond, where the stacking is of ABC type. The space group symmetry of this structure is $O^7{}_h$,
- Hexagonal diamond, where the stacking is of AB type. The space group symmetry of this structure is $D^4{}_{6h}$.

It is characterized by equal 1.54 Å length and 109.54 ° angle between sp^3-type covalent bonds. Therefore diamond exhibits extreme properties. It is the hardest known material on Earth with 10 points on Mohs scale of mineral hardness. Diamond is also one of the toughest among the gems with 3.4 MN $m^{-3/2}$ stress resistivity and a semiconductor with 5.45 eV gap, thus having a number of applications in electronics and engineering.

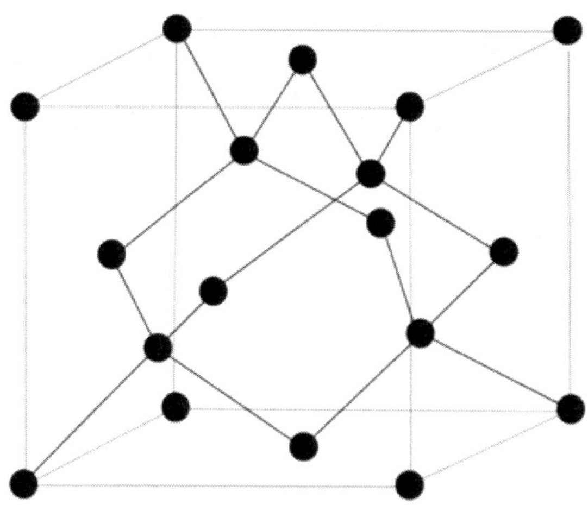

Figure 2. The diamond structure. The bond distance is 1.54 Å and the angle between the bonds is 109,45°.

Diamond however is mostly known due to its total inner reflection property and wide use in jewelry. Its formation requires specific conditions - very strong pressure between 45 and 60 kBar with relatively low, 900-1300 °C temperature. In nature these condition are known to be met in two places on Earth. One is in the lithosphere mantle below stable continental plates and at a meteorite strike site. Therefore natural diamonds are rare and supporting its high price in the global market.

The discovery that graphite and diamond are crystal structures formed with the same element stimulated numerous attempts to perform transformation from one to another.

Initially those efforts were concentrated on treating graphite pills with high pressure shocks after heating to high temperature when finally in 1954 T. Hall reported direct graphite-diamond conversion under static pressure. His method, continued under the name HPHT (High Pressure High Temperature) [7], involved treating carbon pills with high pressure (8-15 GPa) and high temperature (3000° C).

Many possibility of industrial production of diamond started to be achievable, the method possessed many drawbacks. The most important one was that the resulting diamond volume was surrounded with much greater volume of impurities. This method, although able to produce diamonds, was far from being efficient.

Recently a novel method was introduced via Chemical Vapor Deposition (CVD) [13] where carbon atoms in gas are provided with necessary conditions to settle on substrate in crystalline form. In contrast to HPHT method CVD require much lesser pressure (1-27 kPa). The method involves energizing various gases in the chamber and creating conditions for diamond growth on the substrate.

The biggest advantages of the method are ability to grow diamond over large areas and controlling properties of the diamond product.

It should be mentioned that there are alternative methods of manufacturing synthetic diamond using ultrasound cavitations or explosive detonation [14, 15].

2.1. Graphite-Diamond Photo Induced Phase Transition

The paper suggests possibility of an alternative process of Graphite-Diamond conversion i.e. photo induced transition. The key concepts behind the idea are multistability property and proliferation process.

Photoinduced phase transition scheme bases on the assumption that materials, besides true ground state, also possess a false ground state (FGS) [8, 10]. Such a FGS is separated from the true one by an energy barrier which is assumed to be much higher than the energy of thermal fluctuations.

It is also postulated that the FGS is deep enough to become a metastable one. The conversion from one ground state to another is made via photoexcitation. An electron excited by light creates a lattice distortion around it. The distortion triggers proliferation process which may lead to a new FGS.

This scenario may be employed in graphite-diamond transition. One may observe that graphite and diamond structures are nearly degenerated. Following the idea presented in previous paragraph one may assume that graphite is its true ground state whereas diamond is a false one as shown in Figure 3.

A photoexcited electron is transferred from one graphite layer to another one. The process triggers exciton formation leading to the appearance of Coulomb force between the layers which may be strong enough to overcome the barrier between the two ground states.

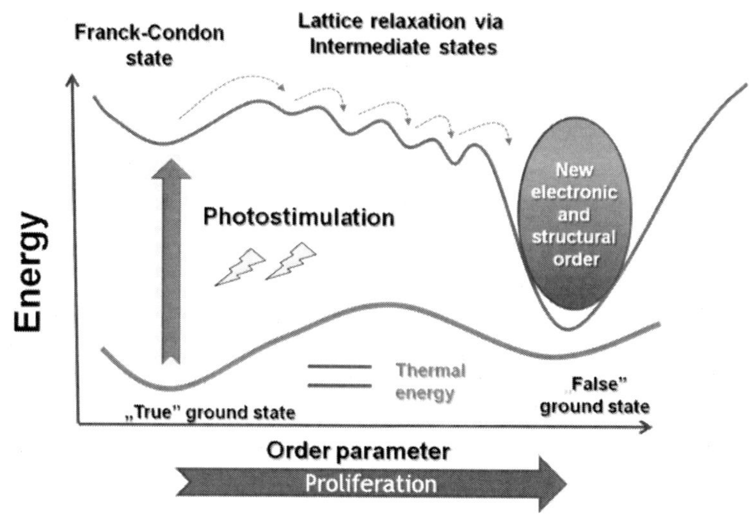

Figure 3. Schematic idea of a photoinduced phase transition [10].

3. Experimental Results

Scenario proposed in previous section is strictly theoretical however, a recent experiment conducted by Tanimura and Kanazaki [3] may bring new light on this problem.

During the experiment graphite sheets were illuminated by laser light. The laser was tuned to 790 nm which corresponds to 1.57 eV energy. Laser's light wave length falls into energy range corresponding to $\pi^*-\pi$ electronic transition. Furthermore the light was polarized perpendicular to the graphite plane. Interesting is that, as a result, structural change occurred. The appearance of the new material formation was investigated using a STM (Scanning Tunneling Microscope). The observations showed that within the structurally changed graphite domain 1/3 of atoms were intruded, whereas the remaining 2/3 were extruded upwards as shown in Figure4.

It was also found that the new domain became stable for up to 10 days long. Further investigations revealed the following properties of transition:

1. The phase transition is a highly nonlinear process of less than 10 photons.
2. The resulting domain is of nanoscopic size up 1000 carbon atoms.
3. The exciting light should be a femtosecond pulse, while a picosecond pulse makes no difference.
4. Too strong irradiation results in ablation of two graphite sheets. However the remaining structure is unchanged hence the conversions is only within two layers.
5. A new peak at around 2 eV appears in the optical reflectivity spectrum which may come from semimetal-insulator transition.

The experimental results suggest that atoms intrusion is a result of $sp^2 \longrightarrow sp^3$ bond type transformation within the space between graphite layers. Nonetheless experiment did not give answer for several crucial questions, namely:

1. What are the electronic and structural properties of the new phase ("diaphite")?
2. What is the physical process governing the transformation?
3. What is the nature of the nonlinear response of the system for the photostimulation?

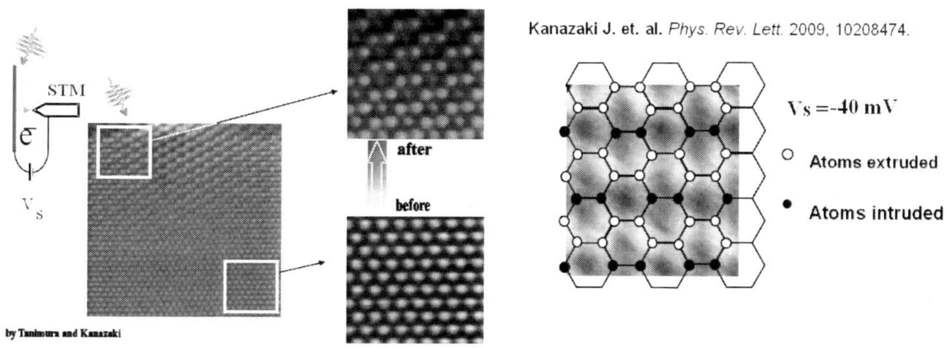

Figure 4. STM image of graphite before and after the illumination by photons (1.6 eV) in Tanimura and Kanazaki experiment [3].

The research presented in the paper gives answer for the above questions.

4. Potential Surface Calculations

4.1. Introduction

Experimental results suggest that the inferred "diaphite" domain is of the size of order of 1000 atoms. In order to acquire correct results the surrounding graphite material has to be of at least of 10 000 atoms size. The Density Functional Theory (DFT) is no longer an efficient method to handle such calculations. Therefore Radosinski et. al. [5] employed the semi-empirical methods like modified Brenner's potential. The idea behind Brenner's potential [16] originates from Abell derivations of binding energies in crystals. Abell proved that the binding energy may be represented as a sum over nearest neighbor interactions which are moderated by the local environment. The idea was further developed by Tersof [17] in large scale silicon structures. In the Abell-Tersof formalism the binding energy is given as a sum over the bonds in the following manner:

$$E = \sum_i E_i \,. \tag{1}$$

where E_i is a cotribution from atom *i*. Following Brenner [16], the sum over the bonds used in our research is represented as:

$$E_i = \sum_{j(<i)} [V_R(r_{ij}) - \overline{B}_{ij} V_A(r_{ij})]. \tag{2}$$

Sum is made over the nearest neighbors of bond i and $V_R(r_{ij})$, $V_A(r_{ij})$ are pair the potentials representing repulsing and attractive coupling respectively.

The key idea in the Abell-Tersof formalism is introducing the B_{ij} factor which represents changes of geometrical parameters of local environment around atom j, such as angles between the bonds and bond destruction-creation process. Abell suggested that B_{ij} may be given as a function of local coordination number Z

$$B_{ij} \propto Z^{\overline{\delta}} \tag{3}$$

The pair potentials are given by:

$$V_A(r_{ij}) = \lambda f_{CC}(r_{ij}) D_{CC}^e S_{CC} /(S_{CC} - 1) \exp\left(-\sqrt{\frac{2}{S_{CC}}} \beta_{CC} \left(r_{ij} - R_{CC}^e\right)\right). \tag{4}$$

$$V_R(r_{ij}) = \lambda f_{CC}(r_{ij}) D_{CC}^e /(S_{CC} - 1) \exp\left(-\sqrt{2S_{CC}}\, \beta_{CC} \left(r_{ij} - R_{CC}^e\right)\right) \tag{5}$$

It is worth to be mentioned, that, by setting the parameter at S_{CC} = 2, the pair terms reduce to the usual Morse potential with a well depth D_{CC}^e. Both the bond coefficient and pair potentials are modulated by the same cutoff function $f_{CC}(r_{ij})$. It simulates the bond formation and the destruction during the transformation, and is given by:

$$f_{CC}(r_{ij}) = \begin{cases} 1 & , \quad r_{ij} \le R_{CC}^1, \\ \cos^2\left[\dfrac{\pi(r_{ij} - R_{CC}^1)}{2(R_{CC}^2 - R_{CC}^1)}\right] & , \quad R_{CC}^1 < r_{ij} < R_{CC}^2, \\ 0 & , \quad r_{ij} \ge R_{CC}^2. \end{cases} \tag{6}$$

In above formula indicates the distance where the bond starts to collapse, and denotes the final interaction distance.

The key element in the Brenner's potential, however, is factor, which reflects the geometry of the local environment of the bond ij. It is defined by

$$\overline{B_{ij}} = \frac{B_{ij} + B_{ji}}{2}, \tag{7}$$

where is given as

$$B_{ij} = \left(1 + \sum_{k(\neq i,j)} G_C(\theta_{ijk}) f_{CC}(r_{ik})\right)^{-\delta_{CC}}. \tag{8}$$

The value of the coefficient depends not only on a bond length relative to the equilibrium disance but also on the angle between the bonds. The angle function $G_C(\theta_{ijk})$ is given by the formula

$$G_C(\theta) = a_0\left[1 + \frac{c_0^2}{d_0^2} - \frac{c_0^2}{d_0^2 + (1+\cos\theta)^2}\right]. \tag{9}$$

The formula favors both the sp^2 and sp^3 hybridizations where the angles between the bonds are equal to and 109.47°, characteristic to the graphite and the diamond structues respectively. According to Brenner the following values for parameters are given as: $R_{CC}^{(e)} = 1.315\text{Å}$, $D_{CC}^{(e)} = 6.325eV$, $\beta_{CC} = 1.5\text{Å}^{-1}$, $S_{CC} = 1.29$, $\delta_{CC} = 0.80469$, $R_{CC}^{(1)} = 1.7\text{Å}$, $R_{CC}^{(2)} = 2.0\text{Å}$, $\alpha_0 = 0.011304$, $c_0^2 = 381$, $d_0^2 = 6.25$. The λ parameter was fitted to be $\lambda = 0.49$, so that it reproduces the energy barrier (0.33 eV/atom) obtained by the DFT calculation of Cohen et al. [14], for the ordinary uniform Graphite-Diamond transition without changing the bond equilibrium distance. Thus, our theory is consistent with the DFT result, if it is used for the uniform phase transition of an infinite carbon system with the translational symmetry. Although Brenner's potential neglects long range (> 2 Å) interactions it does not cause serious mismatch. The transformation occurs only in limited volume (2 layers, 200 atoms each) and the energy difference due to long range interactions is minor. In such a case energy contribution due to long range interactions is several orders smaller then due to inter-layer bond formation. Hence in the calculations we focus our attention to predominant process governing the transformation and use of Brenner potential seems appropriate. Since the two-layer structures are of main interest so information about long range interactions is not necessary. Detailed values of parameters used in the calculations are given in [5].

4.2. Uniform Graphite-Diamond Transition

During recent years a Graphite-Diamond transformation (GDT) has been intensively studied from a theoretical point of view [6, 18, 19]. One of the approaches [6] is based on observation is that both graphite and diamond cells may be described using three parameters as shown in Figure 5. The three parameters are: B - the intralayer bond length, R - the interlayer bond length, θ - the buckling angle. By continuous change of the mentioned values we may travel on the potential energy surface from graphite to diamond structures and vice versa.

Such calculations using DFT method were performed by Cohen et al. [6, 18]. The authors considered R as an independent variable and by adjusting the remaining parameters found minimized value of energy E(R, θ, B). This procedure leads to energy barrier estimation between those two crystals - 0.33 eV per atom.

Following Cohen's idea Radosinski in his Ph.D. thesis [20] described continuous GDT using modified Brenner's potential. The energy barrier in GDT is the same value as obtained by Cohen. Furthermore the method reproduces proper relative positions of graphite and diamond structure with 0.02 eV difference as shown in Figure 6.

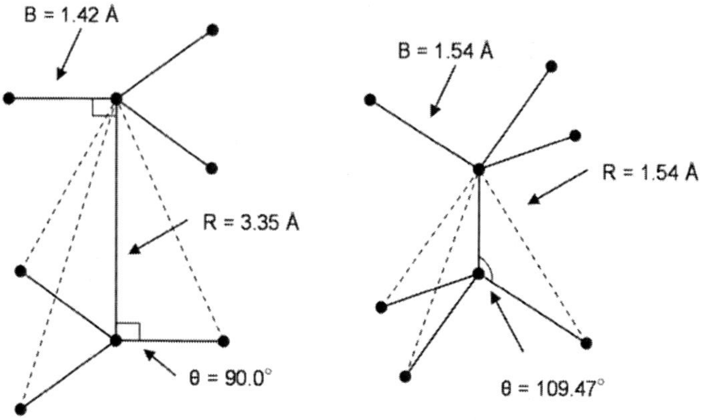

Figure 5. Graphite and diamond structures parameterized by: B – intralayer bond length, R – Interlayer bond length, θ – angle between the bonds [6].

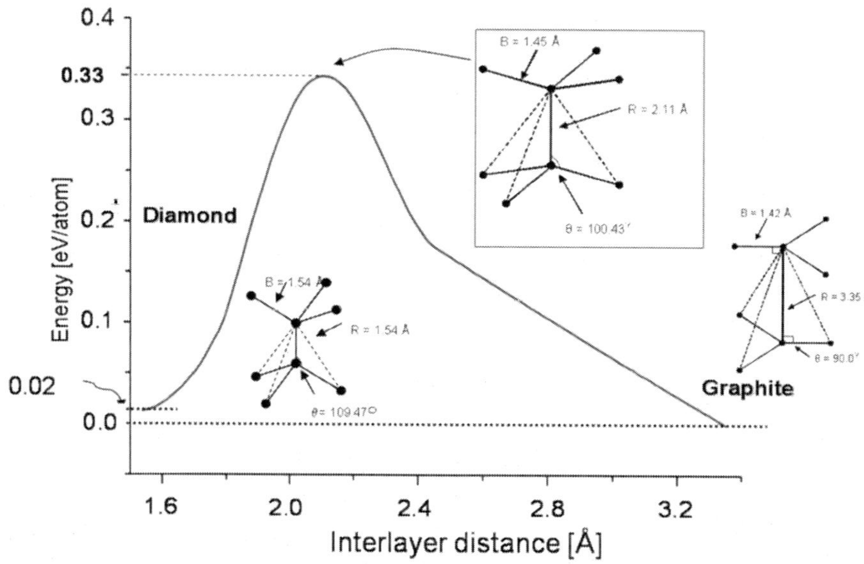

Figure 6. Potential energy surface of diamond-graphite transformation using Brenner potential [20].

The transformation begins with the interlayer distance R=1.54 Å characteristic for diamond Ac structure. With the increase of R energy rises due to the rapid weakening of intralayer bond strength followed by slow energy decrease due to intralayer bond tension release. This competition changes its sign at the distance around 2.1 Å where the barrier reaches its ceiling. After overcoming barrier top energy loss is only due to tension release of intralayer bonds. Finally the transformation ends on graphite structure with R=3.35 Å. Those results (Figure 6) justify the choice of parameter's values.

Analyzing the process of uniform graphite-diamond transformation one finds simple energetic explanation why the energy barrier between two phases is so high. The process is governed by two mutually opposite effects. The first one is the energy gain due to movement of atoms out of the graphite plane resulting in σ bond stretching. The second one is the energy lose due to bond formation between interlayer atoms. The distance between neighboring layers in graphite is much greater than minimal distance when interlayer bond is formed the energy barrier is quite high. This feature will play crucial role in explaining why graphite-diaphite transformation is accessible via photoexcitation.

It has to be mentioned although, that uniform transformation, however theoretically simple, carries one important disadvantage. The main assumption here is that we deal with infinite in size material which is far from reality. Hence values obtained within those type of calculations give us only a lower bound for estimation of energy. Nevertheless, since many parameters values were experimentally confirmed, they are important in process of adjustment potential parameters.

4.3. Graphite-Diaphite Transformation

The next step is to calculate a uniform GDT [20]. Hypothetical "diaphite" structure reflects a pattern which was obtained in Tanimura's and Kanazaki's experiment [3]. As shown in Figure 7 it consists of two regular structures formed by two kinds of atom:

- where atom lies directly above atom from adjacent plane with vertical intralayer bond,
- where atom lies directly above the center of hexagonal from adjacent plane with sheared intralayer bond.

This structure (Figure 7) corresponds to the graphite structure with 1/3 atom intruded inwards and 2/3 extruded upwards as in the experiment. In the calculations however one assumed that the material is infinite without boundaries.

As in the previous, uniform graphite-diamond calculations the system does not reflect a real situation. It provides with a qualitative estimation of the equilibrium distance between atoms from adjacent layers.

The independent variable here is an intrusion depth D. The initial point of the computations is graphite with D = 0.0 Å. Starting from this point one can observe the energy loss due to growth of the intralayer bonds tension. At D = 0.7 Å intralayer bonds start to form resulting additional energy gain.

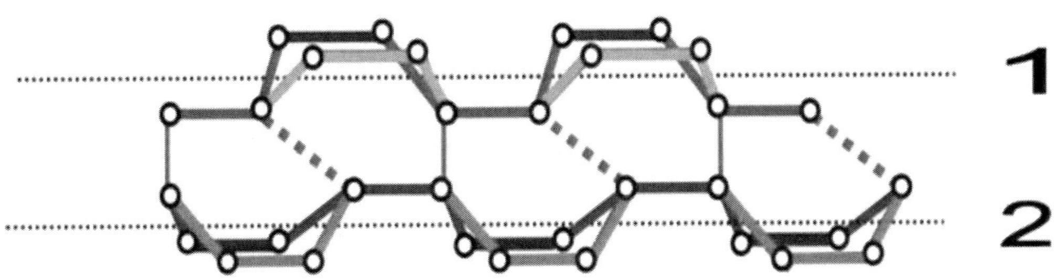

Figure 7. "Diaphite" structure used in graphite-"diaphite" transformation calculations. The structure corresponds to the pattern obtained in Tanimura and Kanazaki experiment [3, 11].

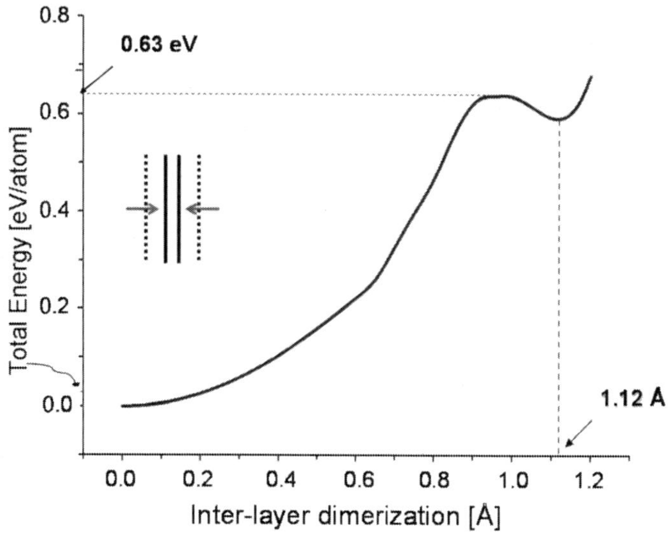

Figure 8. Potential energy surface of a uniform graphite-diaphite transformation obtained using modified Brenner's potential. The structures reaches the equilibrium when interlayer distance D is 1.12 Å. The energy barrier of the transition is 0.63 eV/atom [20].

The competition between those two processes reaches its top at D = 0.81 Å. With further increase of intrusion depth we finally reach an equilibrium distance D = 1.12 Å. The energy barrier between graphite and the "diaphite" structure is estimated as 0.63 eV per atom. Henceforth in the case of uniform graphite-diaphite transformation, when all atoms move simultaneously, the energy barrier is nearly twice higher than in the case of graphite into diamond transition. The experimental results suggest that diaphite is a local structure immersed in the graphite environment.

A more realistic picture should include aforementioned feautures into account. Following the experimental results let us considere two, initially non-interacting, graphite layers (200 x 200 neutral carbon atoms), with the periodic boundrary condition [5]. In eq. (2), the distance r_{ij} is given by the position vector of i th atom as

$$r_{ij} \equiv \left|\bar{\mathbf{r}}_i - \bar{\mathbf{r}}_j\right| , \tag{10}$$

and $\bar{\mathbf{r}}_i$ is defined as a shift from its original position ($\equiv \bar{\mathbf{r}}_{i0}$) in the graphite as

$$\bar{\mathbf{r}}_i \equiv \bar{\mathbf{r}}_{i0} + E_n(\bar{\mathbf{r}}_{i0})[(\Delta q + \delta B_d(\bar{\mathbf{r}}_{i0}))\bar{\mathbf{e}}_z + \Delta X\,\bar{\mathbf{e}}_x] \ . \tag{11}$$

The second part denotes the new diplacement. is the envelope of the new displacement

$$E_n(\bar{\mathbf{r}}_{i0}) = 0.5\left[\tanh\left(\theta\left|\,\left|\bar{\mathbf{r}}_{i0} - \bar{\mathbf{e}}_z(\bar{\mathbf{e}}_z\cdot\bar{\mathbf{r}}_{i0})\right| - L_0\,\right|\right) - 1\right] \tag{12}$$

Here $\bar{\mathbf{e}}_x$, $\bar{\mathbf{e}}_z$ are the unit vectors in the x and z axes, respectively. As shown in Figure 8, the parameters denoted in eq. (12) are as follows: - the amplitude of the planar deflation, - the size of the intruded domain, - the domain boundary width and, ΔX – the magnitude of the shear displacement. While $B_d(\bar{\mathbf{r}}_{i0})$ in eq. (12) denotes the buckling pattern within the six membered ring. In Figure 8, this buckling is symbolically denoted by a zig-zag line

Using pattern described in a previous section and modified Brenner's potential Radosinski et al. [5] performed total cohesive energy calculations. By varying values of chosen parameters one obtains potential energy surface (PES) of conversion with saddle points indicating stable configurations of disordered structure.

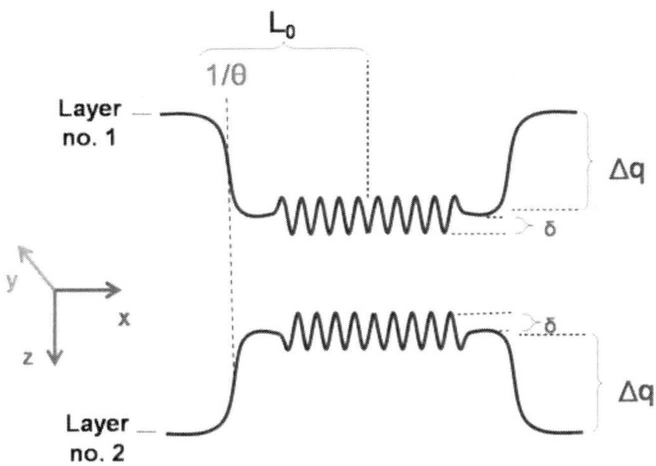

Figure 9. The distortion pattern used in the calculations: Δq - the planar intrusion amplitude, δ-the buckling amplitude, L_0- the system size, θ-the shape coefficient [11].

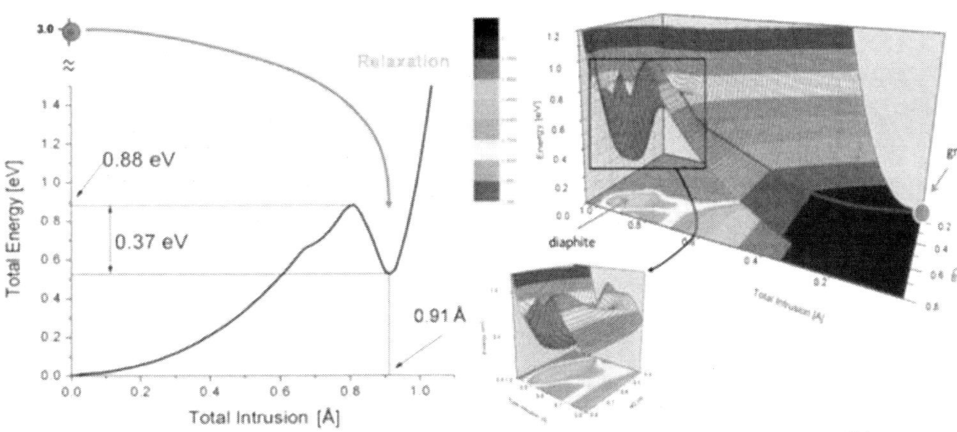

(a) (b)

Figure 10. The PES of graphite-diaphite transformation. The horizontal axis indicates the total intrusion (δ+Δq) of the domain. The second minimum occurs when the total intrusion δ+Δq = 0.91 Å with the energy barrier 0.88 eV. This energy barrier may be easily overcome by visible excitations (≈ 3 eV) and the system may reach the second minimum [5].

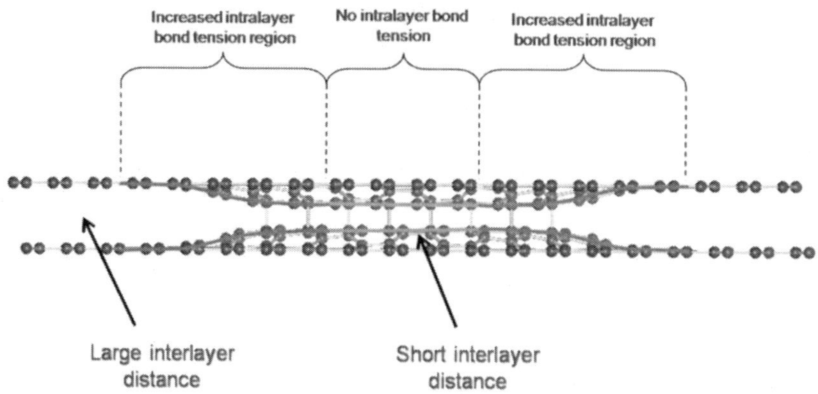

Figure 11. The cross section of diaphite structure. The shift of carbon atoms out of graphite plane is gradually distributed over the diaphite domain causing much less energy loss then in the case of vertical dislocation. As a result the interlayer distance is much shortened enabling sooner interlayer bond formation.

The purpose of the calculation is to find such a values of parameters (Δq, δ, θ, L_0) that minimize the energy barrier between initial graphite structure and a new, stable point. Initial calculation suggested that the energy barrier for minimal diaphite domain is 0.86 eV (Figure 10). One has to stress out that this is a total energy not energy per atom. Thus the diaphite domain formation is possible by a single photon. One may ask why the energy is so low in comparison to graphite-diamond 0.63 eV/atom energy barrier. In the case of graphite-diaphite conversion in the previous paragraph the energy barrier is so high because there is direct, uniform transition performed simultaneously in the whole material. Now however the diaphite is only a local structure immersed in the graphite continuum. The energy loss due to intralayer bond stretching is distributed over the tension on the border of the diaphite domain.

The energy loss due to the tension is rather small since the number of atoms taking part in the transformation is proportional to the diaphite domain circumference. Also carbon atoms of the border are gradually moved from the graphite layer so the angles between the bonds are changing smoothly. On the other hand since the distance between carbon atoms from adjacent planes is much smaller than 3.35 Å in original graphite, the σ bond contributing to the energy gain are created much sooner. Henceforth the energy barrier in the case of local graphite-diaphite transformation is expected to be much smaller than in the case of uniform graphite-diaphite transition.

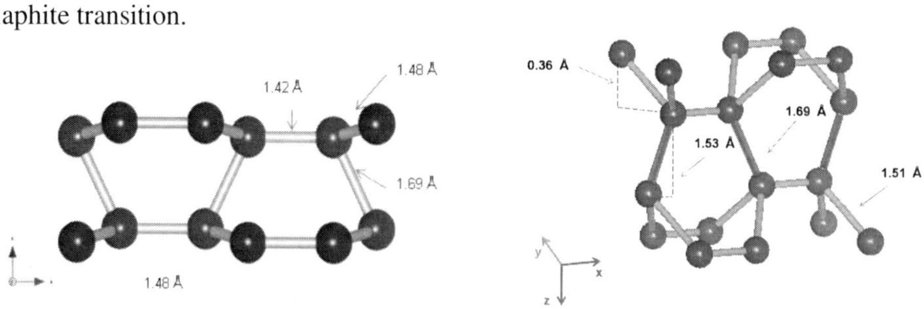

(a) (b)

Figure 12. The diaphite structure (LL type) obtained using modified Brenner's potential. The shear displacement between the layers is 0.35 A. The geometrical configuration of sigma bonds resembles much sp3 hybridization characteristic for diamond [5, 11,34].

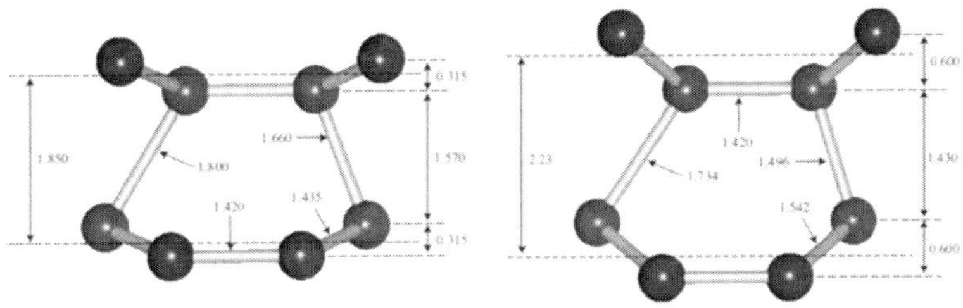

(a) (b)

Figure 13. Two types of diaphite structure [14]: (a) S type structure, (b) L type structure [11, 21-22].

Also the adiabatic energy calculations performed by Ohnishi [21-22] revealed that small diaphite domain appears at 1.7 eV and the energy barrier between graphite and the new phase is about 2 eV. Once the energy barrier is overtaken a number of new stable structures are possible. However we may divide it into two groups: S type with small intrusion and large modulation and L type with large intrusion and small modulation [21-22]. Both structures types are characterized with relatively large shear displacement 0.2 Å and are presented in Figure 13.

The structures somewhat differ from the former one being a combination of features of S and L type i.e. it has large intrusion, large modulation and large shear displacement introducing new symmetry to the system. Nonetheless all structures share same pattern i.e. 1/3 of atoms being intruded and 2/3 atoms being extruded from deformed planes.

Relatively small energy barrier combined with diaphite structure stability is a representation of a fact that this pattern is closely geometrically related with hexagonal diamond. The electronic structure also investigated by Ohnishi [20-22]] revealed that once a diaphite domain appears a pseudo gap is opened in the semi metallic continuum being characteristic for an insulator immersed in the semimetal.

4.4. Size Evolution of the Inferred Domain

Though this paper is mostly concerned with the phenomena occurring only after initially overcoming the transition barrier, one may ask the question what is energetic size evolution of the system? The validity of the question is motivated by the possible scenario of the system growth.

As mentioned before one may treat the energetic evolution of the system from the following perspective. The occurrence and height of the barrier is a result of the competition between the energy loss due to intralayer's bond tension and the energy gain due to interlayer bond formation. Henceforth we may induce that, from qualitative point of view, only the initial transition from graphite to "diaphite" should be energetically expensive since we have to transform number of atoms proportional to πL^2. Then further increase of the system size would require movement of atoms 0only on the circumference of intruded domain.

Hence the number of atoms on the circumference is now proportional only to $2\pi L$ we expect linear increase of energy barrier instead of exponential one, as in the case of direct transition.

The results of the calculations [11] using our method are presented in Figure 14. The black line indicates height of the minimal energy barrier E_b of direct transition from graphite to diaphite with definite value L.

We may observe that with increasing value of L, the value of minimal energy barrier E_b is getting higher rapidly up to 2.57 eV for L = 12 Å and 5.97 eV for L = 24 Å.

On the other hand, the red line indicates evolution of the energy barrier via indirect transition i.e. first overcoming the initial barrier and then increasing the size. The increase this time is much slower with 1.4 eV for L = 12 Å and 2.2 eV for L = 24 Å.

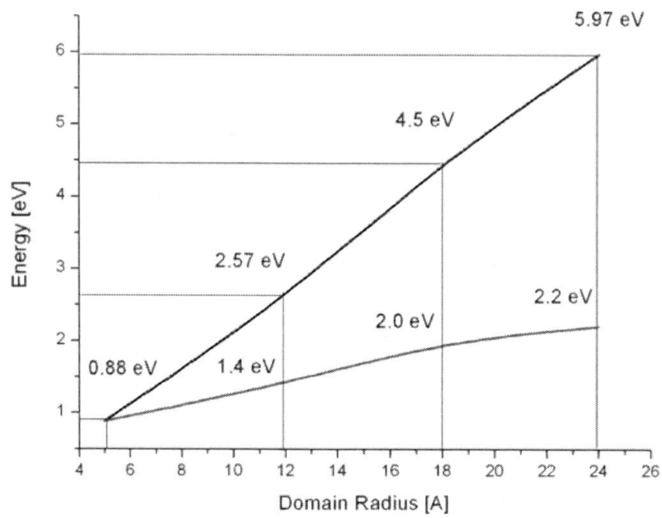

Figure 14. The energy barrier of a graphite-diaphite LL transition using modified Brenner's potential. The black curve indicates direct transition from graphite to diaphite domain of a given radius. The red curve indicates transition from graphite to minimal diaphite LL domain and then expanding to diaphite of a given radius [20].

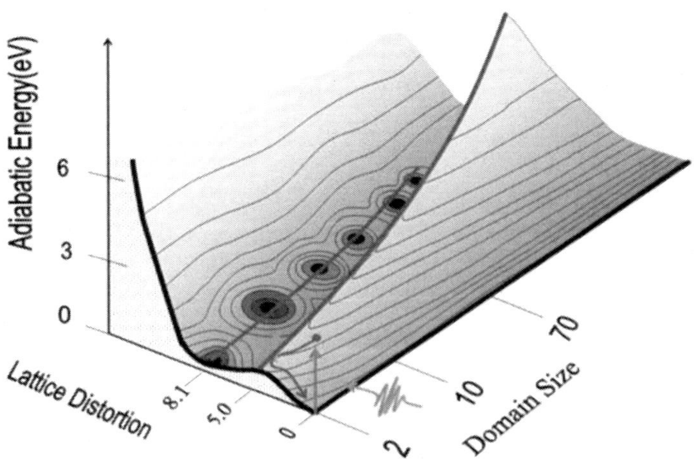

Figure 15. Schematic view of PES of graphite-"diaphite" transition. Large amount of energy is necessary only to overcome the initial barrier. Once the system is in "diaphite" region further increase of the system requires relatively low amounts of energy. Values of the energy, the lattice distortion and the domain size are arbitrary [20].

This calculations conclude the hypothesis of the system evolution and may be connected with graphite-diamond conversion. In the case of heat and pressure transition much energy is required since the transformation occurs simultaneously in large volume of the material. However with the help of photo excitation it may be possible to conduct step by step conversion. As in the case of "diaphite" where only first step is relatively cost full the further increase "diaphite" domain does not require large amounts of energy and it is possible to "jump" from one stable valley into another using small pulses as shown schematically in Figure 15.

5. Photoexcitation and Its Early Dynamics

The phenomenological approach based on the experimental observation of the photoexcited, metastable, phase is to investigate the adiabatic energy landscape by considering deformed phase and identifying for its local minima corresponding to the metastable states. In this case, contrary to the typical conditions of the strongly anisotropic, isolating systems revealing photoinduced phase transitions, graphite is conductor (and also anisotropic but two-dimensional, not one-dimensional system). Therefore the question of proliferation of the photoexcitation appears to be quite nontrivial and below we will discuss the early stage dynamics.

5.1. Model Hamiltonian

Radosinski et. al considered the photoexcitation in the two, initially undistorted graphite layers [5]. During photoexcitation an electron is transferred between the layers and the hole is created. In this way an exciton-like state is formed across the layers.

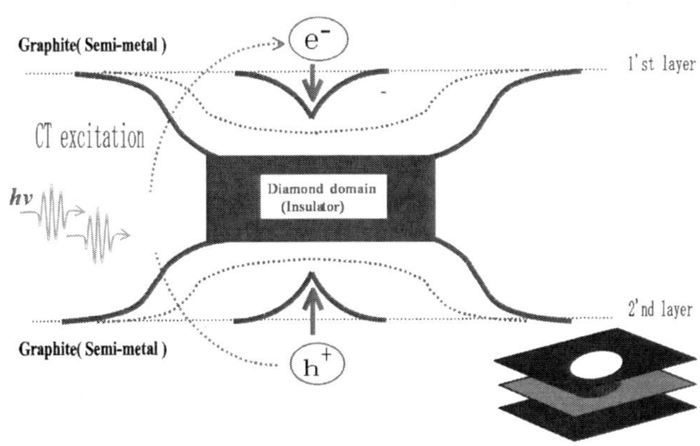

Figure 16. Model used in calculations: after photoexcitation electron from bottom layer is transferred to the top one. Since the absence of electron in the bottom layer, the hole is created. Together with excited electron an exciton-like electron-hole pair is created which spans across the layers. Although graphite is a conductor and most probably electron and hole will disappear we speculate that there is small, though greater than zero, probability that the electron-hole pair will remain at the same site. Hence graphite is neutral, Coulomb interaction between hole and electron enables layers contraction. When the distance between layers is close enough *sp3* type bonds are created leading to "diaphite" domain formation [5].

As the graphite is a conductor, electron and hole would soon disappear as free carriers. However, due to their Coulomb interaction the layers would contract, enabling, σ-type bonds formation. Such a scenario would occur if electron-hole pair remains at the same site. The idea is to estimate whether the probability of such a process would be non-zeroth.

Let us start from a simplified model Hamiltonian of an interacting electron-hole pair, coupled to the two layer lattice with N (>> 1) sites in each layer:

$$H = H_{eh} + H_{ph} + H_c, \quad \hbar = 1 \tag{13}$$

The first term of this Hamiltonian is the energy band of electron or hole, given as:

$$H_{eh} = \sum_{i=1,2}\sum_{j=1,2}\sum_{k} E(k) a^{+}_{kij} a_{kij} \tag{14}$$

where is a usual electron, *i*=1, (hole, i=2) creation operator, index *j* (= 1, 2) is a layer number, and *k* denotes two dimensional wave vector $k = (k_x, k_y)$. E(k) is the energy in a square lattice:

$$E(k) = \sqrt{k_x^2 + k_y^2}, |k_x| \le \pi, |k_y| \le \pi \tag{15}$$

and spin is neglected. As the energy is in the range around the Fermi level one may assume a V-shaped electron band and the hole one centered at the Γ point (k = 0) for simplicity. In the graphite crystal, the Fermi level is at K point but this difference has no effect for our final conclusion. H_{ph} denotes dispersionless, Einstein's phonon localized on a lattice site l with an energy ω and is given as:

$$H_{ph} = \omega \sum_{l} \left(\frac{1}{2} + b_l^{+} b_l \right) \tag{16}$$

where denotes the creation phonon operator. The value of energy of optical excitation ω may be estimated self-consistently. Since the energy barrier height is aproximitely 1eV, corresponding to 0.8 Å (the magnitude of carbon atoms intrusion when interlayer bond is formed), the coefficient of a restoring force is proportional to

$$k \propto \frac{2}{0.64} \frac{eV}{\left(A^{\circ}\right)^2}. \tag{17}$$

This value corresponds to the energy of optical mode of carbon atom,

$$\overline{\omega}^2 \propto 0.4 \times 10^{28}. \tag{18}$$

The corresponding value of energy is then

$$\hbar\omega \approx 0.05 \text{ eV} \tag{19}$$

One may notice that this value is not much different from the one experimentally observed, i.e. $0.025eV$. In our calculations we are taking the value of this parameter, $0.1eV$, according to the level of magnitude.

The H_c part represents the electron-hole Coulomb interaction in the Hamiltonian and is given as:

$$H_c = -\sum_l U(Q_l)\, n_{l11}\, n_{l22}\,, \tag{20}$$

where Q_l is the phonon coordinate and is defined as:

$$Q_l \equiv \frac{b_l^+ + b_l}{\sqrt{2}}. \tag{21}$$

Although there are two square lattices for the electron and the hole, since this Q_l only describe the inter-layer contractions starting from the original graphite, we have only one mode per lattice site l,. In above equation, denotes the number operator on site l and layer j and:

$$a_{lij} = \frac{1}{\sqrt{N}} \sum_k e^{i\mathbf{k}\cdot\mathbf{l}} a_{kij}\,. \tag{22}$$

$U(Q_l)$ is the electron-hole coupling strength. Assuming a linear relation:

$$U(Q_l) = U_0 + U_d\, Q_l\,. \tag{23}$$

The initial value U_0 = 1.5 eV was chosen so it corresponds to the electron-hole coupling in the undistorted graphite system. The energy of the Coulomb attraction between electron and hole was calculated using the Mataga-Nishimoto formula [5]. By relating phonon energy with the energy of the top of the barrier one may compute the value of contraction. At the top, Q_l takes value and is given as

$$\frac{\omega}{2}\overline{Q}_l^2 = 0.8\,eV \tag{24}$$

and gives $\overline{Q}_l = 4$. On the other hand again by using the Mataga-Nishimoto formula [5] the electron-hole Coulomb coupling strength at this point becomes $\overline{U}(\overline{Q}_l) = 2.4\,eV$. Hence U_d is determined, U_d = 0.225 eV.

The calculations were also performed for various values of U_0 ranging from 1.0 eV to 1.6 eV, having no influence on the final conclusions.

5.2. Calculations

Defining the Bloch state [5] composed of an electron at site $l + \Delta$ of the first layer, a hole of site l of the second layer, and m phonons at site l, as

$$\left|\Delta, m\right\rangle = \frac{1}{\sqrt{N}} \sum_l e^{-i\mathbf{k}_{ph}\cdot\mathbf{l}} a^+_{l+\Delta,11} a^+_{l,22} \frac{\left(b_l^+\right)^m}{\sqrt{m!}} \left|0\right\rangle, \quad \left|0\right\rangle \rightarrow vaccum. \tag{25}$$

The expression inside the sum over l, is an electron-hole phonon state localized at site l, wherein $\boldsymbol{k_{ph}}$ is the wave vector of photon that creates this excitation, and we have also assumed that the phonon number m, is zero ($m = 0$), if the electron and the hole are not at a same site ($\Delta \neq 0$) spanning two layers.

The wave vector $\boldsymbol{k_{ph}}$ of a visible photon is almost zero compared to the wave vectors of the electron, hole or phonons in the first Brillouin zone. Henceforth the electron-hole-phonon coupled states which are created by visible photon have the full translation symmetry:

$$\left|\Delta, m\right\rangle = \frac{1}{\sqrt{N}} \sum_l a^+_{l+\Delta,11} a^+_{l,22} \frac{\left(b_l^+\right)^m}{\sqrt{m!}} \left|0\right\rangle \tag{26}$$

The expression inside the sum over l, is an electron-hole-phonon state localized at the site l. One also can see that in eq. (26), Δ stands for electron-hole relative distance. This is just the inter-layer charge transfer and self-localized exciton.

Using this type of state as a base let us formulate eigenvalue problem of H. To do this one can calculate the matrix elements and then diagonalize this matrix.

For practical purpose let us calculate transfer energies in real space rather than wave vector one:

$$T(l) = \frac{1}{N} \sum_k E(k) e^{-i\mathbf{k}\cdot\mathbf{l}} \tag{27}$$

Such $T(l)$ has a long tail in the real space so it may be truncated it at $l = 10$. Finally applying torus type periodic boundary conditions for relative electron-hole distance and using the formula (26) as a basis, matrix elements of the Hamiltonian H may be calculated. Diagonalizing matrix one obtains a set of eigenvalues and eigenfunctions. In the calculations the net consists of 50 x 50 sites with 60 phonons. Having a set of eigenstates and corresponding eigenvalues E_r, one can now calculate the temporal evolution of a coupled electron-hole phonon state using following relation:

$$|W(t)\rangle = \sum_r e^{-itE_r} |r\rangle\langle r | \Delta, 0\rangle . \tag{28}$$

Using as the initial state let us investigate states with various initial electron-hole distance separation Δ while the initial phonon number is zero; $m = 0$. Our main interest is to calculate probability that electron-hole will remain at the same site $\Delta = 0$. This probability is given by $D(t)$, defined as:

$$D(t) = \langle W(t) | [\sum_m |0,m\rangle \langle 0,m|] | W(t)\rangle . \qquad (29)$$

Let us also compute average phonons number as:

$$M(t) = \langle W(t) | [\sum_m |0,m\rangle \, m \, \langle 0,m|] | W(t)\rangle / D(t). \qquad (30)$$

Results for the initial quarter period are presented in figures 17 and 18. In the Figure 17 the time dependence of exciton probability that it will still remain in the same site is shown. One can notice that even if $\Delta = 0$ it decreases rapidly from value 1, which reflects the free carrier relaxation process characteristic for a good conductor. However it does not reach zero but stabilizes at around 0.1, which is nothing else but a self-localization of the electron and hole pair.

At the same time, the number of phonons rapidly increases. It means that the inter-layer contraction has been generated together with this self-localization. One may also calculate largely distant electron-hole evolution by taking as initial state for $\Delta = 10$ and 20. We may observe that these states never localize. On the other hand once self-localization occurs there is a fast response of the net manifested by the rapid increase of average phonon number (Figure 18).

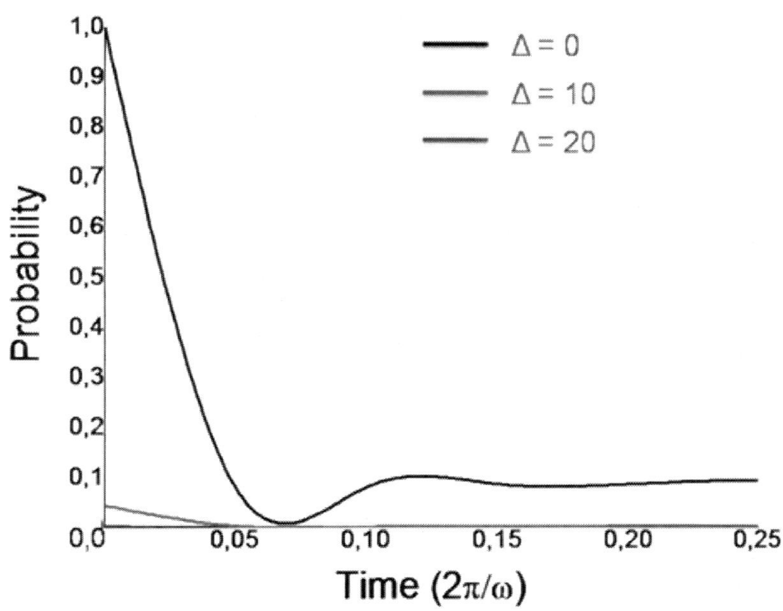

Figure 17. The probability that the exciton remain at the same site with respect to time (in $2\pi/\omega$ units). Different colors indicate different relative electron-hole distances: black $\Delta = 0$, green $\Delta = 10$, blue $\Delta = 20$ [5,11].

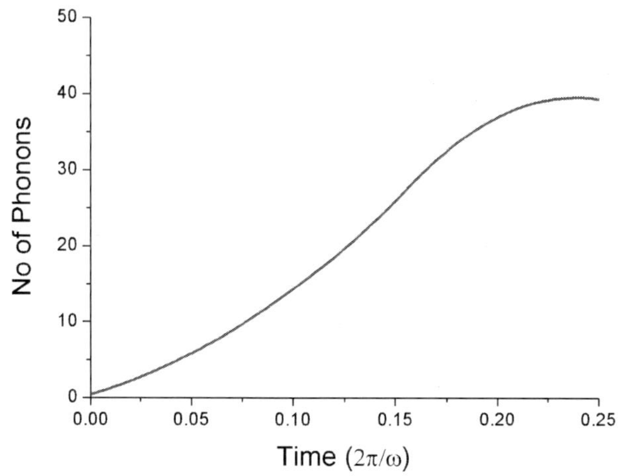

Figure 18. The average number of phonons created during the process of electron-hole localization (in $2\pi/\omega$ units) [2, 11].

Although then graphite is a conductor there exist small but finite probability of self-localization of this pseudo exciton state due to Coulomb interaction. The energy transfer to phonon degree of freedom enables lattice contraction and finally diaphite domain formation.

6. Classical Picture for Interlayer Bond Formation Dynamics

Although the model presented in the previous section [2, 5] explains that self-localization of electron-hole pair induces the motion of the net, it lacks the answer for the subsequent events. After the photoexcitation due to self-localization the quantum phase coherence between the sites is lost and all the sites may be treated separately. Furthermore the distance between electron and hole is small enough so its wave functions are almost overlapped. As a result the graphite-diaphite system may be treated as a classical one without Coulomb coupling. The energy absorbed due to photoexciation is mainly distributed over two channels: the kinetic energy i.e. velocity of the intruded atoms and potential energy i.e. the intrusion. Although the kinetic and potential energy varies over time due to strong nearest neighbor interactions, the total energy is conserved since no heat bath is present. The post excitation evolution of the system is governed by distribution of this extra energy and reaching final equilibrium state. One has to mention that this evolution is fairly complicated due to strong cooperativity between nearest neighbors. The other factor is, as we have seen in former paragraphs, that once the distance between atoms spanning two layers is less that 1.9 Å there are number of stable configurations different from one another energetically.

In order to investigate these phenomena classical molecular dynamics calculations were performed [9, 23-26]. First one has to investigate the interlayer bond formation mechanism. The system consists of two graphite layers without long range interaction. This approach is justified by the fact that long range interaction contribution is only of the 0.02 eV/atom order and the energetic evolution of the system is mainly governed by the formation of σ bonds being 10 times higher. The simulations were performed then using Brenner's potential [16].

The ABAB stacking graphite layer consists of two types of atoms (Figure 19): α atoms are placed directly above each other whereas β atoms are placed above center of hexagonal ring.

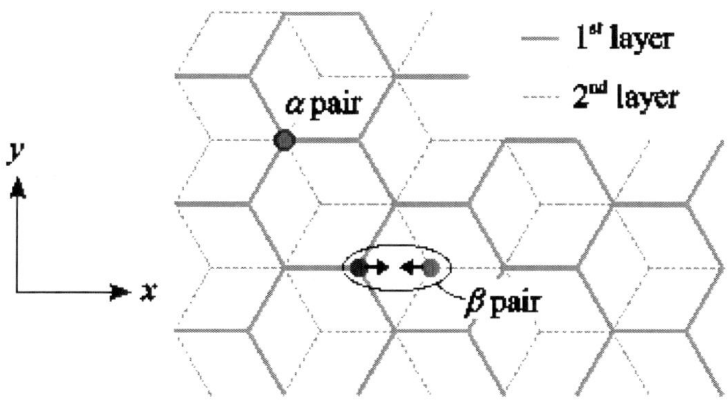

Figure 19. Schematic representation of ABAB stacking of graphite. The net consists of the two types of atoms: the α atoms are located directly one above another whereas β are placed above hexagonal ring from adjacent plane [26].

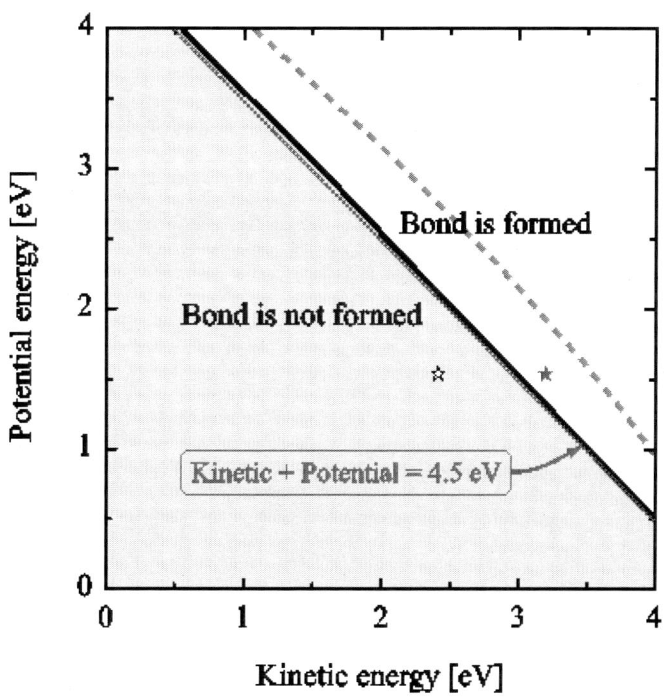

Figure 20. The phase diagram of interlayer bond formation during MD calculations. The potential energy related with initial intrusion (vertical axis) is calculated using Brenner's potential. The kinetic energy (horizontal axis) is calculated as $K=2(m_c v_0^2/2)$ where m is carbon atom mass and v its initial velocity. The interlayer bond is formed when the total energy exceeds 4.5 eV [9].

Since the distance between α atoms is much smaller then between β atoms one initially assumed that contraction appears at one site of α atoms. The initial conditions included various magnitudes of kinetic and potential energy reflecting different values of velocity and intrusion of carbon atoms. The potential energy value was calculated also using Brenner's approach. The initial kinetic energy K_0 is given by $K_0=2(m_c v_0^2/2)$, where m_c is the mass of a carbon atom and v_0 is the initial velocity. The system consists of two graphite layers with the initial interlayer distance of 3.35 Å, and each layer includes 2160 carbons, where a periodic boundary condition is imposed. The stable interlayer bond is created when the total energy (kinetic and potential) is greater than 4.5 eV (Figure 20). As an example, as presented I Figure 21, the interlayer bond is successfully formed when the kinetic energy is 3.2 eV and the potential energy is 1.54 eV, corresponding to the initial intrusion of 0.4 Å,. In Figure 21 the time development of the bond length between the two opposite carbons as a function of time on a logarithmic scale are presented.

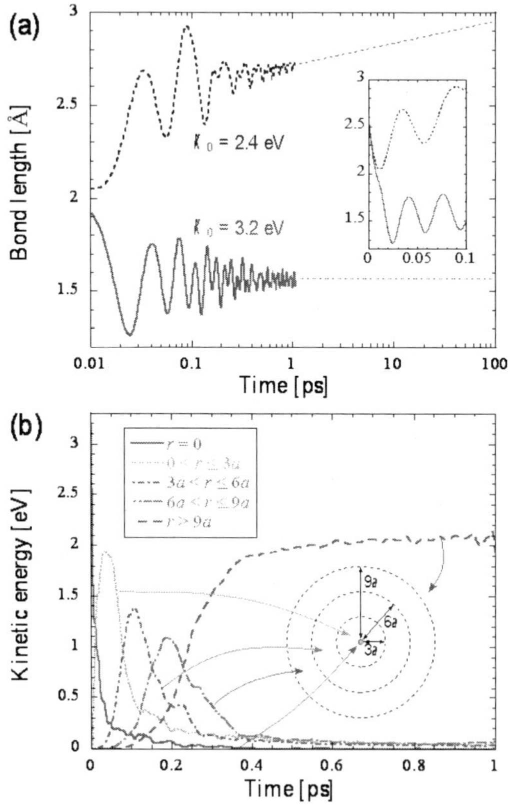

Figure 21. (a) The bond length between the two opposite carbons as a function of time on a logarithmic scale. The inset shows the same plot from 0 to 0.1 ps on a linear scale. A solid line denotes the case of the initial kinetic energy of 3.2 eV and the initial intrusion of 0.4 Å and a dashed line is the case of the initial kinetic energy of 2.4 eV and the same intrusion. (b) The kinetic energies as a function of time. The solid line indicates the sum of the kinetic energies of two central carbons, and the dotted line indicates that of the carbons located within a 3a radius of the self-localized center except the two central carbons, where a is the lattice constant of the graphite (1.42 Å). The one-dotted-dashed and two-dotted-dashed lines are those of carbons located in the radii between 3a and 6a and between 6a and 9a, respectively. The dashed line is that of carbons outside of 9a radius. The corresponding figure about these radii is shown in the inset [9, 11].

The inset shows the same plot from 0 to 0.1 ps on a linear scale. The bond length decreases from 2.55 Å, rapidly, and it settles down gradually in about 1.56 Å, repeating vibration. This bond length almost corresponds to that of the diamond structure (1.54 Å). For comparison, we show the case of the initial kinetic energy of 2.4 eV, which does not result in the bond formation, by a dashed line in Figure 21. In this case, the system does not overtake the energy barrier to form an interlayer bond due to the lack of energy and returns to the graphite without the bond formation.

The key factor however in the photoinduced phase transition phenomena is the proliferation – the process of exponential response of the system for the given stimuli. The experiment indicates nonlinear dependence between number of photons and number of atoms transformed. In order to explain this phenomena one has to investigate the distribution of energy on the given atom surrounding.

Figure 21 presents the transfer of energy as a function of distance from the central atom. The solid line denotes the evolution of kinetic of the excited atom pair energy versus time. The remaining lines indicate the kinetic energy of pair of atoms within 3a, 3a < 6a distance from the central atom. One may observe that due to strong nearest neighbor interaction the energy is distributed from the central atoms to the outer regions. Since the energy is conserved the total energy initially submitted to the central atom is transferred to the reservoir as indicated by the dashed line. The effect plays crucial role in proliferation of the initial disturbance and formation of the diaphite domain.

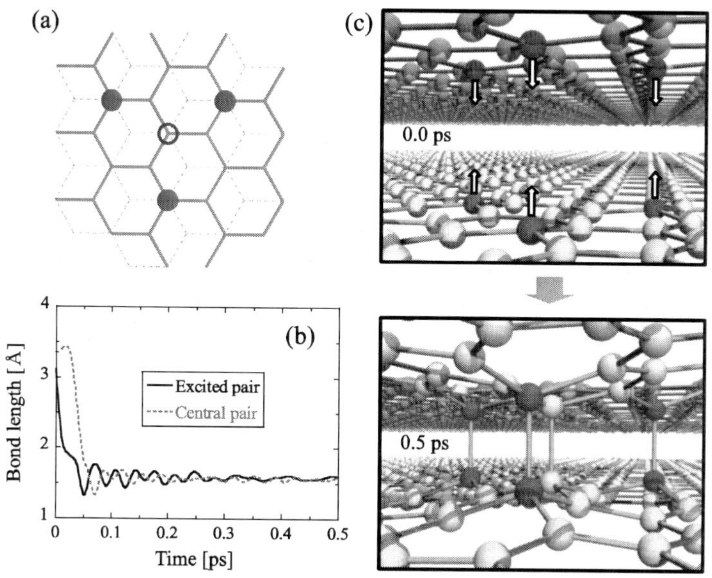

Figure 22. (a) The interlayer bond formation when nearest atoms (red circles) of the central atom (empty circle) are excited (b) snapshots of net during MD simulations. Red spheres indicate excited atoms whereas white arrows its initial velocities. After 0.5 ps the interlayer bonds are formed (c) The interlayer bond length during MD simulations. The red dotted curve indicates interlayer bond length of excited atoms whereas black solid curve indicates interlayer bond length of the central atom. Initially not excited central atom forms interlayer 0.05 ps excitation of nearest neighbors [9].

In order to further investigate this effect one has to verify the response of the surrounding atom on the initial excitation i.e. whether nearest neighbors are able to form interlayer bonds using energy distributed from the excited neighbor. Two kinds of scenarios were investigated. Firstly three α type pairs where exited. Initially each pair was excited with 4.08 eV kinetic energy and 0.14 eV potential energy (corresponding to the 0.12 Å intrusion). One may expect that the energy to the central atom should be sufficient to form interlayer bond. Figure 22 presents the response of such a system. One may observe that in fact the central atom acquires enough energy from its neighbors to form stable bond (the interlayer distance shrinks to 1.56 Å). Somewhat surprising fact is that although each initially excited atom did not have enough energy to form stable interlayer bond by itself after 0.05 ps the bonds are also created. This may be explained by the strong cooperativity over the neighbor atoms. As shown in previous paragraphs the π bond energy strongly depends on the surrounding geometry. Due to energy distribution once the central atom is put in motion the geometry is dramatically changed lowering the activation energy for the interlayer bond formation. This effect is strongly present in the second investigated scenario. In this case each α pair is initially excited with 1.04 eV kinetic energy and 1.47 eV potential energy (corresponding to the 0.39 Å intrusion).

Although total energy 2.51 eV is much lower than 4.5 eV the central pair forms stable interlayer bond as shown in Figure 22. Both scenarios indicate that cooperativity is essential in diaphite domain formation. An excess energy is spread around the graphite increasing probability of forming intralayer bonds between surrounding pair of carbon atoms. In Figure 23 the dependence between lowest excitation energy necessary to form stable intralayer bond and the distance from the excited pair is presented.

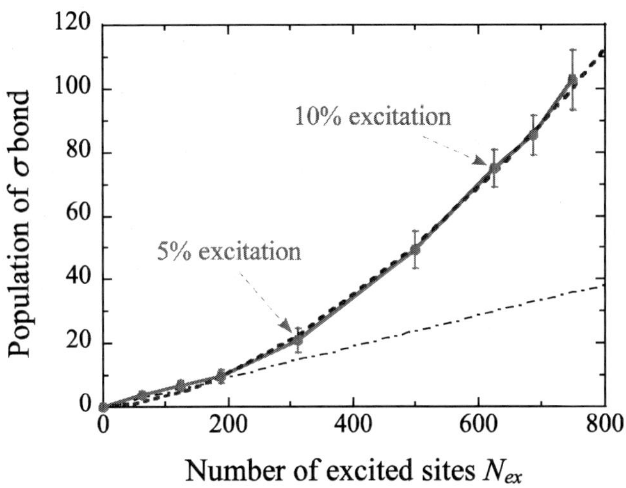

Figure 23. Population of sigma bonds with respect to number of excited sites. The solid red curve indicates exponential growth with the factor 1.7 in contrast to linear one indicated by dashed line [26].

In the previous paragraphs we have proven that there exists small but finite probability of self-localization of electron-hole pair [5]. One has to notice that although the electron-hole pair energy is 3.3 eV due to Heisenberg uncertainty principle there is large energy variance ΔE = 1.8 eV. Thus only a fraction of photoexcited states lead to interlayer bond formation. Nonetheless as presented in Figure 23 the efficiency of diaphite domain formation is exponential with the number of excitations. The excitation energy E_{ex} of each site was chosen randomly according to a Gaussian distribution $G(E_m,\Delta E) = \exp[-\{(E_{ex}- E_m)/ \Delta E\}^2]$, where E_m and ΔE are the mean energy and the width, and are set to be 3.3 eV and 1.8 eV, respectively

[26]. Then the initial potential energy P and the kinetic energy K were also chosen randomly under the condition that $P + K = E_{ex}$. The average number of interlayer bonds was calculated over 1.0 ps. The calculations indicate the exponential dependence with the factor of 1.7 (Figure 23).

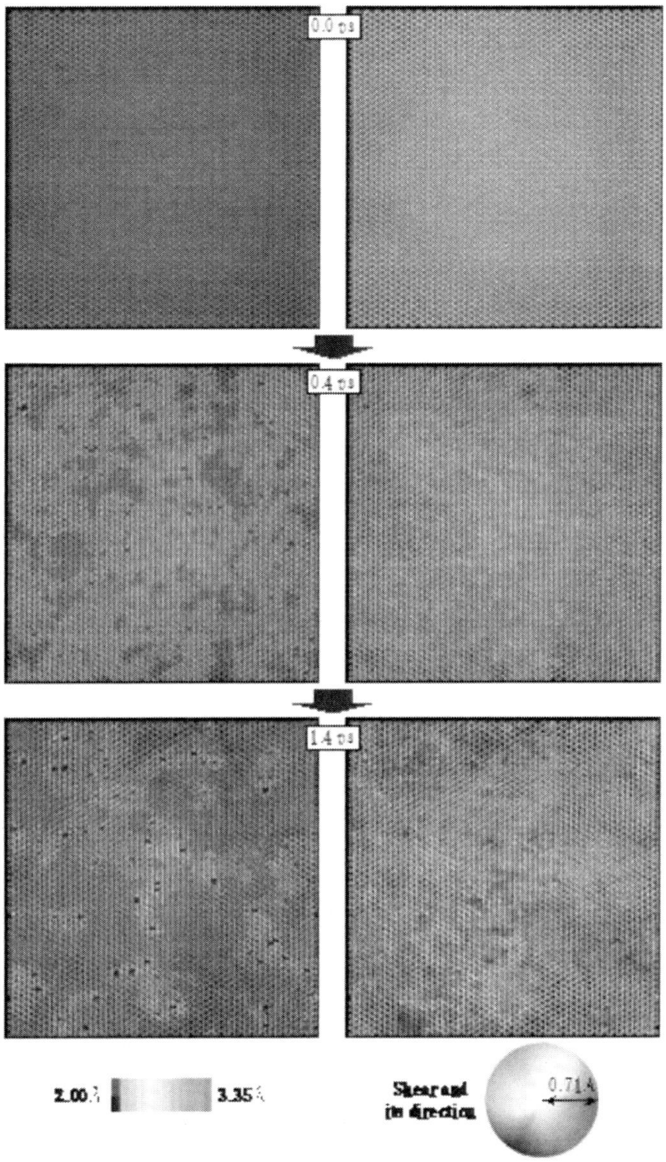

Figure 24. The snapshots of the MD calculation for the excitation density 6 % at t=0, 0.4 and 1.4 ps. The snapshots at the left side represent the distance between two graphite layers by the color palette at the bottom left, wherein the red means that an interlayer bond is formed. Those at the right side represent the shear of the upper layer to the lower layer and its direction by the circular color palette at the bottom right, wherein the white at the center of the circle means no shear [26].

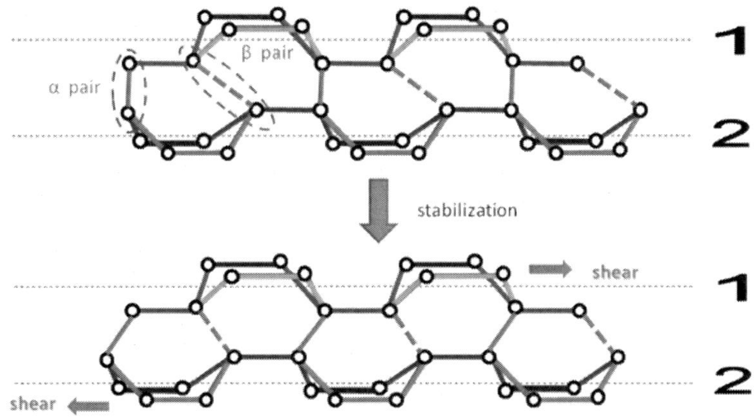

Figure 25. The shear displacement optimizes mismatch between α and β bonds shifting the structure towards sp^3 hybridization (LL structure).

The experiment suggests that after photoexciation there is approx. 0.3 Å shear displacement between atoms from adjacent graphite planes as shown in Figure 25. The shear mode is to be expected since there is a clear mismatch between α and β bonds. There are two favorable geometrical configurations in C-C structures namely the sp^2 and the sp^3. While sp^2 tends to evenly span the carbon atoms in the plane the sp^3 is evenly distributing carbon atoms in space.

Henceforth the diaphite structure with only α bonds are energetically unfavorable and the system eventually evolves so the mismatch is optimized. In order to clarify this fact molecular dynamics calculations were performed simulating a two pulse excitation [25]. Initially a 6% excitation is introduced to the system. After time Δt another 6% excitation is put and the number of α and β bonds is counted. As one may see in figures in Ref. 25, the number of α bonds decrease with Δt and for Δt > 0.9 ps the number saturates. On the other hand when Δt is between 0.3 and 0.5 ps the β bonds are created most efficiently. Comparing these results with average shear evolution with time one may conclude the shear displacement is essential for efficient β bonds formation.

6.1. 3-Dimensional Propagation of Diaphite Phase

So far our consideration where focused on the creation and evolution of diaphite phase. In such a case short ranged interactions take predominant effect. However one may ask the question what is the influence of the diaphite domain on the graphite surrounding? In particular is there any change in equilibrium distance between graphite layers? This question is valid in the light of graphite-diamond transformation. The diamond has greater density then graphite and the desirable direction of the photoinduced graphite-diamond transition would be to reflect this fact. In order to investigate this matter one has to take into account a long range interaction.

Thus the LCBOP [10, 11] (long-range carbon bond order potential) potential was applied of the form:

$$E_b = \frac{1}{2}\sum_{ij}^{N} V_{ij}^{tot} = \frac{1}{2}\sum_{ij}^{N}\left(f_{c,ij} V_{ij}^{SR} + S_{ij}V_{ij}^{LR}\right) \quad (31)$$

where $f_{c,ij} V_{ij}^{SR}$ represents a short range interactions part and $S_{ij}V_{ij}^{LR}$ represents a long range interactions part.

The short range interactions are of similar fashion like a Brenner's one namely:

$$V_{ij}^{SR} = V_R(r_{ij}) - B_{ij}V_A(r_{ij}) \quad (32)$$

where

$$V_R(r_{ij}) = Aexp(-\alpha r) \quad (33)$$

and

$$V_A(r_{ij}) = B_1 \exp(-\beta_1 r) + B_2 \exp(-\beta_2 r). \quad (34)$$

The long range interactions part is of a double-Morse potential type:

$$V^{LR}(r) = \theta(r_0 - r)V_1^M(r) + \theta(r - r_0)V_2^M(r). \quad (35)$$

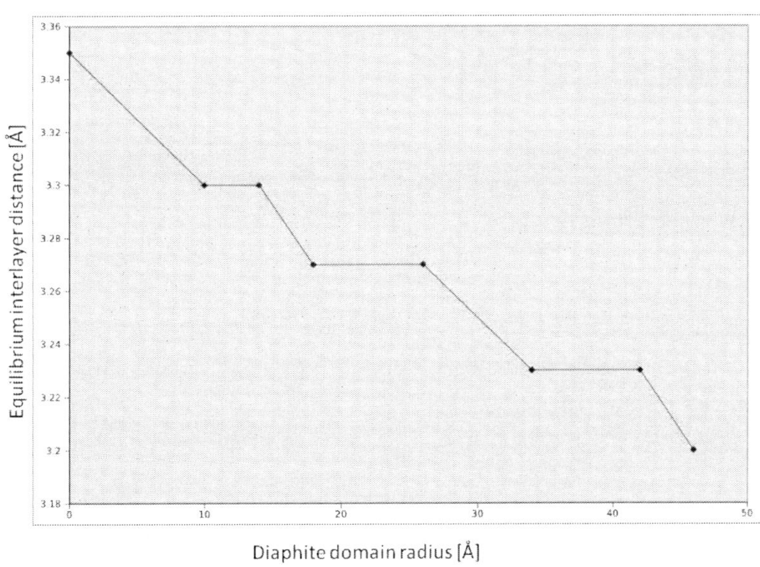

Figure 26. Interlayer equilibrium distance [Å] with respect to diaphite (LL) domain radius [Å]. Calculations were performed using LCBOP [10, 11] potential on the 100x100x12 graphite slab with two layers of the diaphite domain.

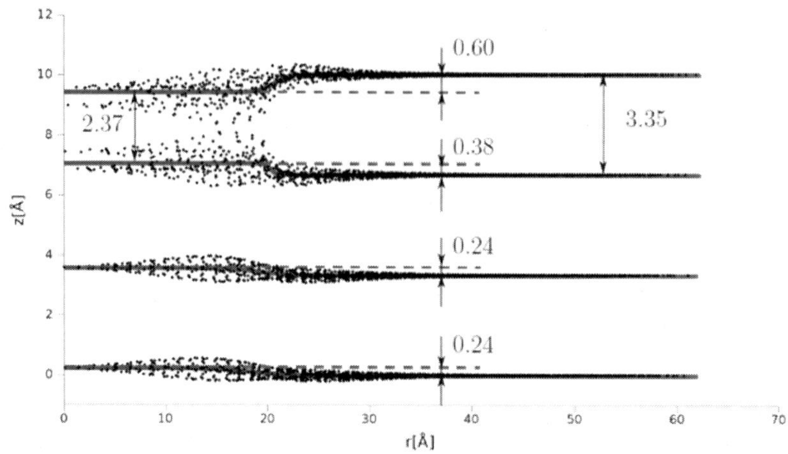

Figure 27. The first 4 layers of graphite with the diaphite structure minimized in 100x100x12 atoms slab using LCBOP potential. The size of the diaphite structure is 20 Å. Each black dot indicates carbon atoms. The red solid line indicates average position in z axis. The long range interaction result in exponentially decaying deformation of bottom layers [34].

The $f_{c,ij}$ is a smooth cutoff function:

$$f_c(x) = \Theta(-x) + \Theta(x)\Theta(1-x)\exp\left(\frac{\gamma x^3}{x^3-1}\right), \tag{36}$$

where $\Theta(x)$ is Heaviside step function. The S_{ij} factor is a switching functions excluding first neighbours and defined as:

$$S_{ij} = 1 - f_{c,ij}. \tag{37}$$

A significant numerical advantage over typical two body potentials is due to integration the long range interactions over 6 Å range. The detailed description of an LCBOP potential one may find in [10].

The simulation box includes a slab of 100x100x12 carbon atoms slab [11]. With the increase of the diaphite domain radius from 0 up to 45 Å a significant change of the interlayer distance up to 0.15 Å (Figure 26) was observed.

7. Temperature Stability

All previous simulations were performed in zero temperature, zero pressure conditions. An interesting question is what is the behavior and a stability of graphite-diaphite system with the increase of the temperature.

Graphite without pressure is transforming into vapor at around 4000 K. This value corresponds to the weak Van der Waals interactions between graphite layers (0.025 eV/atom). On the other hand the appearance of diaphite and σ type interlayer bonds should stabilize top two graphite layers. In order to investigate this matter a molecular dynamics simulations with Nose-Hover thermostat were performed.

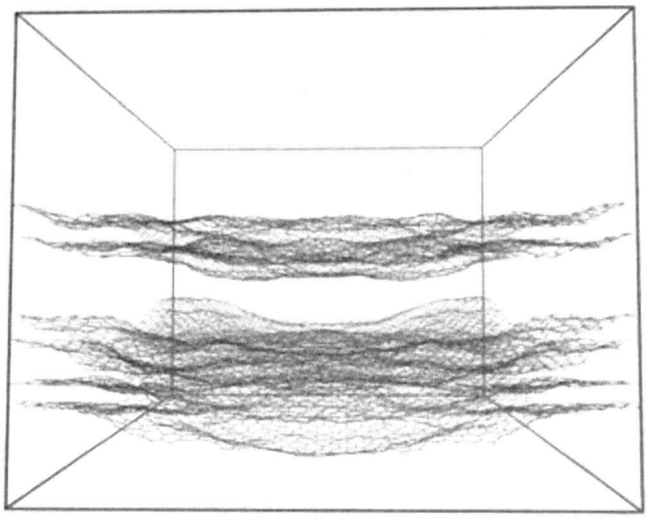

Figure 28. A snapshot of MD simulations using LCBOP [10, 11] potential in 1200 K. Two top graphite layers with diaphite structure separate from remaining graphite. Although the distance between two top layers is increased up to 4.5 Å the cohesion is preserved. The remaining graphite structure also remains undisturbed [34].

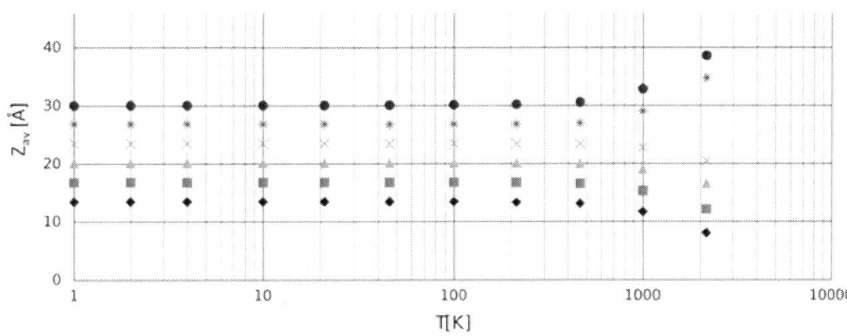

Figure 29. The average distance in the z axis between graphite layers with diaphite structure vs. the temperature of the system using LCBOP potential [10,11]. Each color indicates the average position in the z axis of graphite layers. Once the temperature is over 800 K the distance between graphite layers increases however the cohesion of the system preserved. Over 1000 K temperature two top layers ablate [34].

The system for the simulations consists of 6 layers of [..ab..] stacked graphite, 2800 atoms per layer. The size of the diaphite domain was set to 20 Å. The domain is then relaxed using LCBOP potential and the kinetic energy is introduced gradually with the use of the Nose-Hoover thermostat. Simulation time is 5 ps with a time step of 0.1 fs. With the increase of the temperature up to 1000 K the average interlayer distance is gradually increasing.

This may be expected since graphite layers tend to oscillate in shallow well of potential Van der Waals energy surface. However above 1000 K the two top graphite layers with immersed graphite phase separate from the remaining graphite. Although the average distance between two top graphite layers is increased up to 4.5 Å the diaphite phase inside stabilizes the structure. The diaphite phase itself may be simply quantified by the counting of the interlayer bonds. With the further increase of the temperature more and more bonds tend to be broken and finally and around 5000 K the two top layers loose its cohesion.

8. Possible Applications – Hydrogen Storage

Hydrogen storage is a key technology for the advancement of hydrogen and fuel cell power technologies in transportation, stationary, and portable applications. The main efforts focus primarily on the research and development of on-board vehicular hydrogen storage systems that will allow driving range of greater than 500 km while meeting packaging, cost, safety, and performance requirements to be competitive with current vehicles. While automakers have recently demonstrated progress with some prototype vehicles travelling more than 500 km on a single fill, this driving range must be achievable across different vehicle models and without compromising space, performance or cost.

The research in this field is currently dominated by US projects supported by Department of Energy. This challenge continues to defy all scientific effort and is one of the most technically challenging barriers to the widespread commercialization of hydrogen-fueled light-duty vehicles. The research initiative started in July 2003 by the US Department of Energy (DOE) called a "Grand Challenge" [27] has not achieved the goals that have been proposed for the year 2010 and nothing shows that the next one, planned for 2015, will be achieved (see Table 1). One need a breakthrough in the porous adsorbing materials design and synthesis to propose the structures which will be able to store the desired quantity of hydrogen and serve as a tank for hydrogen fueled cars.

There are many existing structures which have been synthesized and studied during the last 10 years. Many of them have been considered as potential candidates for hydrogen storage [28, 29]. However, the main disadvantage of the existing options is their structure based on very small hydrocarbons and having much too low energy of adsorption of hydrogen. A number of them are not stable enough.

On the other hand recent theoretical research [30, 31] suggests a possibility of increasing graphite storage capabilities by changing its curvature. In [31] authors prove that by tuning the curvature of the grapheme layers the binding energy between hydrogen and carbon changes accordingly.

Table 1. New High Level Storage System Targets for Light-Duty Vehicles (September 2009)

	2010	*2015*	*Ultimate Full Fleet*
Gravimetric Density in g H_2 per kg C (H_2 weight %)	45 (4.5 wt %)	55 (5.5 wt %)	75 (7.5 wt %)
Volumetric Density in g/L	28	40	70
System Fill Time for 5-kg fill, min (Fueling Rate, kg/min)	4.2 min (1.2 kg/min)	3.3 min (1.5 kg/min)	2.5 min (2.0 kg/min)

Source: http://www1.eere.energy.gov/hydrogenandfuelcells/storage/.

It was shown that the binding energy increases of about 2 eV passing from the concave to convex regions, as an effect of the local distortion of the sp^2 hybridization in favor of the sp^3. Furthermore by inverting the curvature of graphene sheets the hydrogen spontaneously desorbs. In the other work [30] a hydrogen chemisorptions is analyzed on various structural modifications of graphene and bilayer graphene is analyzed.

In previous works [30, 31] the authors imply that sp^3 hybridization configuration is favorable for the increase of hydrogen binding energy. In particular they show that two deformations of graphene sheets i.e. an alternating buckling with the amplitude of modulation of 0.4 Å is the most favorable one. They find that for the fully hydrogenated graphene the mass percentage of hydrogen is 7.8 being over the target of DOE. Both works suggest that:

1) Local curvature of carbon based sheets (graphene) affects significantly the hydrogen-carbon binding energy. The change of the curvature may induce the increase or the decrease of the hydrogen adsorption rate.
2) The increase of the hydrogen-carbon binding energy is related with the transformation of local bonds geometry from sp^2 hybridization to sp^3. Breaking the π bonds and producing the additional σ bond, and thus transition from sp^2 to sp^3 is the main mechanism of chemisorptions on graphene.

Although both works [30, 31] analyze chemisorptions on graphene a similar results would be expected on locally distorted graphite surface. Graphite layers or bound by weak van der Walls coupling having no effect storing hydrogen on the surface properties. The binding energy of the carbon-hydrogen systems is mainly due the properties of highly localized, directional depended σ bonds. However question remains what would be an effective mechanism of hybrdization modulation of graphite layers. The key idea of graphite surface curvature/hybridization modfication is to utilize the multistability of the graphite in the context of the PIPT. The MD simulations where performed [33] to confirm the hypothesis that diaphite structure increses binding energy between carbon and hydrogen. The system consists of 400 on 400 on 2 carbon atoms with periodic boundary condittions. It is then minimized using AIREBO [32] potential to form starting diaphite structure. The hydrogen atoms where then placed on each extruded atom on the top plane simulating full coverage of hydrogen on diaphite.

Figure 30. Hydrogen (baige spheres) on diaphite structure. Red spheres indicate carbon atoms extruded from graphite planes whereas blue spheres indicate carbon atoms intruded form graphite planes [33].

The entire structure was minimized again. By substracting the cohesive energy of diaphite and hyrogen combined system from the cohesive energy of diaphite without hydrogen one obtained the hydrogen-carbon binding energy. The preliminary calculations indicate that there is a significant increase up to 0.91 eV/atom corresponding to 87.8 kJ/mol. The energy is , nearly twice higher than the energy recommended by DOE. The hydrogen carbon bond length was estimated on 1.1 Å, whereas the distance to the nearest carbon atoms is now 1.54 Å (same as in diamond). Although the simulation was performed in zero temperature, zero preassure regime it opens new perspective for diaphite as a possible adsorbent for hydrogen. Further calculations will be performed in the close future.

Discussion and Conclusion

In this review paper a detailed discussion of graphite-diaphite transformation was presented. The experimental results by Kanazaki et. al suggested that a new stable phase appears in the photoexcited graphite by a femtosecond laser pulse. The phase is called diaphite. Theoretical considerations presented in the this section described the properties of diaphite and mechanism of its creation in the context of photoinduced phase transitions. The initial transformation is governed by the charge transfer process initiated by the femtosecond pulse. As a result a pseudo exciton state is created spanning two layers. There exists a small but finite probability of self-localization of such a state. Then due to coulomb interaction and electron-phonon interaction two layers are contracted resulting in stable diaphite domain formation. The cohesive energy calculations using Brenner's empirical potential revealed that once the interlayer distance shrinks up to approx. 1.9 Å there are number stable configuration with pseudo sp^3 bond geometry. The energy barrier between graphite and diaphite is of order of few eV being able to overcome by photoexciation. The DFT calculations proved that the domain has a pseudo gap immersed in semmimetallic continuum. Cooperative phenomena play crucial role in the transition enabling exponential growth with excitations number. Molecular dynamics studies suggest that due strong nearest neighbors interactions characteristic for sp^2 systems the energy of the initial excitation is distributed among neighborhood. Once the interlayer bond is created the energy required to form another one in its proximity is much lower. Those two effects result in nonlinear response of the material for the photoexciation. Since there is a mismatch between bond created between α and β atoms a shearing mode appears due to multiphoton excitation. The average value of shear displacement is 0.3 Å. Using the LCBOP potential including long range interaction the effect of diaphite domain on the surrounding graphite was investigated. As a result of diaphite domain formation the interlayer distance between graphite layers shrinks from 3.35 Å up to 3.15 Å. Furthermore a temperature stability of diaphite was investigated.

The molecular dynamics calculations using the LCBOP potential proved the stability of diaphite in the room temperature regime. Above 1000 K however two graphite layers with the diaphite domain ablate leaving undisturbed graphite. Lastly diaphite structure was investigated as a possible substrate for hydrogen adsorbtion. This problem is especially important in the context of hydrogen storage for mobile applications. It was shown that photoinduced curvature of diaphite significantly increase hydrogen adsorption heat beyond

DOE expectations namely 86 kj/mol. This result opens possibility of studying diaphite as a new interesting surface for hydrogen adsorption.

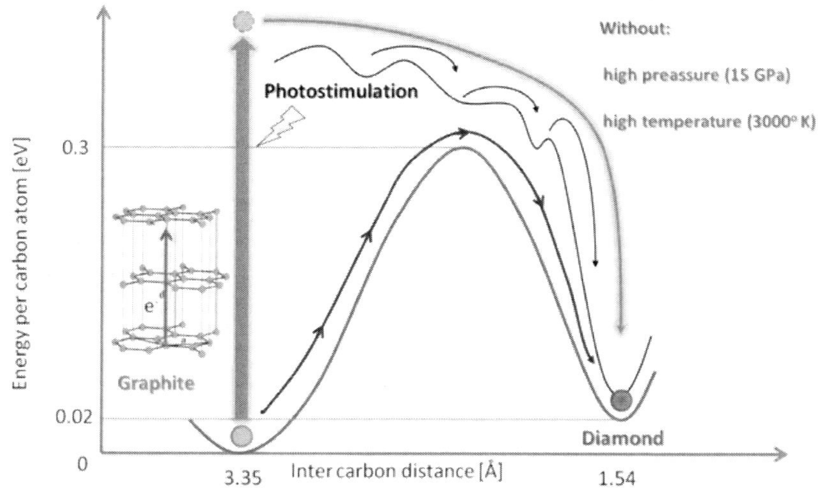

Figure 31. Graphite diamond phototransformation concept via intermediate metastable states.

Since “diaphite” structure resembles diamond’s one in many aspects it would be interesting and justified to ask a question whether this transformation may be further continued? Instead of supplying previous methods of graphite-diamond transformation i.e. via direct transition one could consider a multistep transformation via intermediate metastable states (Figure 31). One of the principal prototype structures is diaphite. In particular one may ask if diaphite is only a local, two-layer, stable structure or maybe it is one of the intermediate phases located on the graphite-diamond phototransformation path? In order to answer those questions one need to analyze the possibility of evolution of the diaphite to macroscopic, volumic (multi-layer) structure. The key element of the investigation will be clarification of the stability of such volumic structures.

It would also be necessary to prove the existence of other stable phases, separated by energy barriers in photoexciation energy regime. Finally it would have to be shown that diamond is the final phase on this path and to characterize processes responsible for those transformations. Although above considerations are purely hypothetical one has to notice that diaphite are very important step on the graphite-diamond path since interlayer bonds are already created. Further structure optimization is energetically relatively cheap. Nonetheless we should not forget that this sort of transformation are most likely possible in a nano scale where no atoms is far less then Avogadro number. However the possibility of uncovering a physical process of transforming graphite into diamond only in the photo stimulated manner by light is most likely as tempting as changing lead into gold by our predecessors.

ACKNOWLEDGMENTS

This work is supported by the Ministry of Education, Culture, Sports, Science and Technology of Japan the Grant-in-Aid for Specially Promoted Scientific Research; Contract

No. 24000006, FY2012 and funds of Polish National Science Center decision no. DEC-2011/03/B/ST5/02677.

References

[1] Novoselov S. et. al. *Science.* 2004, Vol. 70, p. 8.
[2] Radosiński L., Nasu K., Tanimura K., Kanazaki J. book auth.] Naito T. *Molecular electronic and related materials - control and probe with light.* Kerala : Research Network Publisher, 2010.
[3] Kanazaki J. et. al. *Phys. Rev. Lett.* 2009, Vol. 102, p. 08474.
[4] Raman K. et. al. *Phys. Rev. Lett.* 2008, Vol. 101, p. 077401.
[5] L., Radosiński, et al. *Phys. Rev. B.* 2 2010, Vol. 81, p. 035417.
[6] Fahy G. S., Louie S., Cohen M. L. *Phys. Rev. B.* 1986, Vol. 34, p. 1191.
[7] F., Bundy. *The Journal of Chemical Physics.* 1963, Vol. 38, p. 631.
[8] K., Nasu. *Photoinduced Phase Transitions.* Singapore : World Scientific, 2004.
[9] Nishioka K. and Nasu K., *Phys. Rev. B*, 80 (2009) 235420.
[10] Los J. H., Fasolino A. *Phys. Rev. B.* 2003, Vol. 68, p. 0241107.
[11] Radosiński Ł., Luty T., Radosz A., Nasu K., Nishioka K., Ohnishi H. and Wójt D., *Acta Phys. Polon. A*, 121 (2012) 359.
[12] Dresselhaus G., Saito R., Fuijita M. and Dresselhaus M. S., *Phys. Rev. B,* 46(3), 1992.
[13] Koizumi, S.; Nebel, C. E. and Nesladek, M. Wiley VCH 1992, p. 50. ISBN 3-527-40801-0.
[14] Decarli P. and Jamieson J., *Science* 1961, 133 (3467): 1821–1822.
[15] Khachatryan, A.Kh.; Aloyan, S.G.; May, P.W.; Sargsyan, R.; Khachatryan, V.A.; Baghdasaryan, V.S., *Diam. Relat. Mater.* 2008, 17 (6): 931.
[16] W., Brenner D., *Phys. Rev. B.* 1990, Vol. 42, p. 9458.
[17] Tersoff J., *Phys. Rev. Lett.* 1988, 61(25),
[18] Cohen M. L., Fahy S., and Louie S. G., *Phys. Rev. B,* 1987, 35(14).
[19] Katayama-Yoshida H, Nakayama H., *Journal of Physics: Condensed Matter* 2003, 15(3).
[20] Radosinski L., Nanoscale sp2 – sp3 conversion by visible light irradiation in graphite, *Ph.D Thesis*, 2008.
[21] Ohnishi H., Nasu K. *Phys. Rev. B.* 2009, Vol. 79, p. 054111.
[22] Ohnishi H., Nasu K. *Phys. Rev. B.* 2009, Vol. 80, p. 014112.
[23] Nishioka K. and Nasu K., Cooperative, *Phys. Rev. B,* 82 (2010) 035440.
[24] Nishioka K. and Nasu K., *Phys. Status Solidi B,* 248 (2011) 495.
[25] Nishioka K. and Nasu K., *Acta Phys. Polon. A*, 121 (2012) 350.
[26] Nishioka K., Nasu, and Yonemitsu K., *J. Phys.: Condens. Matter* 24 (2012) 205402.
[27] http://www1.eere.energy.gov/hydrogenandfuelcells/2003_storage_solicitation.html.
[28] Kuchta B., Firlej L., Pfeifer P. and Wexler C. *Carbon* 2010 48, 223 –23.1.
[29] Tylianakis E., Klontzas E. and Froudakis G. E., *Nanoscale,* 2011, 3, 856–869.
[30] Tozzini V. and Pellegrini V., *J. Phys. Chem. C*, DOI: 10.1021/ jp208262r.
[31] Boukhvalov D. W., MKatsnelson . I. and Lichtenstein A. I., *Phys.Rev. B* 2008, 77, 035427.
[32] Stuart S. J. et al., *J. Chem. Phys.* 2000, 112, 6472.

[33] Radosinski L., D. Wójt, Wroclaw University of Technology Group Seminar, *unpublished material.*

[34] Wojt D, Hybrid structures with sp2 and sp3 bonds in graphite as a possible transient state on a path of the photoinduced graphite-diamond transformation, *Msc. Thesis*, 2012.

In: Graphite
Editor: Quinton C. Campbell
ISBN: 978-1-62618-576-0

Chapter 4

GRAPHITE POWDER AND RELATED MATERIAL AS THE PRINCIPAL COMPONENTS OF CARBON PASTE ELECTRODES

Ivan Švancara[1*], Tomáš Mikysek[1], Matěj Stočes[1] and Jiří Ludvík[2]

[1]Department of Analytical Chemistry, Faculty of Chemical Technology, University of Pardubice, Pardubice, Czech Republic

[2]J. Heyrovský Institute of Physical Chemistry ASCR, Prague, Czech Republic

ABSTRACT

In this chapter, graphite in powdered form is presented as the material of choice for the preparation and use of the so-called *carbon paste electrodes* (CPEs) representing one of the most popular types of electrodes in electrochemical measurements for more than five decades.

Starting with a short historical reminiscence, when highlighting also the importance of carbon materials in electrochemistry as such, powdered graphite is presented as particularly suitable material for an endless variety of electrodes, sensors, and detectors; the individual forms having found the widespread use in theoretical electrochemical studies, practical electro analysis, as well as in other instrumental measurements. Special attention is paid to the graphite powder with respect to its applicability in carbon paste mixtures and, subsequently, in the configuration of CPEs.

In this respect, classification of traditional types of graphite powders is also given, completed with a survey of related and/or alternate carbonaceous materials whose applicability to the preparation of CPEs has undergone a real boom in the new millennium and, in fact, resulted in a new renaissance of the field. Furthermore, graphite powders (together with related materials) are discussed via physical, physicochemical, and principal electrochemical properties and of similar interest are also the corresponding graphite/carbon paste mixtures, including the experimental know-how on their laboratory preparation and way(s) of housing in specially designed holders and similar assemblies.

*Corresponding author: e-mail: Ivan.Svancara@upce.cz.

At the end of this section, one special example is included, summarizing for the first time the recently performed studies on less well-known phenomena within the electrochemistry with CPEs – the ohmic resistance of carbon pastes in relation with their structure and behavior.

Last but not least, the electrochemistry and electroanalysis with CPEs are briefly surveyed, documenting the versatility of these electrodes in modern instrumental analysis, their adaptability to the present day's trends, as well as great potential in the future, regardless whether one considers traditional graphite powder-based configurations or mixtures from new alternate materials.

4.1. Introduction – Carbon as the Material of Choice in Electroanalysis

For lengthy decades, carbonaceous materials belong amongst the most essential accessories of electrochemical laboratories. Predominantly, they are being selected as the electrode material, offering a reliable performance with widespread use in theoretically oriented electrochemical research, as well as in routine electroanalysis [1-5]. Apart from favorable mechanical and physical properties, often appreciated features of various forms of carbon are electrochemical characteristics and the expressions like "glassy or vitreous carbon (GC; first reported in [6])", pyrolytic graphite" (PyG; see *e.g.* [7])", "carbon fiber (CFi [8])", and now also "carbon nanotubes (CNTs [9])" are respected and already classical terms in the field.

Regarding graphite, representing the most frequent allotrope of the elemental carbon, it exists naturally in three different types — as (i) *crystalline flake graphite*, (ii) *amorphous graphite*, and (iii) *lump/vein graphite* [10] — but, this material of semimetal character can also be produced synthetically.

Nevertheless, in electrochemistry, both natural and synthetic graphites in solid form are employed only occasionally. This is, for instance, the case of dispensable electrode substrates made of pencil lead and in combination with metallic films applicable in electrochemical stripping analysis (see e.g. [11]) or some home-made prototypes of simple graphite sensors [4,12]. Otherwise, as the electrode proper, these "crude" graphite materials are not normally used and the above-mentioned special forms of carbon, *i.e.*, GC, PyG, CFi or CNTs are highly preferred. The reason is often insufficient purity of common graphites, their lesser compactness (compared to GC or PyG, respectively) and considerable adsorption capabilities of typically untreated surface.

However, there is one particular form of graphite, which had become popular as early as it was introduced into the electrochemistry during the late 1950s by Adams [13]. It is a fine graphite powder with particles in the range of micrometers, or tens of micrometers, that had unleashed the rapid development and later even real boom of the so-called ***carbon paste*** electrodes (CPEs [1,14]). Despite this, when the respective area of electroanalysis with CPEs covers to date maybe 3,000 scientific papers [15,16], most of review articles or monographs dedicated to carbon or carbon electrodes in electrochemical measurements (*e.g.* [7,17-20]) have mentioned graphite powder and CPEs only briefly and without deeper characterization of typical properties and requirements to act in the role of the carbon moiety in carbon paste mixtures. Furthermore, these texts usually do not reflect the empirical experience that each powdered form of graphite is a very specific material that may significantly differ in properties from each other, as well as form graphite in compact form, including the same material of which the powder can also be made. (Perhaps, the Adams' monograph is an exception [1], but this excellent book got somewhat old from today's point of view; mainly, with respect to knowledge coming afterwards.)

Thus, when omitting the monographic material dealing exclusively with electro-analysis at CPEs [14, 16], there is still missing a concise report on the role of graphite and related carbon forms on the physicochemical and electro analytical characteristics of the carbon paste-based electrodes. After being invited by the Publisher to contribute into this book, the authors have gladly agreed and prepared such hitherto missing text being focused primarily on graphite/carbon in the CPE configurations, but considering also the importance of the second moiety in the carbon paste material – the agent binding mechanically the graphite particles into uniform ensemble, but being necessarily reflected in the overall behavior of each CPE.

4.2. Graphites As the Principal Component of Carbon Pastes and Related Mixtures

4.2.1. Basic Configurations of Carbon Pastes and Carbon Paste Electrodes

From traditional point of view, the carbon paste is a mixture of carbon (graphite) powder and a suitable binder [1,7,16]; the latter being chemically and electrochemically inert organic liquid, insoluble in aqueous solutions as the most common media in electroanalysis. The most widely used pasting liquids are paraffin /mineral oils —see later, par. x.3.1— or various silicone fluids of higher density. The far prevailing way of preparation of carbon pastes is *manual homogenization in laboratory*, when using porcelain or agate pestle and mortar. A minority of users then prefers the ready-to-use mixtures (*e.g.* [21]) that are supplied by some dealers of electrochemical instrumentation.

The resultant carbon paste issoft and non-compact mass, needing to be fixed in a glass/plastic tubing [22, 23] or specially constructed electrode bodies [24-28]; see also Figure 1. Such configurations are consensually called *"**carbon paste electrode**"* (***CPE*** [7,16] or, alternatively but seldom, "*graphite paste electrode*" (GPE; *e.g.* [23,29]). In carbon pastes and the respective CPEs, both *main components*; *i.e.* (i) carbon / graphite powder and (ii) (liquid) binder may be altered and the history of the field has seen many attempts of this kind; especially, when exchanging / altering the carbon moiety.

This can be well documented on research activities within the first decade of the new millennium, spawning the whole series of brand new carbon paste-based electrodes, sensors, and other detection units; for details, see [16,30] and below, in par. 4.2.3.

NOTE: The alteration of traditional binders in carbon pastes is less frequent, but principally feasible and — in some cases — surprisingly effective; see *e.g.* [31] and other refs in [16]. The mixtures undergoing these variations often lose the typical paste-like character, forming either rather compact and hard composites [32] or, *vice versa*, quite fluidic and soft matter [33].

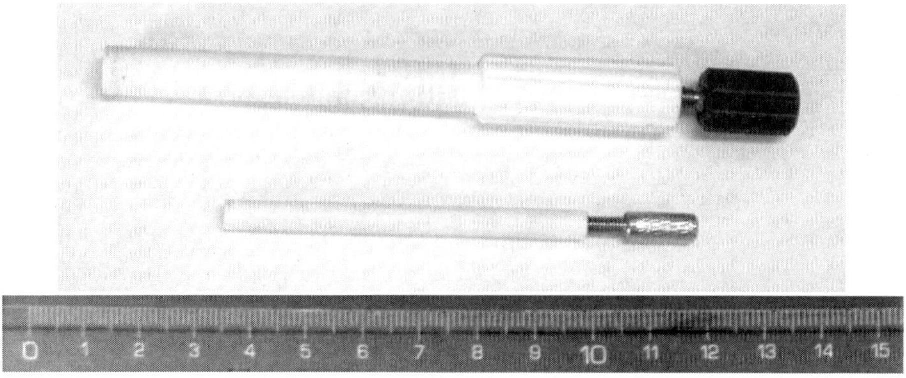

Figure 1. Two size variants of the home-made piston-driven electrode holders for carbon pastes (on bottom: a ruler with cm scale; from authors' archives; for further details, go into [27]).

There is a wide choice of interesting examples showing such arrangements; however, they already are beyond the scope of this chapter focused on the role of the carbon moiety in the CPE and its graphitic form. Thus, the basic facts with many useful details on CPEs with alternate binders cannot be given in further text and the interested reader(s) are advised to consult the relevant literature; best, contemporary reviews on CPEs [15,30] or similar compilations [14,16].

When going back to graphite in CPEs, it must be emphasized again that its powdered form is indeed a specific matter bringing along a synergistic, minimally triple effect [16], when assuming the overall behavior of a CPE. At first, it is (i) quality and type of carbon/graphite and the way of synthetic production, including possible treatments of the particles during the proper technological process [34]. Secondly, it is (ii) the actual distribution of the individual particles, when both the average size and uniformity play key roles. Third, there is also the aspect of (iii) manual preparation requiring certain skills and experience to obtain the final mixture in reasonably reproducible way [16,35,36].

Apart from deductions which factor is more important, it is evident that each carbon paste represents *an individual* and very specific electrode material that possesses a number of attractive properties, together with some unique features.

4.2.2. Classification and Typical Properties/Features of Graphite Powders Applicable in the Carbon Paste Mixtures

Spectroscopic (Spectral) Graphite. This material is undoubtedly dominant among the carbon powders used for preparation of CPEs. Usually, this graphite is of synthetic origin being produced via a *thermal degradation* of suitable hydrocarbon-based compounds [10]. Alternately, the source for artificially made graphite powder may be a soot from incomplete combustion of special coke, which is the case of "RW-B" spectral carbon [33,34,41].

As the title suggests one, its main use is to be for spectroscopic measurements as a special conductive substrate. Spectral graphite powders are commonly included as a fine chemical in annual catalogues of renowned companies and global dealers, such as *Aldrich, Fluka, Merck, Sigma,* or *Riedel-de-Haen.* Relatively frequent are also some local products being supplied under the respective trademarks and/or abbreviations.

From this group, one can quote the products "RW" (in two types, "RW-B" and "RW-C", both available in Germany and Austria), "BDH" (U.K. and Scandinavia), "SMMC" (China), "Acheson", "UCP", and "SP" (U.S.A.); the last two named being popular in the early era of CPEs (see *e.g.* [1,7] and *Table I* in [37]). Unquestionable advantage of spectral graphite powders is the fact that renowned commercial products have guaranteed high purity.

Graphite Gear Lubricants. As alternate source of powdered graphite for carbon pastes, this product is attractive not only from financial point of view, when the price of the respective product(s) may be even in lower than those described above.

NOTE: It is worthy of mentioning that graphite lubricant may also be of natural origin; namely, in the Czech Republic, such a powder — under the trademark abbreviation "CR" and specified with a number declaring the average mesh in μm — is manufactured from the piece material unearthed from graphite mine in Southern Bohemia; see Figure 2. The production of gear lubricants then takes place nearby – in a local factory [38], where the exploited graphite is finely grinded into the micrometer particles and ultimately chemically purified. Along with, the same company manufactures also other special graphites, including the fully synthetic products.

Despite apparently lesser purity compared to spectroscopic graphites, such graphite lubricants can be recommended for the preparation of carbon pastes and — according to our recent experience — its performance in the carbon paste mixtures is surprisingly good (see *e.g.* [40,41] and refs in [16]).

Herein, it can be concluded that both types of graphite powders presented in previous paragraphs exhibit the characteristics and features that must be considered when selecting the proper carbon material for making of carbon paste. Among such criteria and guaranteed properties, one should definitely include:

- ***Particle Size***. As already mentioned, the particles of common spectroscopic graphites range in micrometers or in tens of micrometers, respectively, when most of manufactures are able to provide an official certificate specifying the overall size distribution. Otherwise, relatively reliable information about the average distribution can also be obtained with the aid of microscopy [34,41] and, in perspective, also using the sieve analysis [42].
- ***Particle Distribution***. Common spectral powders or gear lubricants have the particle size distribution satisfactorily uniform, typically ranging within 5-20 micrometers with more than 75% of the average diameter (see *e.g.* [34] and other refs in [16]). Renowned manufacturers usually offer a certificate and/or distribution diagram of their actual products; for instance, the above-quoted graphites can be characterized by the following average mesh: "*CR-5*": 5 μm, "*RW-B*": 10-25 μm, "*Acheson G-38*": 10-20 μm, "*UCP-1-M*": 1 μm, and "*Graphon GP*": 25 μm (for other specification, see again [37]).

  Empirical experience says that the distribution of graphite particles should not differ too much and the more uniform the grain size of graphite, the better properties of the resultant carbon paste can be anticipated [36].
- ***Graphite Powder Microstructure***. Since the invention of CPEs [13], carbon powders used for preparation of carbon pastes were of permanent interest with respect to the microstructure (see *e.g.* [21,34,43,44] and refs. therein). It can be stated that neither spectroscopic graphites nor graphite lubricants exhibit any special structural pattern

or any typical morphology. Nevertheless, thanks to modern microscopic techniques like SEM, one can reveal some differences concerning synthetic and natural graphite powders. As seen in Figure 3, by comparing the images "A" and "B", spectral carbon forms geometrically irregular pieces (3-D objects), resembling small stones with relatively rough and scratched surface, exhibiting also visible relief of hexagonal graphitic structure (see "A").

Figure 2. Touristic entrance into the graphite mine in Český Krumlov (Czech Republic) – an unique source of natural graphite used for manufacturing traditional pencil leads and newly also gear lubricants; the latter applicable as carbon powder for CPEs (from free-accessible website [39]).

A B

Figure 3. Typical microstructures of two different graphite powders. Legend: A) spectral graphite, "RW-B" (see [34]); B) natural graphite, "CR-5" [38]. Experimental conditions: SEM, magnification: A) 1 : 1 000 (1 cm = 15 μm) and B) 1 : 3 000 (1 cm = 5 μm). [From authors' archives] In contrast, the cleaved graphite lubricant is formed by units with more-or-less planar architecture, looking like a portion of tiny corn flakes. Also, the surface of grinded natural graphite is fairly smooth without any visible pattern or relief upon – again, compare both SEM microphotographs featured in Figure 3.

NOTE: During authors' experimentation, the "RW-B" spectral graphite (shown in Figure 3A) was found to be one of the best and most suitable carbon powders among many other tested in the configuration of carbon pastes over lengthy decades (see *e.g.* [31,34,45] *vs.* [36, 39]). And this evaluation applies despite relatively rough surface of the "RW-B" product and its atypically large particles with average diameter of *ca.* 25 μm; some achieving up to 50 μm; see again Figure 3 and image "**A**".

Besides ordinary µm-sized graphite powders, there have been some reports on special carbon paste mixtures containing the carbon moiety with extremely small particles, from 100 nanometers [46] down to 30 nm [47]. Such fine materials can nowadays be purchased from specialized dealers, but they are also obtainable on one's own in laboratory – by thoroughly grinding commonly marketed spectroscopic graphite powder in a miniature mill [41,46].

- ***Low(er)Adsorption Capabilities***. Normally, only specially proposed assays can prove that a graphite powder intended for carbon pastes exhibits undesirable adsorption activity [35,36]. During measurements, this can be revealed easier; usually, *via* the enhanced content of oxygen entrapped in pores of graphite or absorbed additionally during mechanical homogenization in the mortar.

  By using CPEs from spectral graphites or graphite lubricants in faradic measurements —*i.e.*, current flow-based and electron transfer involving redox transformations — higher concentrations of oxygen are then registered as characteristic plateau-like responses, in fact, the increased background within the potentials from –0.2 to –0.6 V vs. Ag/AgCl during the scanning in cathodic direction (see *e.g.* [16,30,43]).

NOTE: In the past, the problems associated with the presence of oxygen in graphite powder were tried to be overcome by pre-treating with the aid of some special procedures. One of such purifying operations had *e.g.* utilized thermal desorption of oxygen at *ca.* 400°C in the presence of inert gas passed over a small portion of graphite. Reportedly, such a process taking several hours had allowed the experimenters to remove significant part of adsorbed oxygen [48]. A similar operation, improved yet by impregnating the refined graphite with ceresin wax, was found to be effective as well [49]. Nevertheless, further attempts on purification of graphite powders from oxygen appeared only rarely [50, 51]; the latter procedure having also had different goal – improvement of reaction kinetics at the treated surface [51].

Scarce applications of such procedures can be seen in their unease and time consumption. Moreover, carbon pastes are still being prepared mainly by manual hand-mixing and therefore, the powder may absorb back a part of previously removed oxygen during this operation made at open air. Finally, from the present day's of view, it seems that additional deoxygenation is not so necessary since graphite powders produced by modern technologies do not exhibit strong adsorption capabilities; oppositely, they can be specially maintained against this effect [36].

- ***The Remaining Criteria.***.Just to complete the *survey of essential properties* of the carbon / graphite moiety, there are yet further, more or less obvious features that can be quoted:
  (i) *Sufficient Chemical Resistance.* It is a typical property of both natural and synthetic graphites, when solely extreme conditions and/or highly aggressive reagents may transform the elemental carbon into some oxidation products (*e.g.*, quinone and hydroxyl derivatives; see discussion in [16, 28, 43]).
  (ii) *Excellent Conductivity.* Generally known and one of the most valuable properties of carbon in graphitic form. As highlighted below, almost zero resistance is also

characteristic for the carbon pastes themselves, apart from the presence of binder, typical insulator.

(iii) High Purity. An essential demand when the powder of choice must not contain any electroactive traces that may interfere with current-flow measurements. (However, in potentiometry or conductometry, where the changes in the equilibrium electrode potential or conductivity of a solution are indicated, the high purity is not a strict limitation and, in principle, even relatively raw graphites may be fairly applicable [52].)

4.2.3. Alternate Carbonaceous Materials for Carbon Paste(-like) Electrodes

Despite the main focus of this book featuring various achievements in technology and chemistry of graphite, the association of this material with carbon pastes for electrochemical measurements requires to consider also some related materials—mainly, the other forms of carbon—because, at present, they may represent a half among all newly introduced carbon paste (-like) mixtures and related heterogeneous carbon-based composites [16,29,53].

- Acetylene Black (see *e.g.* [54, 55]) ... Our survey can be opened by this product obtainable by total combustion of acetylene in inert atmosphere or, better, by controlled chemical decomposition. According to the propagators of this alternate material, acetylene black has a semi-crystalline consolidated structure, contributing to more reproducible behavior of the corresponding carbon paste mixtures and offering also specific surface characteristics with enhanced adherence to the accumulated analyte species [55].
- Carbon Black ... Of similar nature is also an amorphous mass obtainable by the incomplete combustion of heavy petroleum fractions, introduced into the realm of CPEs by Kauffman et al. [56] as the constituent of alternate "home-made" carbon paste mixture.
- Colloidal Graphite ... Next in the list is a hexagonal carbon with extremely fine flakes and enhanced conductivity, having been presented by the same group [56] and recommended for similar purposes – electroanalysis of selected pharmaceuticals.
- Soot ... In contrast to all previous items, the product of incomplete combustion of diverse flammable materials of organic origin was found totally inapplicable for making of carbon paste already by their inventor, Adams [43].
- Activated Charcoal ... Nearly the same can be stated about graphitic carbon widelyused in medicine against flatulence in digestion tract or as the active substance in the filters of gas masks. In this case, the tested specimen had revealed the anticipated enormous absorption, reflected in undesirably high background signal of the respective CPE [31]. Nevertheless, one type of commercially available activated charcoal — an "USP" powder [57]— was reported as quite well functioning in a CPE tested and recommended for clinical analysis.
- Coke (Snow) ...In a certain defiance with the momentarily massive preference of all new forms of carbon (see below), yet another common variety of carbon has been chosen and, very recently, also examined by our research group as a particularly

cheap carbon alternative [41]. The first results are surprisingly good; however, it should be pointed out that these still continuing experiments are carried out with specially purified specimens.

- Natural Coal ...With similar motivation and in an effort to complete the spectrum of widely available carbonaceous materials, also the most common form of natural carbon and worldwide used solid fuel is now under investigation as a possible alternative of traditional graphite powders [58]. So far, in electrochemistry, "black coal" has *e.g.* been used as the modifier of an ordinary carbon paste [59]. On the contrary, a CPE made of another naturally occurring carbon, lignite or "brown coal", has not been yet reported [16].

NOTE: In our new experiments [41, 58], there are some difficulties with grinding rather soft pieces of common black coal, whereas "**anthracite**" (the oldest and hardest modification of natural coal) appears to be more promising. Also, in this case, each specimen is carefully selected and purified in mineral acids before being applied as a carbon powder.

- Diamond Powder ... Last but not least, as indeed unexpected representative of carbonaceous materials, pure (non-doped) diamond has also been recommended for making of carbon pastes in the early 2000s [60]. Both natural and synthetic diamonds in a form of fine powders with particle diameter of about 1-5 μm were extensively employed in the configuration of the so-called ***diamond paste electrode*** (***DPE***) by the van Stadens, when *ca.*dozen of reports — see a sub-chapter in [16] and refs therein — were to convince the electrochemists that one of the most typical electric insulators can be used as the true replacement of traditional graphite powders.
- New Forms of Carbon. Almost all these materials are wholly synthetic and exhibit a number of attractive physicochemical and electrochemical properties, such as distinct catalytic capabilities, enhanced conductivity, or specific mass/charge transfer at the electrode/solution interface [16,29]. Undoubtedly, these features are given mainly by *specific microstructure* of the individual types that are being classified as "new carbons" or "new forms of carbon", respectively, although their appearance in chemical laboratories goes back to the late 1980s (see [9] and refs. therein).

One can start the brief survey of new carbons with (i) glassy carbon powder (GC), having been for the first time been tested in the mid-1990s [31, 34] and, later, successfully used in practical analysis (see *e.g.* [61-64]). The GC powder is specially treated graphite with spherical particles produced by temperature controlled degradation of highly molecular resins [34], when the respective forms may differ each from other in the final surface treatment, which is, for instance, the case of two *Sigradur*® powders.

Chronologically, the next new material is (ii) fullerene "C-60", a representative of ***molecular carbons*** forming — together with related "carbon nanotubes" and "carbon nanohorns" (see below) — three major classes of carbon nanomaterials that can be titled as "*carbon of the new millennium*" [29]. In the carbon paste mixtures, fullerene "C-60", whose structure of a hollow sphere from hexagonal aromatic rings resembles a soccer ball, was firstly used in 1999 [65], followed soon by some other reports (*e.g.* [66,67]). In all cases, fullerene "C-60" was applied as the additional modifier with distinct electrocatalytic properties, whereas a binary mixture of the "C60" with an oil has not been yet reported. Similarly, within the electrochemistry with CPEs, there is no mention on the use of an ellipsoidal fullerene, "C-70" [16, 29].

The second carbon nano-material, (iii) carbon nanohorns (CNHs) have so far been examined as a CPE in one study only [68], whereas (iv) carbon nanotubes (CNTs [9]), in both single-walled (SW) and multi-walled (MW) variants, clearly dominate among all new forms of carbon as documented by a wide collection of more than 150 scientific reports concerning either CNT-modified CPEs (firstly described in 1996 [69]) or the so-called ***carbon nanotube paste electrodes*** (***CNPEs***; with a debut in 2003 [70,71]), where the CNTs totally replace the powdered graphite.

Their widespread applicability has already been the subject of special reviews [16,29,53] and it can be stated that both CNT-CPE and CNTPE variants utilize fully the unique properties of this still very popular material.

NOTE: Finally, the last carbonaceous nano-material is (**v**) ***carbon nanopowder*** which has also been used in a configuration of CPE (see *e.g.* [47]), including the newest studies, where it acted as a particularly fine CP-proper of some biosensors [58]. In fact, this nano-material is more or less common carbon powder, but with extremely small particles (20-50 nm) and with its predecessors in specially grinded carbon powders used already in the late 1980s for the preparation of ensembles of CP-ultramicroelectrodes (UMEs [46]).

The order of new carbons continues with (v) template carbon, (vi) porous carbon foam, (vii) porous carbon microspheres (all tested as new CPEs by Nossol's group [72], (viii) ordered mesoporous carbon (omC [73]), (ix) carbon nanofibers (CNFi) obtainable either by electrospinning [74,75] or with the aid of chemical vapor deposition [76], and (x) graphene(Gr or "graphite sp^2" [77]) plus graphene oxide (GrO), with planar structures of an indefinite polyaromate and both applied as binary GR(O)-binder mixtures. To date, this special carbon allotrope and its oxidized form are the last newcomers within the large family of carbon pastes.

4.3. Brief Characterization of Carbon Paste-Based Electrodes

4.3.1. Pasting Liquid as the Binder and Second Moiety in Carbon Pastes

Since this text is focused predominantly on the carbon/graphite moiety in carbon pastes and CPEs, all the aspects associated with the presence and functioning of the binder cannot be herein discussed in detail. Instead, a concise survey will be given based on information material summarized ***in special literature***, such as the already *cited reviews* [15,29,36, 37,43,52,53] and some other *similar articles* [1,14,16,78-85], or even *principal original papers*—namely: [13,21,24,28,30,34,70,86-98]— dealing exclusively with the characterization of carbon paste and/or CPEs and appearing during the entire half-a-century of existence of CPEs in electrochemistry and electroanalysis.

The primary role of the binder/pasting liquid in the carbon paste mixture is to *connect mechanically* the individual carbon particles into a ***uniform mass***. Nevertheless, the presence of binder is more or less reflected in the resultant properties of each paste – starting with (i) the actual consistency (given mainly by the mutual carbon-to-binder ratio), via (ii) the effect on the mass/charge transfer at the particles coated with lipophilic binder (resulting in typical hydrophobic character), up to some (iii) nuances in electrochemical characteristics which can be attributed to the particular type of pasting liquid and its specific physicochemical properties, including chemical structure.

- Important Criteria for a Carbon Paste Binder ... The choice of a suitable binding agent for the preparation of carbon pastes and the respective CPEs should take into account the following requirements: (i) *chemical inertness*, (ii) *electroinactivity*, (iii) *low volatility*, (iv) *minimal solubility in water or in a medium used for measurement*, (v) *controlled miscibility with organic solvents.* In context with still rising effort to direct modern electrochemistry towards the principles of environmentally friendly ("green") analysis; see *e.g.*[99,100], one can also add: (vi) *low toxicity* and (vii) *replacement of highly harmful substances.*
- Pasting Liquids and Related Binders ... Although none of pasting liquids obeys all the criteria for ideal binder, there is a wide offer of organic compounds or mixtures that form carbon pastes of good quality. Specifically, these are as follows:
  - ***Paraffin (Mineral) Oils***. These liquids are the most frequently chosen carbon paste binders and the may represent *ca.* 75% from all pasting liquids ever proposed and then used in the configuration of carbon paste mixtures. The most popular product of this kind is ***Nujol***® (oil), a mixture of liquid aliphatic hydrocarbons, and a product declared as mineral oil of ***Uvasol***® purity grade which very similar in nature and basic properties.
  - ***Silicone Oils and Greases***. The binders of this kind ranking the second position with respect to popularity in use for CPEs, belong among polymerized siloxanes with organic side-chains and usually with high molecular weight.
    In general, both SO and SG share a majority of valuable properties highlighted above for paraffin oils; moreover, they are non-flammable, forming more stable mixtures with less pronounced ageing effect, and generally more resistant to some organic solvents (see e.g. [16,36] and refs. therein).
  - ***Other Pasting Liquids***. From rather wide palette of hitherto examined/used liquid binders, one can highlight: (i) *aliphatic, aromatic, and halogenated hydrocarbons,* (ii) *lipophilic organic esters*, (iii) *natural and synthetic oils or greases* (namely: castor oil, vaseline), and (iv) *new types of carbon paste binders*, such as *polycationic electrolytes* and, mainly, various*ionic liquids*(e.g.), having contributed — together with CNTs (see par. 4.2.3) — to a stormy development of numerous new types of carbon pastes being reported in nearly hundred reports within the archived literature on electroanalysis with CPEs.

4.3.2. Physicochemical and Electrochemical Properties of Carbon Pastes and Carbon Paste Electrodes in a Concise Survey

Similarly as in the previous section, also this paragraph offers a survey of the most important observations and facts, gathered and discussed ***in special literature*** (*monographs and chapters*, *reviews* and *similar texts* [1,14-16,29,36,37,43,52,53,78-85], or *principal experimental studies* [13,21,24,28, 30,34,70,86-98]).

Despite numerous specific features, ***traditional carbon pastes*** consisting of two main components (carbon and binder moieties), as well as their ***chemically and biologically modified variants*** [14-16,78-81, 84,85] still belong to the *solid or solid-like carbon electrodes* and therefore exhibiting the properties typical for carbonaceous materials. Herein, one can quote the following:

- Basic Chemical Properties. Compared to mercury or noble metals, carbon paste is almost ***inert electrode material*** with substantial resistance against unwanted chemical or electrolytic transformations. Nevertheless, like with carbon alone, also the surface of carbon paste can be *oxidized under extreme conditions*, which usually leads to principal changes in the electro-chemical behavior (see below).
- Consistency of Carbon Pastes. Although common mixtures resemble in *softness and plasticity* a peanut butter [1,7], the consistency may also significantly vary, which principally depends on the mutual adherence of both main components. In extreme cases, the resultant CP-mixture may be muddy or, *vice versa*, quite hard; both being difficult to handle.
- Structure and Microstructure. Carbon paste mixtures can also be described as a ***solid dispersion*** of graphite particles in the liquid binder used. In all principal microscopic studies (see *e.g.* [34,44] and *Table I* with refs in [101]), it has been repetitively proved that typical morphology of carbon paste surface is a *random array of carbon particles linked together with pasting liquid* whose molecules form a *very thin film of nm-dimensions* covering almost each solid particle in the mixture.

A number of interesting microscopic images featuring different types of carbon pastes have been obtained in our laboratories since the mid-1990s [34], when using scanning electron microscopy (SEM). The newest experiments performed on the verge of 2012/ 2013 are then premiered in this text [58]; see Figure 4 below with a quartet of images "A-D".

The individual photos document pretty well notable differences between the pastes from the two different graphite powders (images "A" and "B"), as well as completely dissimilar structure of two mixtures based on spherical particles of glassy carbon powder (photo "C") and carbon nanotubes ("D").Concerning the former, the carbon moiety was either spectral graphite "RW-B" or gear lubricant from refined natural graphite (again, see above and refs. [34,38,39]). As can be seen, the overall pattern of both CP-surfaces is rather similar, when the structure copies the original texture of graphite particles (Figure 3 A,B), including the grainy appearance of the CP-mixture from RW-B graphite whose surface is markedly rougher.

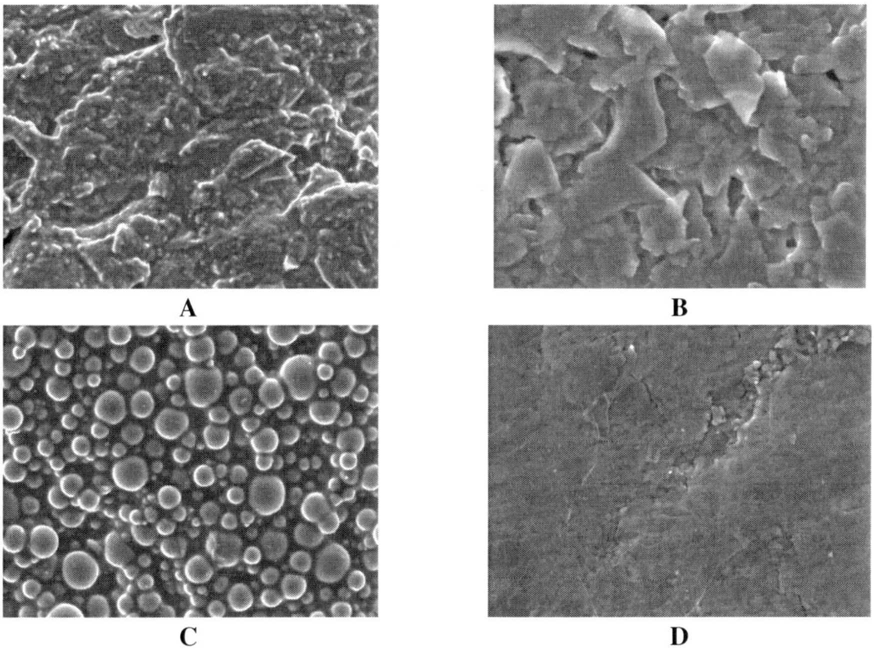

Figure 4. Typical surface microstructures of four different carbon pastes. Legend: Carbon paste mixture made of spectral graphite + silicone oil (**A**), graphite lubricant + silicone oil (**B**), glassycarbon powder + silicone oil (**C**), single-wall carbon nanotubes + mineral oil (**D**). Actual carbonpaste ratio = 1gr : 0.5 mL. Exp. cond.: SEM, 1 : 1 000, 1 cm = 15 μm. [from authors' archives].

Similarly, the globules of the glassy carbon powder (*Sigradur*® product, specified in [34]) are tightly obscured by oil, forming picturesque design in Figure 4C. In this case, it is quite interesting to compare this new photo with a shot of sample obtained already two decades ago [34].) Quite unique is also the structure of SW-CNT-based carbon paste (image "D"), manifesting a very compact surface, which corresponds to the already published reports (see [16] and refs therein) and is being attributed to strong adhesiveness at CNTs.

Finally, a carbon paste from multi-walled CNTs was also found distinctly compact, but its surface exhibited somewhat less homogeneous surface with abundant scratches and small craters (not shown here).

- Minimal Ohmic Resistance. As already mentioned during characterization of graphite powders, carbon pastes and the respective CPEs exhibit a very low ohmic resistance; typical values ranging below 10 Ω, although some "wetter pastes" or mixtures from dense greases may exceed even 100 Ω; both limits enabling the comfortable use of CPEs in potentiometry [52]. The ohmic resistance of carbon pastes and some accompanying phenomena are altogether so interesting that these features are discussed again below (see par. x.3.4), also in association with some recently performed studies that have not yet been reviewed.

NOTE: Extremely low ohmic resistance is the most distinct difference between carbon paste and otherwise very similar carbon inks used for screen-printed electrodes that, after hardening, exhibit significantly higher resistances; typically, in hundreds of ohms [14].

- Ageing of Carbon Pastes and Effect of "Self-Homogenization". This rather less known phenomenon can be observed for *CP-mixtures from more volatile binders*, such as some organic esters. Regarding common mineral and silicone oil-based carbon paste, the ageing effect is much less pronounced, but some freshly made CPEs may exhibit notable instability in behavior shortly after preparation — manual mixing of both main components — and requiring one or two days for settling and getting into sufficiently homogeneous state.
- Low (or No) Stability of Carbon Pastes in Organic Solvents. Due to the presence of non-polar binders in carbon pastes and their certain miscibility with organic solvents, the employment of CPEs in such media or aqueous solutions containing these solvents is considerably limited. In these cases, one can expect *disintegration of the paste* and, in fact, its total destruction. The stability of carbon pastes can be improved by using more resistant pasting liquids; *e.g.*, highly viscous silicone fluids or — quite surprisingly — powdered graphite with spherical particles (similar to that forming the eye-catching relief in Figure 4C) and their specially treated surface [16,28].
- Polarization Characteristics. In faradic measurements associated with the electron transfer and current flow, common types of CPEs can be polarized within a *potential range of–1.0 V and +1.0 V vs. SCE*; the respective potential limits being given either by obvious hydrogen evolution at cathodic side or by oxidation of the electrode in anodic direction. By proper choosing of the carbon paste constituents or via modification, the resultant *potential window* may be somewhat extended in both anodic and cathodic direction. Such achievement may have the reason for the hydrophobic surface of a CPE, which is a feature commented next.
- Hydrophobic Character of Carbon Pastes with the Effect on Kinetics of Electrode Reactions. Definitely the most typical and most highlighted characteristics of carbon pastes and CPEs is lipophilicity of the electrode material, giving rise to *decelerated/moderated reaction kinetics* of numerous electrode processes, resulting in partially or totally irreversible behavior of redox-systems and complicating the corresponding reactions.
- Interactions at the Carbon Paste Surface and in the Bulk. CPEs are undoubtedly appreciated for a wide variety of interactions and mechanisms in which either CP-surface or even CP-bulk may participate. Here, without going into detail, one can name the following processes and principles: (i) ***electrolytic redox transformations***, including the involvement of the CP-surface after the so-called *surface activation/hydrophilization* by its partial oxidation; (ii) a***dsorption at the CP-surface***with enhanced affinity towards lipophilic molecules;(iii) ***extraction/penetration onto the CP-bulk***as most valuable and almost unique feature of carbon paste (comparable only to membrane-devised electrodes); (iv) ***ion-exchange and ion-pair formation***, *i.e.* electrostatic interactions between the charged counter-ions and often employed for extremely high selectivity; (v) ***electrocatalysis***often enabled by suitable modifier(s) contained in the CP-paste and, in general, leading to a marked enhancement of the final response; (vi) ***indication of some specific signals or states***, such as the *equilibrium electrode potential*, *conductivity*, electrochemiluminescence, or temperature-dependent *diffusion*, or *(ultra)sonication effect*. Finally, the whole survey can be ended by (vii) ***synergistic***

mechanisms combining some of the above-surveyed processes and, in general, improving the overall electroanalytical performance of the corresponding measurements.

4.3.3. Some Notes on the Making of Carbon Pastes and Their Experimental Testing

- Carbon-to-Pasting Liquid Ratio. Usually, graphite powder and a pasting liquid are mixed together in quantities chosen accordingly to empiric experience; the respective "*carbon-to-pasting liquid ratio*" varying in an interval of ***1.0 gr.: 0.4-1.0 mL*** [14-16,36,43,49,50]. Some mixtures require longer homogenization when mixed manually and may contain a higher percentage of liquid binder since the optimal ratio of the two main components depends also on their mutual adherence.
  Thus, it can be generally advised to seek and "tune" the suitable ratio of the two paste constituents experimentally, when using empirical experience and, best, in the way, as shown in the following section.
- Testing of Carbon Pastes. In the era of CPEs, as well as later on, numerous authors — often pioneers of the field in their countries and geographical regions — had found highly useful the sets of specially proposed experiments for electro-chemical characterization of the individual types of CPEs. Still inspiring works of this kind were first published by Adams' team [86,87], followed by electrochemists from Middle Europe [24,88-90], Scandinavia [91,92], or later again from the U.S. [93-95]. In the late 1990s, the substantial material from these very valuable reports has been gathered into a review [36], which can be recommended as a true guide through the inevitable experimentation with CPEs and related sensors.

Such experimental characterization or shortly "testing" is consensually the most straightforward way of how to define the basic properties and behavior of freshly made carbon pastes, including mixtures made from brand new or hitherto unused components. The strategy of the corresponding tests involves sequentially running assays, comprising ***cyclic voltammetry with model systems***; usually, the redox pairs like $Fe(CN)_6^{3-}/Fe(CN)_6^{4-}$; I_2 / I^-; *quinine/hydroquinone, or ascorbic acid/dehydroascorbic acid*, for which the electrode behavior at common solid or solid-like electrode is well known and described in literature. Then, in ***advanced testing procedures***, some further assays follow, allowing one to define specific features of carbon pastes, such as the *degree of hydrophobicity* and the phenomena associated with (*i.e.*: moderated reaction kinetics and possible surface reactivation), *adsorption, extraction and ion-pairing* capabilities, *the effect of oxygen* (contained in the paste), *ageing*, or actual *ohmic resistance.*

Regarding the last named parameter, it has been shown recently in a series of reports by our group [102-105] that the knowledge on the fundamental properties and structure of carbon paste can be extended yet nowadays – more than five decades since CPEs had came into this world. Moreover, some relations with the ***ohmic resistance*** of the individual types of carbon pastes that had hitherto been of interest in a few reports only (namely: [52,89,91]) were found advantageous as new testing procedures, where one might test, in parallel, a set of mixtures differing in the actual carbon-to-pasting liquid ratio [35,36,102].

Although experiments of this kind still continue [41,58], the summary of the already published material with numerous practical advices [102-105] is given below – in a special paragraph included as an example of *importance of experimental testing* with carbon paste-based electrodes and the role of both main carbon paste components on the overall structure of a carbon paste mixture and, subsequently, upon the resultant properties and behavior of a CPE in electrochemical and electroanalytical measurements.

4.3.4. New Contribution to the Characterization of Carbon Pastes by Means of the Ohmic Resistance Measurements

Generally, the commonly used carbon paste mixtures have a very low ohmic resistance, usually ohms or tens of ohms only [14,16,28,36,56].

With growing proportion of the binder the resistance increases [29] but until now, this feature has not yet been in CPEs satisfactorily investigated, described or explained. When plotting experimentally the resistance data acquired for a CPE versus the carbon-to-binder ratio, the graph is not regularly increasing, but it has a shape of an "ice-hockey stick" with two linear segments.

When the paste is rich in the carbon moiety (the so-called "drier" or "dry" composition [16,29,43]), the resistance is low and almost constant. Once the amount of binder increases above the critical ratio, the resistance raises abruptly exhibiting a "break-point". This characteristic shape was observed in all our studies with various combinations of the CP-composition; hence, there should be a general principle behind this phenomenon.

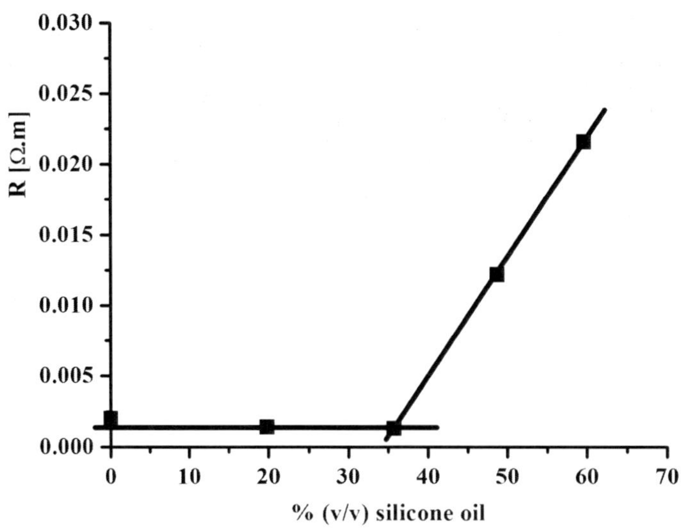

A

Figure 5. (Continued).

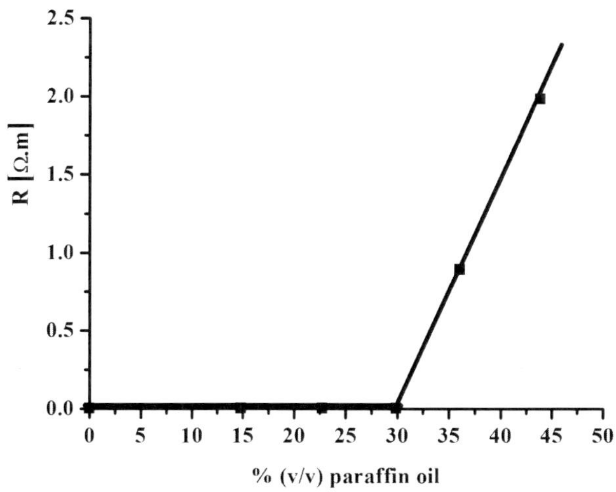

B

Figure 5. Typical dependence of the resistivity upon carbon paste composition for two different carbon paste mixtures: C/SO type (consisting of graphite powder and silicone oil; plot "A"); GC/PO type (with glassy carbon powder and paraffin oil; plot "B"). [From authors' archives].

The best explanation can be given applying the basic principle based on "close packing of spherical particles" known in crystallographic sciences. Accordingly to this theory, the space filled with spheres represents theoretically 52–74% of the total volume depending on the type of packing-space groups (of the *Fm3m, P63/mmc, Im3m*, and *Pm3m* type, respectively [106]). Evidently, only some carbon materials can be considered as spheres (*e.g., Sigradur®* glassy carbon powder); but, still, even irregularly shaped particles obey this principle and exhibiting the break point at the different carbon-to-pasting liquid ratio.

When the graphite particles in a carbon paste mixture are in high excess over the binding liquid, they are in maximal permanent contact with the full conductivity attained between them. Thus, the non-conducting binder/pasting liquid represents only a filler of the free space between the graphite particles and has practically no effect on the overall conductivity up to a critical carbon-to-binder ratio displayed in itself as a break point in the respective plot mentioned above and shown in Figure 5.

When, however, a proportion of the pasting oily liquid further increases, the carbon particles start to "float" in oil phase and then, the permanent contact among the individual particles is not fully ensured and solely the random touches of the graphite units in such oily phase cause a rapid increase of the electrode resistance with the higher fraction of the pasting liquid. This model resembles the *Percolation Theory for Rigid Lattices* [107], seeming to be more practical and qualitatively applicable also to the viscous "moving" systems.

In our studies [102-105], various combinations of both carbon paste constituents, including newly [41,58] also *"CP-components of the new millennium"*, such as CNTs, carbon fibers, or some ionic liquids were examined and characterized. The revealed differences in break-points were discussed and elucidated based on characteristic properties of the respective material (shape, porosity, hydrophobicity, adsorptivity, viscosity, molecular size, atomic distances *etc.*). In addition to this, problems connected with homogeneity and stability of carbon pastes, their handling, storage, or aging effects were also discussed.

The reliability and applicability of such characterization of CP-mixtures utilizing the resistance measurements have been confirmed using a standardized evaluation *via* cyclic voltammetry (see *e.g.* [1-3]) with a one-electron reversible model system, where the anodic-to-cathodic peak difference, the ΔE_P parameter, follows the "ice-hockey-stick" shaped graph (see again Figure 5). As a result, it could be postulated that the optimal carbon-to-pasting liquid ratio for a CPE suitable for electroanalytical use can be found close to the break-point, where the given CP-mixture exhibits still full conductivity and sufficient stability with minimal effect of undesirable ageing and similar variations in the actual CDP-composition.

A number of users are preparing the CPEs intuitively and empirically, doing hope that the paste mixtures with the same consistency/plasticity share the same electrochemical properties. Similarly, it is believed — mainly, by new and inexperienced users — that the same proportion of both main carbon paste components ensures the same electrochemical properties even for different materials used.

Of course, neither of these assumptions is correct and, as emphasized in the previous section x.3.3, some experimental examination is always needed. And although such testing of each CPE, including the above-recommended resistance measurements, can be somewhat time consuming, the results and data gained from such experiments usually bring the desired benefit – the information on the proper carbon paste composition, without losing too much effort in a random choice of both key components and/or their mutual ratio, apart from quite difficult "tuning" of the resultant properties of the carbon paste mixture tested and the respective CPE as late as during electroanalytical measurements.

4.4. Carbon Paste Electrodes in Electrochemical Measurements: Facts and Numbers

Vast archives of all electroanalytical applications and representing maybe 2000 or even 2500 original methods and procedures [15] have been reviewed recently [14,16] and hence, only a brief information suffices here, assembled as selected surveys and summarizing data.

NOTE: If instructive and unless stated otherwise, the individual items are specified by a number with the ***approximate percentage*** (from the whole database or the corresponding part/category involved in the actual chart / comparison).

- ***Survey of the Overal Employment of Main Types of CPEs ...*** unmodified: 15%, chemically modified: 50%, biologically modified and enzymatic biosensors: 30%, others: 5%.
- ***The Most Frequently Used Configurations of CPEs and Related Sensors:***
  (i) *common disc electrodes* ... 40% (from which: in quiescent solution ... 40%, in rotated or stirred solutions ... 55%, other arrangements ... 5%)
  (ii) *membrane-coated disc electrodes* ... 3%,
  (iii) *ring-disc and other dual electrodes* ... 1%,
  (iv) *electrodes with atypical shape and/or geometry* ... 1%.
  (v) *(bio)sensors of specific construction* ... 35%,
  (vi) *flow-through detectors* (in FIA and HPLC) ... 15%,

Table I. Instrumental Techniques in Combination with Carbon Paste-Based Electrodes

Direct voltammetry (DV)	**20**[1]	**Coulometry**	**1**
• cyclic, DC-, AC-voltammetry	45[2]		
• differential pulse or square-wave voltammetry	52	**Direct potenciometry + titrations**	**4**
• voltammetry of 1,5; 2,5-order	2		
• voltammetrie with rotated eld.	1	**Stripping (chrono)potenciometry**	**2**
		PSA mode	30
		CCSA mode	70
Stripping voltammetry (SV)	**40**		
• anodic stripping voltammetry	40	**Conductometry** ...	**< 1**
• adsorptive stripping voltammetry	50		
• catalytic stripping voltammetry	10	**Other techniques**	**3**
		• microscopic observations	90
Amperometric detection (AD)	**30**	• impedance measurements	2
Batch and/or hydrodynamic AD	25	• spectroelectrochemistry	1
AD in FIA, SIA, CE, and HPLC	75	• electrochemiluminescence	7

Legend and abbreviations used: 1) total percentage, 2) partial percentage (in the corresponding ategory);
DC ... direct current, F(S)IA ... flow (sequential) injection analysis, CE ... capillary electrophoresis, P(CC)SA ... potentiometric (constant currect) stripping analysis; eld ... electrode(s).

(vii) *large (pool) electrodes* ... << 1%,
(viii) mini-, *micro-electrodes, and UMEs* ... 5%

- ***The Most Important Areas of Electrochemistry with CPEs and Related Sensors:***
  (i) *electrochemical research* (*e.g.*, studies on electrode reactions) ... 10%,
  (ii) *development and testing of new types* (of electrodes and sensors) ... 10%
  (iii) *analysis of inorganic ions and molecules* ... 20%,
  (iv) *analysis of organic substances and environmental pollutants* ... 20%,
  (v) *analýza of biologically important compounds* ... 25%,
  (vi) *pharmaceutical and clinical analysis* ... 10%
  (vii) *others* (*e.g.*, solid state electrochemistry, voltammetry *in vivo*, investigation on new materials and technologies) ... 5%.
- ***Instrumental techniques employed in combination with CPEs*** can be surveyed as well, which shown in the Table I, incorporating all the techniques that ever been in used since the late 1950.
- ***Electroanalysis of inorganic ions, complexes and molecules at CPEs, CMCPEs, and CP-biosensors***:
  - *heavy metals [**40**[1]]:* Zn(5[2]), Pb (40), Cd (30), Cu (15), Tl(5), In (1), Sn (1), Sb (1), Bi (2);
  - *metals from the V. to VIII. group of the periodical system [**20**]:* V (1), Cr (5), Mo(1), W (1), Mn (2), Co (40), Ni (30), and Fe (20);
  - *precious and platinum metals [**10**]:*Ag (30), Au (20), Hg (40); Pt (7), Ir (2), Pd (< 1), Os(< 1), Ru (<< 1), and Rh (<< 1);
  - *alkaline metals, alkaline earths [**5**]:*Li (80), Na (2), K (1), Cs (1), Be(1), Mg (5), Ca(10);
  - *other metals [**5**]:*Al (40), Ga (5), Ge (5), Ti(30), Zr (10), Sc (3), U (5), Tc (1), and Me^{RE} (1);

- *semimetals and non-metals [5]:* As (80), Se (15), and Te (5);
- *cations [5]:* H^+ [pH-measurement] (5), NH_4^+ and NH_3(15), $N_2H_5^+$ (50), and NH_3OH^+ (30);
- *anions [7]:* Cl^- (5), Br^-(5), I^- (40), IO_3^- (1), CN^-(5), SCN^- (1), N_3^-(1), NO_2^-(20), NO_3^-(7), SO_3^{2-}(3), SO_4^{2-}(2), S^{2-}(2); HPO_4^{2-} (3), ClO_4^-, F^-, SiO_4^{4-}, BF_4^-, $[B(O_2)(OH)_2]^-$(all ... 1%);
- *molecules [3]:* H_2O_2 (60), O_2 (30), NO_X(5), Cl_2 (2), SO_2 (2), and H_2S (1),

where: 1) total percentage, 2) percentage in the given category; CMCPEs ... chemically modified carbon paste electrodes, CP- ... carbon paste-based, RE ... rare earths.

- ***Analysis of organic substances, including environmental pollutants***
  - *N-compounds [**40**[1]]:* amines (alifatic/aromatic): $5^2/65^2$; nitro-componds (alif./arom.): 5/20; nitroso- der. (arom.): 5; hydrazo-der. (arom.): 5; hydroxylamines (arom.): 2; azo-compounds: 1; others (e.g., synthetic dyes, natural alkaloids): 2;
  - *O-compounds [**40**]:* alcohols (MeOH/EtOH/arom.): 1/15/4; PhOH/other phenols: 30/35; aldehydes (alif./arom.) 1/4; quinones: 5; organic acids (except AAs): 4; organic peroxides: 1;
  - *Compounds with heteroatom(s) [**20**]*: X- (50), S- (35), P- (10), As- (3), Me(2);
  - *Environmental Pollutants:* NPAHs and APAHs(*ca.* 10 items), pesticides (100 it.), and surfactants (5 it.),

  where: 1) total percentage, 2) percentage in the respective category, except the last one; X ... halogen (Cl, Br, I), AA ... amino acid; N(A)PAHs ... nitro- (amino-) polyaromatic hydrocarbons.
- ***Pharmaceutical and clinical analysis of commercial preparatives, formulations, and abuse drugs:***
  - *pharmaceuticals and vitamins* (*ca.* 150+50 items, of which AA: 30), *drugs* (5 it).
- ***Analysis of biologically important compounds, including macromolecules:*** *aminokyseliny*: 25%; *nucleic acids* (DNA, RNA): 20%; *purines and pyrimidines* (UA + ostatní): 5%; *sugars* (GLU + monosaccharides + polysaccharides): 25%; *enzymes and co-enzymes* (NADH/others): 5%; *hormons and steroids*: 1%; *neurotransmitters:* DOPA/other catechols: 10%; *flavanols and flavonoids*: 2%; *lipids and fatty acids*: 2%; *lactic acid and lactates*: 1%; *miscellaneous* (org. macro-molecules like creatine, creatinine or chitosan; miniature cells, viruses, and bacteria): 3%;

  where: DNA ... deoxyribonucleic acid, AA ... ascorbic acid, UA ... uric acid, GLU ... glucose, NADH ... nicotine adenine dinucleotide, DOPA ... dopamine.

4.5. Other Possible Applications of Graphite Powders in Electrochemistry

Although the role of graphite powder as the carbon moiety in CPEs is undoubtedly dominant, involvement of this versatile material in electrochemical measurements is yet wider and some special applications can be traced up even beyond the borders of electrochemistry or electroanalysis.

The first category—*i.e.*, another applicability of graphite powders in electrochemistry and electroanalysis—can be represented by ***special constructions of carbon solid electrodes***, where a *small portion of graphite powder* serves as the inner electrical contact. In this way, powdered graphite enables highly *conductive connection* of a metal wire inserted into a glass tube with the surface of a graphite piece immersed in the lower end of tubing and fixed by paraffin wax [4,12].

Such a contact may also be accomplished via a dispersion of graphite in liquid mercury [108]. Other examples of possible use of graphite powders already represent more or less ***curious applications of special carbon pastes***, which is a topic comprehensively reviewed recently – in a separate chapter prepared for the very first monograph on CPEs [16] and some examples can also be repeated herein – just for illustration. We can start with a report on *carbon paste briquettes* as continually smelting cathode in industrial production of Al, Si, and CaC_2 [109]. Another CP-mixture contained a Cu-alloy as a catalyst to reduce CO_2 [110]. Various carbon pastes are being used as the *conductive support for samples in SEM* [34,101] and related CNT-composite in construction of a field emission display [111].

Finally, there were two successful attempts on *solar cells*, where the photocurrent was released from carbon paste-like substrates [112,113]; the latter configuration based on the mixture from carbon black.

Conclusion

The main goal of this contribution was to present a *wide-spread use of* ***powdered graphite*** *and related materials* in the form of the so-called *carbon paste electrodes* for electrochemical measurements. To show a reality which — up until now — might not have been so obvious to wider scientific audience. Indeed, the applicability of graphite in CPEs is a rich history lasting more than half a century, it is the exciting present, as well as the promising future. When looking at the web and brand new publications about CPEs, it can be stated that prospects for the up-coming years are wide open. Even a random selection of seven "fresh" contributions [114-120] confirms this statement. Thus, there is so far a space for traditional graphite-based pastes employed either as such [116] or in the form of chemically modified electrodes [115,117]; all being applicable in still attractive biological and clinical analysis. Of continuing interest are also "green" bismuth-based electrodes, represented here by a CPE made of mixed binder with ionic liquid [114], which is — together with nanomaterials (used in [117]) — the "*carbon paste component of new millennium*" [16,29]. Our short survey can be symbolically completed with such "carbon pastes of new millennium"; namely with *carbon ionic-liquid* and *carbon nanotube ionic-liquid electrodes* (CILEs and CNT-ILEs, respectively); both variants being described within this small collection of newest contributions (see [118-120]).

Acknowledgments

Financial support from the Ministry of Education, Youth, and Sports of the CzechRepublic (project "CZ.1.07/2.3.00/30.0021") is gratefully acknowledged.

References

[1] Adams, R. N. (1969). *Electrochemistry at Solid Electrodes*, M. Dekker: N.Y., US.

[2] Bard, A. J. and Faulkner, L. R. (1980). *Electrochemical Methods: Fundamentals and Applica-tions*. J. Wiley: N.Y., US.

[3] Wang, J. (1985). *Stripping Analysis: Principles, Instrumentation, and Application*, Orig. Ed. VCH Publishers: N.Y., US.

[4] Brainina, Kh. Z., Neiman, E. Ya. and Slepuschkin, V. V. (1993). *Инверсионные електроаналитические методы*, Orig. Ed. (in Russian), Химия: Moscow, SU; 1988; or: Brainina, Kh.; Neiman, E. Ya. *Electroanalytical Stripping Methods*, Eng. Ed., J. Willey: N.Y., US.

[5] Wang, J. (1994). *Analytical Electrochemistry*, 1st Ed., VCH Publishers: N.Y., US.

[6] Cowlard, F. C. and Lewis, J. C. (1967). *J. Mater. Sci.*, *2*, 507–512.

[7] Kinoshita, K. (1988). *Carbon: Electrochemical and Physicochemical Properties*. J. Wiley: N.Y., US.

[8] Inagaki, M., Yang, Y. and Kang, F. -Y. (2012). *Adv. Mater.*, *24*, 2547–2566.

[9] Iijima, S. (1991). *Nature*, *354*, 56–58.

[10] anonymous (2013). Graphite, http://en.wikipedia.org/wiki/Graphite.

[11] Demetriadis, D., Economou, A. and Voulgaropoulos A. (2004). *Anal. Chim. Acta*, *519*, 167–172.

[12] Brainina, Kh. Z. (1990). Personal communication.

[13] Adams, R. N. (1958). *Anal. Chem.*, *30*, 1576–1576.

[14] Kalcher, K., Švancara, I., Metelka, R., Vytřas, K. and Walcarius, A. (2006). in: *Encyclopedia of Sensors, Vol. 4* (Grimes, C.A.; Dickey, E.C.; Pishko, M.V.; Eds.), 283–430. American Scientific Publishers: Stevenson Ranch, US.

[15] Švancara, I., Vytřas, K., Kalcher, K., Walcarius, A. and Wang, J. (2009). *Electroanalysis*, *21*, 7–28.

[16] Švancara, I., Kalcher, K., Walcarius, A. and Vytřas K. (2012). *Electroanalysis with Carbon Paste Electrodes*. CRC Press: Boca Raton, US.

[17] Štulíková, M. and Štulík, K. (1974). *Chem. Listy*, *68*, 800–835.

[18] Dryhust, G. and McAllister, D. L. (1985). in: *Laboratory Techniques in Electroanalytical Chemistry*, 1st Ed. (Kissinger, P.T.; Heineman, W.R.; Eds.); M. Dekker: N.Y., US, 294–301.

[19] McCreery, R. L. and Cline, K. K. (1996). in: *Laboratory Techniques in Electroanalytical Chemistry*, 2nd Ed. (Kissinger, P.T.; Heineman, W.R.; Eds.); 311–332. M. Dekker: N.Y., US.

[20] McCreery, R. L. (2008). *Chem. Rev.*, *108*, 2646–2687.

[21] Digua, K., Kauffmann, J. -M. and Del Plancke, J. L. (1994). *Electroanalysis*, *6*, 451–458 and 459–462.

[22] Hernández, L., Hernández, P., Blanco, M. H. and Sánchez, M. (1988). *Analyst(London)*, *113*, 41-43.

[23] Bobbitt, J. M., Colaruotolo, J. F. and Huang, S. -J. (1973). in: *J. Electrochem. Soc.*, *120*, 773–773.

[24] Monien, H., Specker, H. and Zinke, K. (1967). *Fresenius Z. Anal. Chem.*, *225*, 342–351.

[25] Lindquist, J. (1968). in: *J. Electroanal. Chem., 18*, 204–205.
[26] University of Pardubice, Švancara, I., Metelka, R. and Vytřas, K. (2010). *CZ Patent registered at the Industrial Property Office*, Czech Rep.; No. 301714/22.4.2010.
[27] Švancara, I., Metelka, R. and Vytřas, K. (2005). in: *Sensing in Electroanalysis*, Vol. 1 (Vytřas, K.; Kalcher, K.; Eds.). University of Pardubice: Pardubice, CZ, 7–18.
[28] Zima, J., Barek, J. and Muck, A. (2005). *Rev. Chim. (Bucharest), 55*, 657–662.
[29] Rice, M. E., Galus, Z. and Adams, R. N. (1983). *J. Electroanal. Chem., 143*, 89–102.
[30] Švancara, I., Walcarius, A., Kalcher, K. and Vytřas, K. (2009). *CEJC– Cent. Eur. J. Chem., 7*, 598–656.
[31] Švancara, I. and Vytřas, K. (1993). *Anal. Chim. Acta, 273*, 195–204.
[32] Nagy, G., Feher, Zs. and Pungor, E. (1970). *Anal. Chim. Acta, 52*, 47–54.
[33] Hvízdalová, M. (1994). *A Study on Applicability of Carbon Paste Electrodes in Voltammetry*. MSc dissertation (in Czech), University of Pardubice, Pardubice, CZ.
[34] Švancara, I., Hvízdalová, K., Vytřas, K., Kalcher, K. and Novotný, R. (1996). *Electroanalysis, 8,* 61–65.
[35] Švancara, I., Zima, J. and Schachl, K. (1998). *Sci. Pap. Univ. Pardubice, Ser. A*; *4*, 49–63.
[36] Švancara, I. and Schachl, K. (1999). *Chem. Listy, 93*, 490–499.
[37] Kalcher, K., Wang J., Kauffmann, J.-M., Švancara, I., Vytřas, K., Neuhold, C. and Yang, Z.-P. (1995). *Electroanalysis, 7*, 5–22.
[38] anonymous (2013). "Hi-Tech in Graphite"; http://www.graphite.cz/ index.php?stranka=onas.
[39] anonymous (2013). "Cesky Krumlov: Unesco World Heritage/ Graphite Mine"; http://www.ckrumlov.info/docs/cz/atr18.xml
[40] Sopha, H., Baldrianová, L., Tesařová, E., Grincienė, G., Weidlich, T., Švancara, I. and Hočevar, S. B. (2010). *Electroanalysis, 22*, 1489–1493.
[41] Rosecká K. (2013). *Physicochemical andElectrochemical Characterization of Some New Types of Carbon Pastes*. MSc Dissertation (in Czech), University of Pardubice: Pardubice, CZ.
[42] Ramirez, M., Díaz, A., Luna, E., Figueroa, Y., Larsen, T. and Bjarte, P. (2005). in: *AADE 2005 – National Technical Conference and Exhibition*; presentation No AADE-05-NTCE-30. Wyndam Greenspoint, Houston (TX), April 5-7, or also: anonymous (2013). http://www.scribd com/doc/47714011/Successful-Application-of-Synthetic-Graphite-to-Overcome-Severe-Lost-Circulation.
[43] Adams, R. N. (1963). *Rev. Polarog. (Kyoto, Jpn), 11*, 71–78.
[44] Wang, J., Martinez, T., Yaniv, D. R. and McCormick L. (1990). in: *J. Electroanal. Chem., 286*, 265–272.
[45] Švancara, I., Kalcher, K., Vytřas, K. and Diewald, W. (1996). *Electroanalysis, 8*, 336–342.
[46] Cheng, I. F. and Martin, C. R. (1988). *Anal. Chem., 60*, 2163–2165.
[47] Hočevar, S. B. and Ogorevc, B. (2007). *Talanta, 74*, 405–411.
[48] Zakharchuk, N. F. (1991). Personal communication.
[49] Lindquist, J. (1973). *Anal. Chem., 45*, 1006–1008.
[50] Ali Qureshi, G. (1978). *Libyan J. Sci., 8A*, 37 41.
[51] Johansson, E., Marko Varga, G. and Gorton, L. (1993). in: *J. Biomater. Appl., 8*, 146–173.

[52] Vytřas, K. and Švancara, I. (2007). in: *Sensing in Electroanalysis,* Vol. *2*, (Vytřas, K.; Kalcher, K.; Eds.). University of Pardubice: Pardubice, CZ; 7–22.
[53] Zima, J., Švancara, I., Barek, J. and Vytřas, K. (2009). *CRAC – Crit. Rev. Anal. Chem.*, *39*, 204–227.
[54] Li, G., Ji, Z. -M. and Wu, K. -B. (2006). *Anal. Chim. Acta*, *577*, 178–182.
[55] Yang, C. -H., Zhao, J., Xu, J. -H., Hu, C. -G. and Hu, S. -S. (2009). *Int. J. Environ. Anal. Chem.*, *89*, 233–244.
[56] Kauffmann, J. -M., Préte, M. P., Viré, J. -C. and Patriarche, G. J. (1985). *Fresenius Z. Anal. Chem.*, *321*, 172–176.
[57] Schmid, G. M. and Bolger, G. W. (1973). *Clin. Chem.*, *19*, 1002–1005.
[58] Mikysek, T., Stočes, M., Švancara, I. and Kalcher, K. (2013). Unpublished results.
[59] Navrátilová, Z. (2009). in: *Electroanalysis*, *21*, 1758–1762.
[60] Stefan, R. I. and Bairu, S. G. (2003). *Anal. Chem.*, *75*, 5394–5398.
[61] Wang, J., Kirgoz, U. A., Mo, J. -W., Lu, J. -M., Kawde, A. N. and Muck. A. (2001). *Electrochem. Commun.*, *3*, 203–208.
[62] Varma, S. and Mitra, C. K. (2002). *Electrochem. Commun.*, *4*, 151–157.
[63] Rodriguez, M. C. and Rivas, G. A. (2002). *Anal. Chim. Acta*, *459*, 43–51.
[64] Ricci, F., Goncalves, C., Amine, A., Gorton, L., Palleschi, G. and Moscone. D. (2003). *Electro-analysis*, *15*, 1204–1211.
[65] Kureishi, Y., Tamiaki, H., Shiraishi, H. and Maruyama., K. (1999). *Bioelectrochem. Bioenerg.*, *48*, 95–100.
[66] Miranda Hernández, A., Rincón, M. and González, I. (2005). in: *Carbon*, *43*, 1961–1967.
[67] Pruneanu, S., Ali, Z., Watson, G., Hu, S. -Q., Lupu, D., Biris, A. R., Olenic, L. and Mihailescu, G. (2006). *Sci. Technol.*, *24*, 311–320.
[68] Zhu, S. -Y., Fan, L. -S., Liu, X. -Q., Shi, L. -H., Li, H. -J., Han, S. and Xu, G. -B. (2008). *Electro-chem. Commun.*, *10*, 695–698.
[69] Britto, P. J., Santhanam, K. S. V. and Ajayan, P. M. (1996). *Bioelectrochem. Bioenerg.*, *41*, 121–125.
[70] Rubianes, M. D. and Rivas, G. A. (2003). *Electrochem. Commun.*, *5*, 689–694.
[71] Valentini, F., Amine, A., Orlanducci, S., Terranova, M. L. and Palleschi, G. (2003). *Anal. Chem.*, *75*, 5413–5421.
[72] Nossol, E. and Zarbin, A. J. G. (2008). *Electrochim. Acta*, *54*, 582–589.
[73] Zhu, L., Tian, C., Zhu, D. and Yang, R. (2008). *Electroanalysis*, *20*, 1128–1134.
[74] Tang, X. -F., Liu, Y., Hou, H. -Q. and You, T. -Y. (2010). Talanta, *80*, 2182–2186.
[75] Liu, Y., Wang, D. -W., Huang, J. -S., Hou, H. -Q. and You, T. -Y. (2010). *Electrochem. Commun.*, 12, 1108–1111.
[76] Wu, Y. -H., Mao, X. -Y., Cui, X. -J. and Zhu, L. -D. (2010). *Sens. Actuators, B*; *145*, 749–755.
[77] Wang, L., Zhang, X. -H., Xiong, H. -Y. and Wang, S. -F. (2010). *Biosens. Bioelectron.*, 26, 991–995.
[78] Kalcher, K. (1990). *Electroanalysis*, *2*, 419–433.
[79] Švancara, I., Vytřas, K., Renger, F. and Smyth, M. R. (1992/93). *Sb. Ved. Pr., Vys. Sk. Chemicko- technol., Pardubice*, *56*, 21–57.
[80] Ulakhovich, N. A., Medyanceva, E. P. and Budnikov, G. K. (1993). *Zh. Anal. Khim.*, *48*, 980–998.

[81] Gorton, L. (1995). *Electroanalysis*, *7*, 23–52.
[82] Zhang, Z. -Q., Liu, H., Li, Y. -F. (1998). *Fenxi Kexue Xuebao*, *14*, 80–86; *Chem. Abstr. 125*, 175403x.
[83] Švancara, I. and Vytřas, K. (2000). *Chemija (Vilnius)*, *11*, 18–27.
[84] Švancara, I., Vytřas, K., Zima, J. and Barek, J. (2001). *Crit. Rev. Anal. Chem.*, *31*, 311–345.
[85] Kalcher, K., Švancara, I., Buzuk, M., Vytřas, K. and Walcarius, A. (2009). *Monatsh. Chem.*, *140*, 861–889.
[86] Olson, C. and Adams, R. N. (1960). *Anal. Chim. Acta*, *22*, 582–589.
[87] Olson, C. and Adams, R. N. (1963). *Anal. Chim. Acta*, *29*, 358–363.
[88] Farsang, Gy. (1965). *Acta Chim. Acad. Sci. Hung.*, *45*, 163–176.
[89] Beilby, A. L. and Mather, B. R. (1965). *Anal. Chem.*, *37*, 766–768.
[90] Neeb, R., Kiehnast, I. and Narayanan, A. (1972). *Fresenius Z. Anal. Chem.*, *262*, 339–343.
[91] Lindquist, J. (1974). in: *J. Electroanal. Chem.*, *52*, 37–46.
[92] Söderhjelm, P. (1976). *J. Electroanal. Chem.*, *71*, 109–115.
[93] Ravichandran, K. and Baldwin, R. P. (1981). *J. Electroanal. Chem.*, *126*, 293–300.
[94] Ravichandran, K. and Baldwin, R. P. (1984). *Anal. Chem.*, *56*, 1744–1747.
[95] Wang, J. and Freiha, B. A. (1984). in: *J. Chromatogr.*, *298*, 79–84.
[96] Danielson, N. D., Wangsa, J., Targove,M. A. (1985). *Anal. Chem.*, *61*, 2585–2588.
[97] Wang, J., Gründler, P., Flechsig, G. -U., Jasinski, M., Rivas, G., Sahlin, E. and Paz, J. L. L. (2000). *Anal. Chem.*, *72*, 3752–3756.
[98] Maleki, N., Safavi, A. and Tajabadi, F. (2006). *Anal. Chem.*, *78*, 3820–3826.
[99] Wang, J. (2002). *Ass. Chem. Res.*, *35*, 811–816.
[100] Pingarrón, J. et al. (2012). in: *Handbook of Green Analytical Chemistry* (De la Guardia, M.; Garrigues, S.; Eds.). J. Wiley: N.Y., US,262–268 & 282–284.
[101] Švancara, I., Baldrianová, L., Tesařová, E., Vlček, M., Vytřas, K. and Sotiropoulos, S. (2007). in: *Sensing in Electroanalysis, Vol. 2* (Vytřas, K.; Kalcher, K.; Eds.), 35-58. Univ.PardubicePressCenter, Pardubice.
[102] Mikysek, T., Švancara, I., Bartoš, M., Kalcher, K., Vytřas, K. and Ludvík, J. (2009). in: *Anal. Chem.*, *81*, 6327-6333.
[103] Mikysek, T., Stočes, M., Švancara, I. and Ludvík, J. (2010). in: *Sensing in Electroanalysis*, Vol.5 (Vytřas, K.; Kalcher, K.; Švancara, I.; Eds.). Univ.PardubicePressCenter, Pardubice, CZ,69-75.
[104] Mikysek, T., Jovanoski, V., Sopha, H. -I., Švancara, I. and Ludvík, J. (2011). in: *Sensing in Electroanalysis*, Vol. 6 (Kalcher, K.; Metelka, R.; Švancara, I.; Vytřas, K; Eds.), Univ.PardubicePressCenter, Pardubice, CZ, 157-166.
[105] Mikysek, T., Stočes, M., Švancara, I. and Ludvík, J. (2012). *RSC Advances*, *2*, 3684-3690.
[106] Hahn, T. (Ed.) (1983). *International Tables for Crystallography,* Vol. *A*. Reidel Publishing: DordrechtandBoston, NL and US.
[107] Grimmet, G. (1999). *Percolation*, 2nd Ed. Springer: Berlin, DE.
[108] Efron, C. and Ariel, M. (1979). *Anal. Chim. Acta*, *108*, 395–399.
[109] Fitt, A. D. and Howell, P. D. (1998). *J. Eng. Math.*, *33*, 353–376.
[110] Pettichini, S., Boggetti, H., Lopez de Mishima, B. A., Mishima, H. T., Rodriguez, J. and Pastor, E. (2003). in: *J. Argentine Chem. Soc.*, *91*, 107–112.

[111] Kim, T. -H., Kwon, S. -J., Yoon, H. -H., Chung, C.-W. and Kim J.-S. (2008). *Colloids Surf., A, 313*, 448–451.

[112] Kureishi, Y., Shiraishi, H. and Tamiaki, H. (2001). *J. Electroanal. Chem., 496*, 13–20.

[113] Lee, C. -P., Chen, P. -Y., Vittal, R. and Ho, K. -C. (2010). *J. Mater. Chem., 20*, 2356–2361.

[114] Wang, Z. -Q., Gang Liu, G., Zhang, L. -N. and Wang, H. (2012). *Int. J. Electrochem. Sci., 7*, 12326–12339.

[115] Chandrashekar, B. N., Kumara Swamy, B. E., Reddy, S. and Nagendra, P. (2012). *Anal. Bioanal. Electrochem., 4*, 544–552.

[116] Zayed, S. I. M. and Arida, H. A. M. (2013). *Int. J. Electrochem. Sci., 8*, 1340–1348.

[117] Mostafa Najafi, M., Darabi, S., Tadjarodi, A. and Imani, M. (2013). *Electroanalysis*, online: DOI: 10.1002/elan.201200462.

[118] Zhang, X. -Z., Li, M., Zhao, J., Chen, B. -J. and Qu, K. -M. (2013). in: *J. Chinese Chem. Soc.*, on-line: DOI: 10.1002/jccs.201200465.

[119] Ansari, M., Kazemi, S., Khalilzadeh, M. A., Karimi Maleh, H. and Pasha Zanousi, M. B. (2013). *Int. J. Electrochem. Sci., 8*, 1938–1948.

[120] Fouladgar, M. and Karimi Maleh, H. (2013). *Ionics*, DOI: 10.1007 / s11581- 012-0832-7.

In: Graphite
Editor: Quinton C. Campbell

ISBN: 978-1-62618-576-0

Chapter 5

GRAPHITE: SAILING IN A COST-EFFECTIVE ELECTRON SEA

Gabriela da Cruz Mello[1], Edilson Moura Pinto[2], Eduardo Ariel Ponzio[3] and Felipe Silva Semaan[1*]

[1]Departamento de Química Analítica, Universidade Federal Fluminense, Outeiro São João Batista s/n, Niterói, Rio de Janeiro, Brazil

[2]Departamento de Ciências Exatas e Sociais Aplicadas, Universidade Sagrado Coração, Rua Irmã Arminda, Bauru, São Paulo, Brazil

[3]Departamento de Físico-Química, Universidade Federal Fluminense, Outeiro São João Batista s/n, Niterói, Rio de Janeiro, Brazil

ABSTRACT

Carbon is an element easily found in different polymorphic arrangements, existing in at least three main well-established forms (diamond, graphite and fullerenes), each of which can exist in a variety of materials with different electrochemical and mechanical properties; such characteristics lead to many advantages of carbon electrodes such as low cost, wide potential window, relatively inert electrochemistry, and electrocatalytic activity for a variety of redox reactions, specially to organic compounds. The most common allotrope are based on the graphite structure, consisting of ideally infinite sheets of graphene stacked like parallel sheets. In this arrange carbon atoms are all sp^2 hybridized, with an intraplanar and interplanar C-C bond lengths of 1.42 Å and 3.354 Å, respectively. Diamond is entirely sp^3 hybridized and tetrahedral, with a C-C bond length of 1.54 Å, its poor electrical properties requires the use of dopants (such boron) to provide sufficient conductivity for electrochemical purposes; geometrically, graphite and the newer forms of carbon have a strong family resemblance, as they are all built of sp^2-hybridised atoms. The two-dimensional structure of graphite is constructed by hexagons, and, as a result, diamond has a three-dimensional crystal structure (covalent network solid), while graphite consists of carbon layers (with covalent and metallic bonding within each layer) which are stacked in such a sequence different from that found in a hexagonal close packed crystal structure, being linked by a weak van der Walls

* felipesemaan@gmail.com, semaan@vm.uff.br.

interaction produced by a delocalized π-orbital. The anisotropy observed in graphite provides a good electrical and thermal conductor within the layers (due to the in-plane metallic bonding) and a poor electrical and thermal conductor perpendicular to the layers (due to the weak van der Waals forces between the layers), anisotropy in such case also allows these carbon layers to slide with respect to one another quite easily, making graphite a good solid lubricant, besides these graphite is able to undergo chemical reactions by allowing the reactant/intercalate to be located between the graphene layers, forming compounds (called intercalation compounds). Although carbon materials used in electrochemistry share some of the electronic properties of metals, their structures and chemistry differ dramatically from all metallic electrodes; being the development and application of carbon-based composites specially cited.

LIST OF ABREVIATIONS

AA	Ascorbic acid
AFM	Atomic force microscopy
APTES	3-aminopropyltriethoxysilane
ATR	Attenuated total reflectance
CA	Chronoamperometry
CNT	Carbon nanotubes
CRGO	Chemically reduced graphene oxide
CV	cyclic voltammetry
DPV	Differential pulse voltammetry
EDS	Energy dispersive X-ray spectroscopy
EIS	Electrochemical impedance spectroscopy
EQCM	Electrochemical quartz crystal microbalance
ERGO	Electrochemically reduced graphene oxide
FET	Field Effect Transistors
FGGE	Functionalized-graphene modified graphite electrode
FTIR	Fourier Transformed Infrared
GCE	Glassy carbon electrode
GO	Graphene oxide
GOD	Glucose oxidase
GOx	Glucose oxidase
GR	Graphene
HOPG	Highly oriented pyrolytic graphite
IL-Graphene	Ionic liquid functionalized grapheme
MWCNT	Multi-walled carbon nanotube
ORC	Oxidation–reduction cycling
PAN	Polyacrylonitrile
PB	Prussian blue
PBS	Phosfate buffer solution
PDDA-G	Poly (diallyldimethylammonium chloride) functionalized graphene nanosheets
PTBO	Poly(toluidine blue O)
PMB	Poly(Methylene Blue)

QCM	Quartz crystal microbalance
RGO	Reduced graphene oxide
SBS	Poly(styrene-co-butadiene-co-styrene)
SEM	Scanning electron microscopy
NMP	N-Methylpyrrolidone STM – Scanning tunneling microscope
SWCNT	Single-walled carbon nanotube
TEM	Transmission electron microscopy
THF	Tetrahydrofurane
Trp	l-tryptophane
Tyr	l-tyrosine
UA	Uric acid
XPS	X-ray photoelectron spectroscopy

1. Introduction - A Brief Overview

Carbon (represented by "C") has the name related to the latin word "carbo" which means "charcoal". Such element has atomic number "6", and atomic mass 12.01, with 6 protons, 6 electrons, 6 neutrons. It is one of the most extraordinary elements known, being found in numerous forms, like diamond, coal, soot, graphite or more them ten million other compounds extensively present in all living organisms. Proteins, fats, fat acids, carbohydrates, and nucleic acids are some of those examples [1]. This element occurs in the nature in more different forms and structures than any other known element [2].

Recently, many other forms of carbon arrangements such as nanotubes, fullerenes (or *buckyballs*) and graphene were discovered and produced artificially, opening a whole new field of chemistry and analysis, specially applied in sensors development for electrochemistry [3]. Bonding resulted from the hybridization in the single, double or triple bonds, gives its element the possibility to form many allotropes with different physical and chemical properties.

Carbon allotropes are extensively used as electrode material due to their properties, different microstructures lead to completely different mechanical, chemical and physical properties, among these allotropes some may be mentioned e.g. glassy carbon, amorphous carbon species, carbon fibers and nanotubes [4-12], and so one. Among the non-crystalline allotropes of carbon it´s possible to cite coal, lampblack, charcoal, carbon black, coke and others; those allotropes are amorphous, or even without crystalline shape. Diamond and graphite represents two examples of allotropes of carbon with crystalline structures, with atoms arranged in a neat orderly pattern.

1.1. Amorphous Carbon

Among many carbon-derived forms studied it is easy to start discussing relatively disordered ones, in which carbons can display sp, sp^2 and sp^3 bonds, and are classified under the conditions of preparation, leading to distinction in properties. Some can be easily cited and highlighted such as glassy carbon, carbon fibers, activated carbon, porous carbon and amorphous carbon [13-16].

The disordered carbons can be structurally compared considering (1) the size of their microcrystals toward the basal plane of graphite (L_a), determined by Raman spectroscopy; (2) the thickness of microcrystals perpendicular to the planes of graphite (L_c), determined by X-ray diffraction and also by (3) interplanar spacing value respect to the plane 002 (d_{002}), also determined by x-ray diffractometry. In the case of amorphous carbon, an increasing occurs at the value of d_{002} as compared with graphite.

The glassy carbon is widely used as one of the types of isotropic carbon like diamond, e.g. its properties are constant along the material, independent of direction. The name glassy is due to the glossy appearance when polished and for the fracture similar to glass. The preparation of this electrode is made by carbonizing with heating phenol/formaldehyde polymers or polyacrylonitrile between 1000°C and 3000°C, under pressure. However, its preparation is complicated by their hardness and fragility, thus restricting to the use of existing forms of carbon.

The carbon fiber is also used as beams in porous electrodes. The preparation of carbon fibers is made by two methods: from polyacrylonitrile (PAN), which gives a tape of graphite concentric circular rings; or from pitch, which tends to provide a radial structure of laminar graphite. These carbon fibers have a diameter 2-20 μm and hardness greater than the steel in the fiber direction [17].

The different types of porous carbons are materials which are generally synthesized by reaction of carbon fibers (obtained from pitch) with O_2, H_2O and / or CO_2 at temperatures between 800-1200°C [14]. The material formed is called activated carbon fiber, which has fibers randomly distributed along the structure, resulting in the presence of vacancies that give the pores of this material. However, when they are heat-treated, occurs a ordering on the structure and a decrease in pore volume, and these parameters are controlled with the aid of heating temperature. This structure porous control is extremely important because the porosity has influence on physical properties such as density, thermal conductivity and mechanical strength.

Other material also extensively studied is the amorphous carbon, which presents two important parameters, which are the type of bond between the carbon atoms (expressed in terms of the ratio sp^2/sp^3) and hydrogen content in its structure (which can vary up to 65%). When there is a high degree of sp^3 hybridization, the amorphous carbon has properties very similar to those of diamond, for example, high hardness. The amorphous carbon can be formed during neutron irradiation, electron or ion beam in carbon-based materials [18].

1.2. Graphite

Graphite is one of the most famous allotrope of carbon, formed due to the different bonds permitted by carbon electronic arrangement. Carbon may form single, double or triple bonds [19]. Together with fullerene and diamond, graphite is one of the most known carbon allotrope, being the last discovered in 1985 [14]. These compounds may exist in many forms materials with different electrochemical properties [20].

Graphite is widely distributed throughout the world though much of it is of little economic importance. Large crystals or "flake" occur in metamorphosed sedimentary silicate rocks such as quartz, mica schists and gneisses. Crystal size varies from less than 1mm up to about 6mm (average - 4mm) and the deposits form lenses up to 30 meters thick stretching several kilometers across country. Mexico contain up to 95% of carbon microcrystalline graphite (sometimes referred as "amorphous"), this type of allotrope occurs in carbon-rich metamorphosed sediments naturally stored [21].

There is a world shortage of natural graphite which is particularly marked in North America and Europe. Natural graphite is used in refractories (27%), lubricants (17%), foundries (14%). brake linings (12%), pencils (5.3%), crucibles, retorts, stoppers, sleeves and nozzles (4%). Artificial graphite was first manufactured on a large scale by A. *G.* Acheson in 1896. In this process coke is heated with silica at 2500°C for 25-35h:

$$SiO_2 + 2C \rightarrow SiC\ (-2CO) \rightarrow Si_{(g)} + C_{(grafite)}\ (2500°)$$

The industrial processing is made by flotation followed by treatment with HF and HC1, and then by heating to 1500°C in vacuum. Artificial graphite is now made on a large scale, being used mainly for electrodes, crucibles and vessels, and various shapes; specialist uses include motor brushes and contacts and refractories of various sorts [22].

The molecular orbital theory can explain why the graphite allotrope is conducting form of carbon. Taking in to account that, the bonding and anti-bonding molecular orbitals are formed from the adjacent *p-orbitals* that extend over the surface of graphene sheet. The energy is different between these orbitals, but these differences are no so large. Thence, electrons can pass through the planes, moving from one to the other. Looking for the distribution of atoms and there bonds, it is possible conclude simplistically that, the free electrons can easily pass through the planes of graphene sheets, giving to this alotrope, the intrinsic conducting characteristics.

In fact carbon atoms in graphite have sp^2 hybridization, with 120^0 between the atoms of the net, adopting a hexagonal arrangement organized such as planar layers, called graphene sheets (Figure 1).

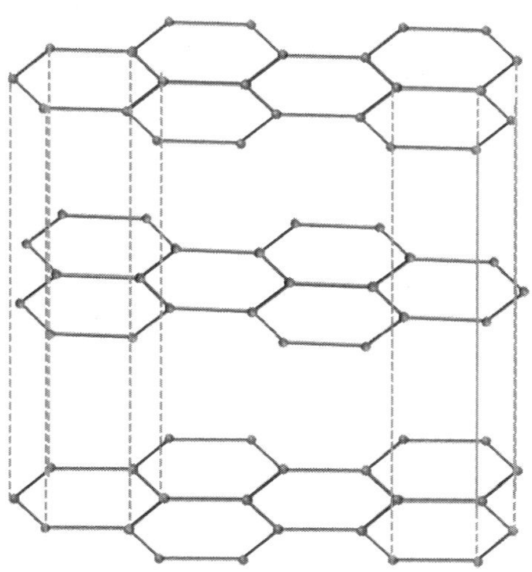

Figure 1. Crystalline structure of graphite [23, 24].

Table 1. Detailed carbon hybridization according to the structure and shape

Hybridization	Number of Hybrid Orbitals	Structure	Shape
sp	2	Linear	
sp^2	3	Trigonal planar	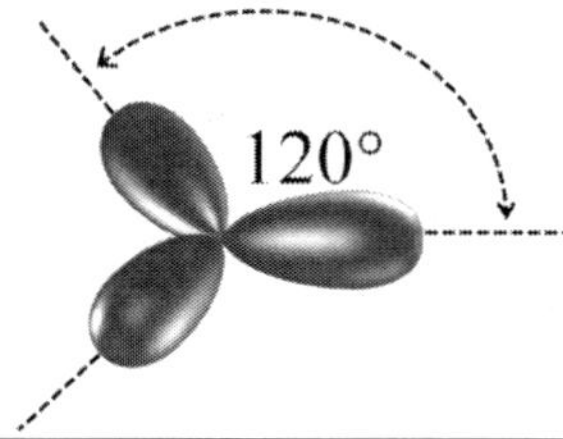
sp^3	4	Tetrahedral	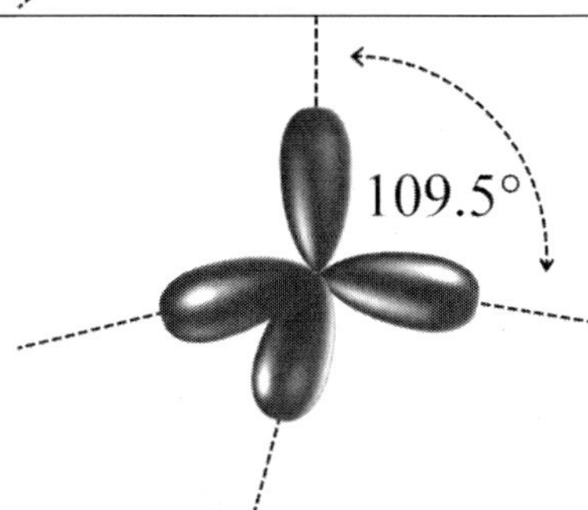
sp^3d	5	Trigonal bipyramidal	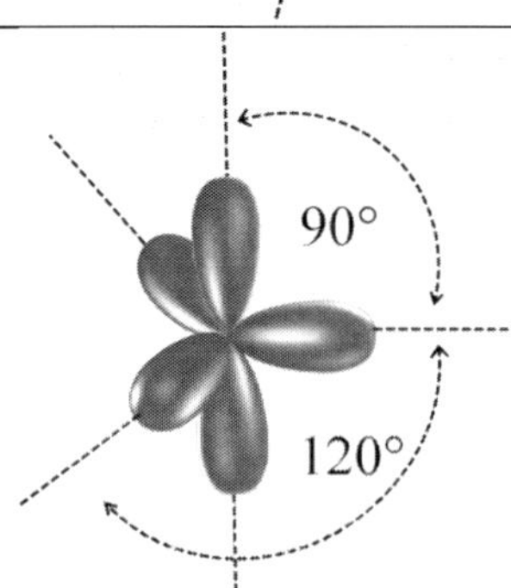
sp^3d^2	6	Ochtahedral	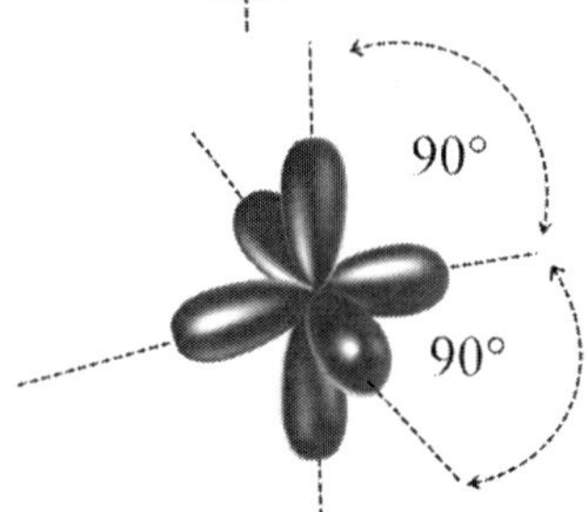

The bond length on the sheet, e.g. between neighboring carbons, is equal to 1.415 Å, and the length between the basal planes is equal to 3.40 Å (c axis). This difference indicates that in the arrangement there are multiple bonds, while plans are bonded by Van der Waals forces and not by covalent bonds [19-20].

Usually the binding behavior of molecular orbitals is the binding with same kind of orbital. Two *s-orbitals* or two similar *p-orbitals* bind together in an anti bonding and bonding way depending on the sign of the orbital. However, for many of the carbon atoms, including atoms of amorphous graphite, it is possible to find bonds by mixing the *s-* and *p-orbitals* and in specific cases, with *d-orbital*. In the following sp^3- and sp^2- hybridization and shape of carbons better represented on the Table 1.

However the conducting character of the graphite can be explained by energy level of the graphite. Tacking in to account that, each carbon atom has four valence electrons and each also four atoms. Hence, there are sixteen energy bands, not counting spin, were twelve are *σ-bands* and four are *π-bands*. There are two groups containing six *σ-bands,* bonded and other six with higher energies anti bonding, both groups, separated by ~ 5 eV.

The four π-bands are positioned between those *σ-bands* groups, similarly with then, two *π-bands* are bonding and two of them are anti-bonding. All *π-bands* are coupled with four of them, strongly coupled. Since graphite has sixteen electrons per unit cell, only eight energy bands are filled. Therefore, the Fermi level it is found just in the middle of the four *π-bands*.The *π-bands*, with highest valence bands, upper to Fermy level, overlaps along the Brillouin zone edges, this overlaps gives to graphite its semi-metal characteristic. The band overlap energy is about 0.03 eV, the same value founded for the valence and conduction bands. The Fermi energy lies within the band overlap region, resulting in pockets of holes and electrons [24, 25].

Graphite can also be produced in the laboratory. An example is pyrolytic graphite, polycrystalline graphite material of high quality. This preparation is made by decomposition of gaseous hydrocarbons at high temperature on a hot surface. This process results in crystals with density greater than 2.2
g cm^{-3} [25]. If the pyrolytic graphite is annealed under pressure and elevated temperature, this becomes highly oriented pyrolytic graphite (HOPG), which is very reproducible and highly anisotropic [17].

In comparison, the diamond is an optically transparent insulator and graphite is a dark solid, opaque, with slightly metallic brightness and a very high electrical conductivity in the direction parallel to the planes, and smaller in the perpendicular direction. The brightness and conductivity of graphite are explained by the various type of bonds on plans, which behaves as a "sea of free electrons", as found in metal bond [19].

Due to its crystalline structure oriented upon planes the graphite has anisotropic, e.g., their properties vary with the direction within the crystal. This differs from the diamond that is isotropic [25]. Graphite is formed by lamella and the choice of plans (basal or perpendicular to this lamella) changes the electrochemical response. The graphite becomes more stable at 25°C and 1 atm and the reaction that transforms diamond into graphite presents a ΔG equal to 2.9 KJ / mol [14, 17, 19, 20].

Graphite has a rich chemistry in which it can participate in reactions as either a reducing agent (electron donor) or an oxidizer (electron acceptor), and form compounds. This is a direct consequence of its electronic structure, which results in both an electron affinity and an ionization potential of 4.6 eV [26].

Graphite stands for one of the most useful material applied in sensors due to it conductivity and malleability, allowing the construction of many geometric forms of commercial, or even lab-made, electrodes. Carbon species vary from 0 to 3D dimensionality In recent years there has been much interest on using carbon-based material as electrodes, being graphite one of the most used material due to many characteristics to be discussed along this text.

This material has chemical stability in different environments from acidic to alkaline buffers [7, 11], organic solvents and ionic liquids, with the possibility of application on a wide range of temperature [27] with activation methods that allow the development of structures with controlled superficial area and distribution of pores, determining, this way, suitable interfacial (electrode/electrolyte) properties for electrochemical purposes [11, 12]. Graphite provides a high surface activity, making connections with hydrogen, hydroxyl and carboxyl and in special cases, quinone groups.

Electrochemical reactions occur more slowly on carbon them metal electrodes, and the transfer kinetics of the electrode depends on the structure and pre-treatment of the electrode surface. These materials provide reduced size-scale electrochemical studies in solutions with high resistivity, such as diluted electrolytes and organic solvents. Furthermore such compounds allow the investigation of kinetic processes that occur in nanometric scale [28].

There are many different graphite-derived compounds, classified into three main groups, namely surface compounds; substitutional compounds and intercalation compounds.

The surface compounds of graphite are formed by the reaction with the graphite surface atoms. Adsorption occurs on the edge atoms of the carbon planes, because of the free valence bonds of the edge atoms, what makes it more reactive. An example of this surface reaction is the oxidation reaction [25]. And, finally, the intercalation compounds of graphite are interstitial compound in which the foreign species is included in the interplanar interstitial sites of the graphite crystal. These compounds are the most well-known of all the compounds of graphite and because of it a special attention must be done for this class.

The large interlayer distance between the parallel planes of C atoms in graphite implies that the interlayer bonding is relatively weak [26]. This accounts for the ready cleavage along the basal plane and the remarkable softness of the crystals [21]. It also enables a wide range of substances to intercalate between the planes under mild conditions to give lamellar compounds of variable composition (Figure 2). These reactions are often reversible and the graphitic nature of the host lattice is retained.

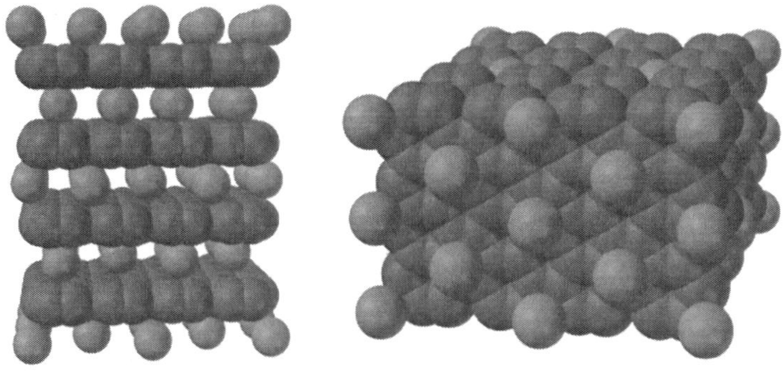

Figure 2. Space-filling model part of the crystal structure of potassium graphite.

The electrical resistance of graphite intercalation compounds is even lower than for graphite itself, resistance along the a-axis dropping by a factor of -10 and that along the c-axis by -100. Moreover, in contrast to graphite, which is diamagnetic, the compounds have a temperature-independent (Pauli) paramagnetism and also behave as true metals in having a resistivity that increases with increase in temperature [21].

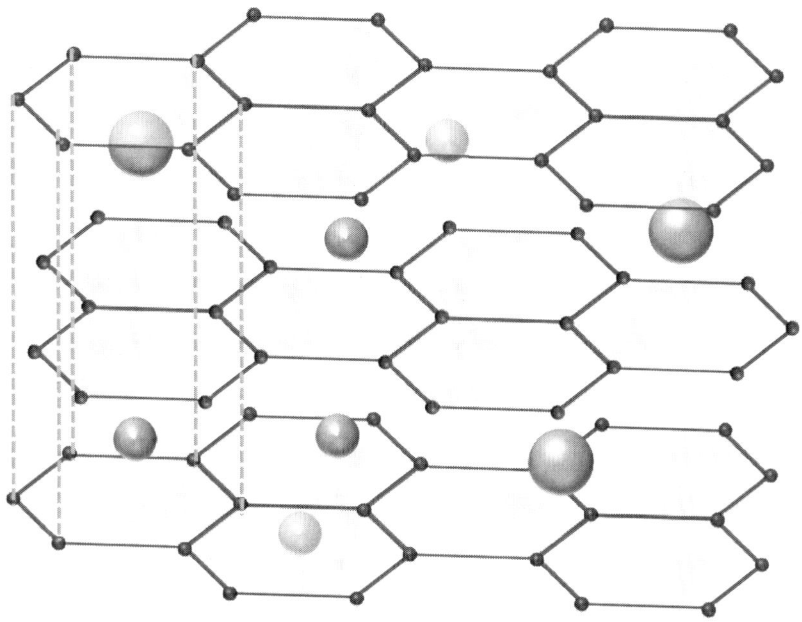

Figure 3. Example of graphite intercalation compounds structure.

Examples of this type are the graphite alkali-metals; the graphite-acid compounds, which have received much attention because of the high electrical conductivities; the graphite-halide compounds, which has halogens intercalated on graphite planes [25].

1.3. Diamond

In diamond structure, carbon atoms have a high hardness and sp^3 hybridization, e.g. in their structure the atoms are arranged so that each carbon atom is covalently bonded to another four ones. Thus, the atoms are at the vertices of a regular tetrahedron (Figure 4). In diamond, the C-C bond has a length of 1.54 Å and binding energy equal to 355 kJ mol^{-1} [19].

Diamond is a rigid, transparent and electrically insulator. It is the hardest substance known and best heat conductor (circa of five times better than copper). The diamond is also used as a protective coating for drill bits. Its high thermal conductivity makes diamond films ideal as the basis for some integrated circuits that cannot overheat.

In nature the diamond is found embedded in a soft rock called *kimberlite*, which rises in columns within the earth and forms diamonds under intense pressure. The diamond can be synthetically produced by compression of graphite at pressures exceeding 80 Kbar and temperatures above 1500°C [29].

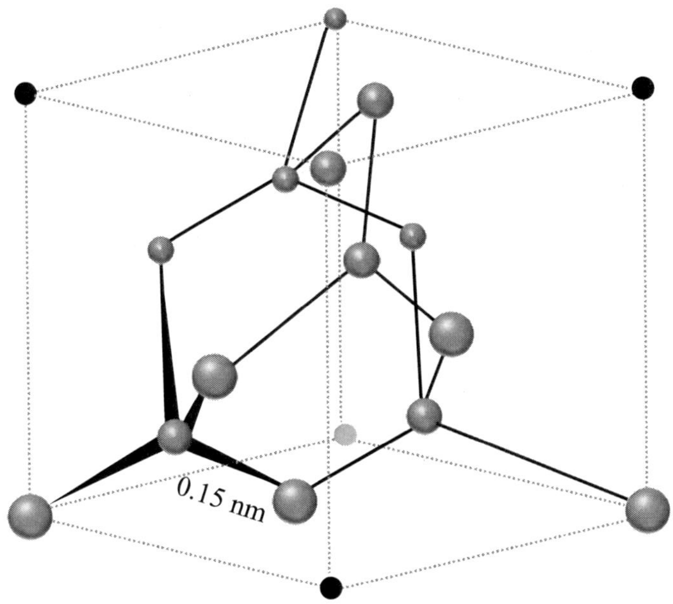

Figure 4. Crystal structure of diamond [28].

Due to such low electrical conductivity, doping elements are inserted in the diamond structure generating conductivity up to obtain different electrochemical properties and interesting to use as electrodes. Some examples are diamonds doped with boron, nanocrystalline diamonds and diamonds doped with nitrogen. The first two have useful electrochemical properties compared to graphite and are much more inert [20].

1.4. Fullerenes

Another interesting applicable form of carbon is found on fullerenes, discovered in 1985 by Kroto et al. The group lead by H. W. Kroto (Sussex,UK) and R. E. Smalley (Rice, Texas) laser-blasted graphite at temperature up to 10°C and showed that the products contained a series of molecules in a wide range of masses from C_{44} to C_{90} [28]. Concentrations of individual molecules varied with conditions but the peak for C_{60} was always by far the strongest, followed by C_{70}. This experiment suggested the existence of new molecular forms of carbon but was not a bulk preparation.

A simple preparation method of fullerenes in bulk consists of striking an arc between graphite electrodes at 3500°C, under low pressure of helium gas. By this way, a mix of different fullerenes is obtained, although separation of individual members proved to be more difficult. Pure C_{60} and C_{70} were obtained for the first time at 1990 by chromatographic separation whit alumina in hexane [30]. Other fullerenes which have been isolated and characterized include C_{70}, C_{76} (chiral), C_{78} (3 isomers), C_{84} (3 isomers), C_{90} and C_{94}, but there is mass spectrometric evidence for all even C_n from C_{30} to $C_{>600}$ [21].

Composed by carbons arranged as vertices of a polyhedron consisting of 20 hexagons and 12 pentagons, C_{60} is an icosahedric soccer-ball-like carbon compound with sp^2

hybridization (Figure 5). Its appearance is a black solid, in appropriate solvents it can form colored solutions/dispersions [31].

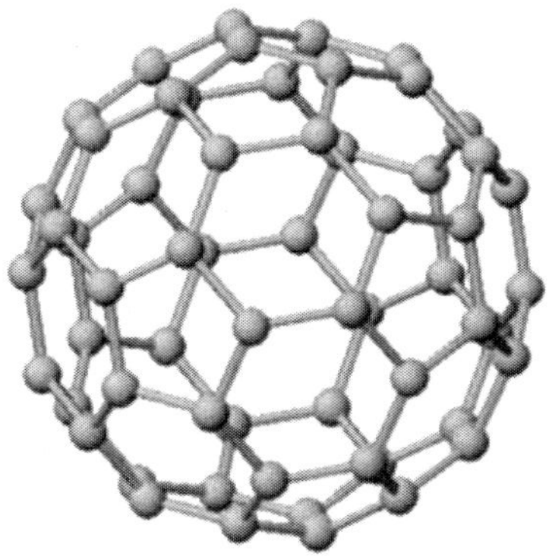

Figure 5. Crystal structure of fullerene [32].

Many properties of fullerenes are related to its high symmetry, for example, its high stability. Furthermore there are also thermal and electrical conductivity, and many others related to the vibrational and optical properties. One of the problems of C_{60} is the low solubility in usual solvents, making its handling a little complex, being this draw-back circumvented by chemical modifications or complexation with other compounds [31, 33].

1.5. Graphene

Since its discovery, graphene has been the target of intense research, having become a candidate to replace silicon technology. Among the various properties that make graphene so interesting, can be cite the high mobility of charge carriers, the higher mechanical strength existing between the material and a thermal conductivity greater than twice the diamond. [22, 34, 35]. All these properties make graphene highly attractive for applications in electronics, ranging from the construction of Field Effect Transistors (FET), passed by memory devices, photovoltaic cells, spintronic devices, until its application in the synthesis of composite materials [36-39].

The planar structures that together form the structure of graphite are referred to graphene (Figure 6). And despite the graphite be known by everyone for over 500 years, the structural design of graphene emerged only in 1947. And only years later, in 2004, it was possible to isolate graphene layers of graphite, work done by a group led by Andre Geim using duct tape to separate the layers until it can be isolate [26].

As previously mentioned, the graphene sheets have only carbons with sp^2 hybridization, forming planar hexagons that make this material highly resistant. Another important property is that the electronic configuration of this structure requires only three of the four valencies of each carbon, generating free electrons which assist in the transfer of current with almost null resistance [40]. For this and several other attractive properties, graphene is currently a material widely studied. In 2010, the Nobel Prize in Physics went to Andre Geim and Konstantin Novoselov for research using graphene as a superconductor of heat.

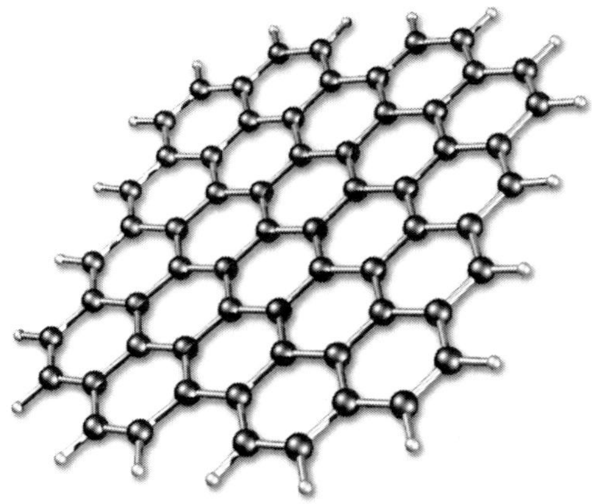

Figure 6. Structure of graphene.

1.6. Nanotubes

In the study of nanoscience, it is possible to work with dimensions of 1 to 100 nm and, thus, and obtained information from materials since its structural composition relative to atoms and molecules. Development of these nanomaterials, based on these initial studies, is performed to obtain the physical and chemical properties required for their use that in macro scale would not be attractive. So the combination of these materials at the nanometer scale with its willingness to generate a high surface area advantages for various cases. One such advantage is the possibility of electrochemical studies on medium that exhibits high resistivity, such as organic solvents [41].

Carbon nanotubes (CNT) consist of graphene sheets arranged in a cylindrical shape and are classified into two forms [42]. The first is the carbon nanotube single wall (SWCNT, single-walled carbon nanotube), which consists of a single sheet of graphene cylinder (Figure 7a). The second classification is the carbon nanotube multi-walled (MWCNT, multi-walled carbon nanotube), which in this case consists of several concentric cylinders spaced (Figure 7b) [43].

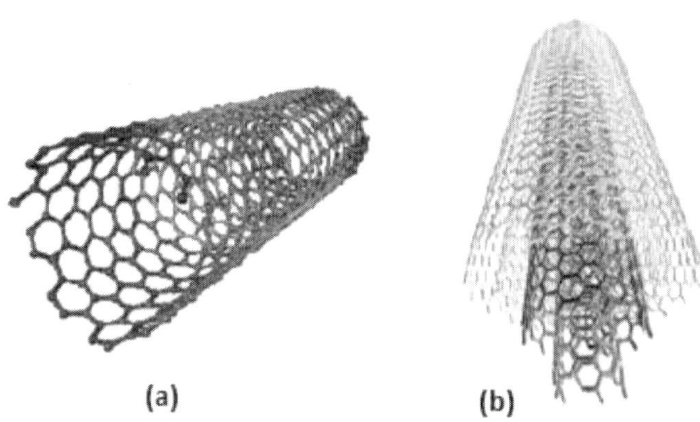

Figure 7. Structural representation of the carbon nanotubes, (a) single wall and (b) multi-walled.

2. General Applications

2.1. Methods of Graphite Exfoliation

The graphene can be produced by various methods, including by chemical vapor deposition, epitaxial growth on silicon carbide, epitaxial growth on metal substrates, Growth from metal-carbon melts etc. These techniques can produce defect-free materials with exceptional properties, but cannot produce significant amounts [44].

The industrial application of graphene depends on large scale production methods. Exfoliation from bulk, stacked graphite is the most economical way to achieve large quantities of single layer grapheme. Many methods have been developed to achieve exfoliation of graphite. The application of mechanical forces to solutions of layered materials via stirring, ultrasonication, or shaking has become one of the most important approaches to exfoliate stacked materials. Further, we describe a brief current exfoliation methods and techniques used to produce single layer materials from graphite precursors. Exfoliated graphene can also be obtained for graphite, graphite oxide (GO), graphite intercalation compounds (GICs) and expanded graphite (EG).

The first successful sonication-based exfoliation of graphite was achieved in the organic solvent n-methyl pirrolidine (NMP) [45]. Graphene dispersions with concentrations up to 0.01 mg ml^{-1}, produced by dispersion and exfoliation of graphite in organic solvents such as N-methyl-pyrrolidone were obtained. This is possible because the energy required to exfoliate graphene is balanced by the solvent–graphene interaction for solvents whose surface energies match that of graphene. The method was then largely improved by several groups. In particular, it was greatly developed at the Department of Chemistry of the University of Sassari by the Italian group of Alberto Mariani. Alzari et al. [46] reached the concentration of 2.21 g L^{-1} in N-methylpyrrolidone.

Once exfoliated from natural graphite under sonication, the graphene sheets can be stabilized by surfactant. Liu et al. [47] was produced high-concentration of graphene sheets non-covalently modified with poly(styrene-co-butadiene-co-styrene) (SBS) in NMP . The graphene sheets can be stabilized by SBS through the π–π stacking with the PS chains. The graphene/SBS composite obtained shows good electrical conductivity with a percolation threshold of ~0.25 % (v). Figure 8 shows an scheme for the exfoliated graphene sheets with poly(styrene-co-butadiene-co-styrene).

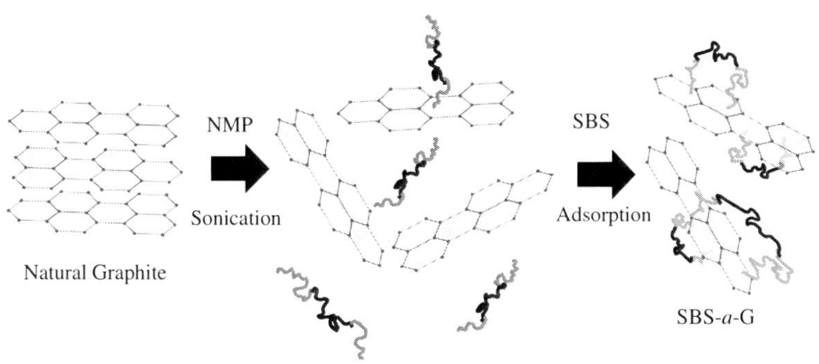

Figure 8. Scheme for the noncovalent modification of the exfoliated graphene sheets with SBS.

Single-, bi-, and multilayer graphenes were prepared by sonicating graphite in distilled water by Asanithi et al. [48], in these studies the effect of different sonication times (3, 6, 12, 36, and 72 hours) on thickness of graphene layer were assessed. Sonication time strongly affected the sedimentation and homogeneity of the graphene suspensions, they shows that the longer time of sonication increased the concentration of exfoliated graphene.

Other process for exfoliation of graphite is using supercritical carbon dioxide ($scCO_2$) [49, 50]. The $scCO_2$ has been used as a solvent intercalation and exfoliation in some layered structured materials, such as clays and was recently used for exfoliation of graphite to obtain grapheme. The method consists of spreading layers $scCO_2$ through the material and subjecting the system to a catastrophic depressurization, CO_2 that has spread between the layers, undergoes rapid expansion and promote the exfoliation of the material. The simplicity, low cost and short processing times make this method very promising for industrial scale production. A further advantage of this method is to obtain a material with high purity because the $scCO_2$ returns to the gas phase after depressurization of the system, leaving no residual solvents [49].

Park et al. [51] was obtained graphene nanosheets by exfoliation of graphite using supercritical carbon dioxide. They show that the repeated $scCO_2$ process reduces the thickness of the exfoliated graphene nanosheets. The thickness and lateral size of the exfoliated GNs produced in supercritical condition at fixed temperature of 45° and pressure of 150 bar were 1.0–6.0 nm and 0.2–1.0 μm, respectively.

For a large scale production of graphene, reduction of graphite oxide (GO) method has been currently used. The graphite oxidation causes an introduction of functional groups which leads to an increase in graphite interlayer distance, producing GO. The GO can be reduced either by chemical or thermal methods.

The reduction of GO can be performed either by the action of reducing agents as by heat treatments. Various reducing agents are reported in literature, such as sodium borohydride [52, 53], hydrazine hydrate [54], hydroquinone [55], among others. When reduction is performed in the GO aqueous medium, the layers repackage irreversibly, making impossible a subsequent exfoliation and dispersion in water or other media [56].

For the reduction by heat treatment, it is necessary to heat the GO at temperatures up to 1050°C using heating rates greater than 2000°C min^{-1}. This rapid heating promotes the rapid release of gases from oxygenated groups that were introduced in the structure of GO during oxidation, causing its expansion and exfoliation [57 - 59].

To increase the size of dispersed flakes of graphene some progress has been development, for example some methods have been demonstrated to separate graphene oxide flasks by lateral size. Coleman et al. [60] describe a new method to take an existing dispersion of graphene in a solvent and to separate flakes by size using controlled centrifugation. Centrifugation at high rates results in small flakes and centrifugation at a lower rate results in a dispersion of slightly larger flakes. Figure 13 shows a schematic separation process. Repeating this procedure a number of times results in the separation of the original dispersion into a number of fractions each with different mean flake length (~1 to ~3.5 μm).

A Graphene-based nanosheet was obtained via chemical reduction of exfoliated graphite oxide, the reduction of a colloidal suspension of exfoliated graphene oxide sheets in water with hydrazine hydrate results in their aggregation and subsequent formation of a high-surface-area carbon material which consists of thin graphene-based sheets [61].

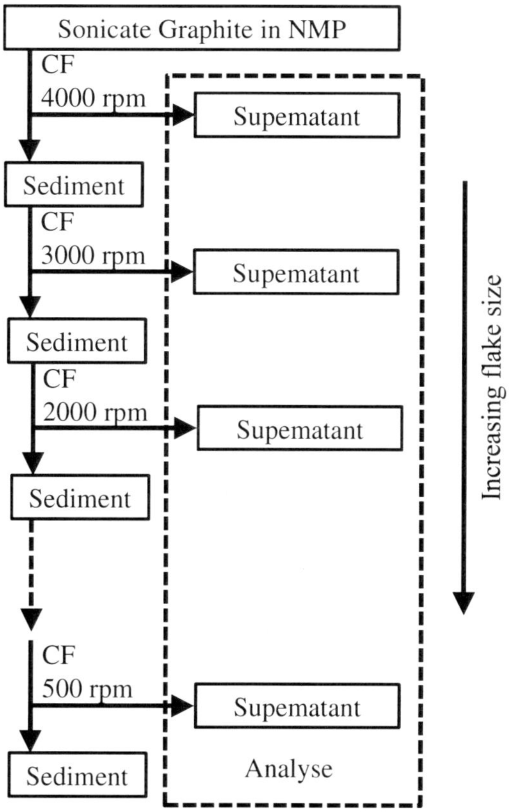

Figure 9. Schematic showing separation process [60].

Shen et al. [62] was produced reduced graphene oxide (RGO) nanosheets by chemical reduction of exfoliated graphite oxide. Atomic force microscopy (AFM) images show that the obtained RGO nanosheets vary greatly in lateral-dimensional sizes, ranging from less than 100 nm to more than 2000 nm. In order to separate these nanosheets, one simple and low-cost method mainly based on magnetic-stirring and centrifugation treatments was proposed. Preliminary statistical analysis of RGO nanosheets, based on AFM images, shows that the dot-like RGO nanosheets (with lateral dimensions less than 100 nm) and leaf-like RGO nanosheets (with lateral dimensions more than 500 nm) were effectively separated by this simple method. Among the different methods for the production of graphene sheets by exfoliation of graphite oxide, graphite, etc., the electrochemical route has been poorly investigated. One of the earliest reported routes of production of graphene by electrochemical exfoliation was published by Liu et al. [63], in this paper the electrochemical exfoliation of graphite represents a simple one-step electrochemical approach for the preparation of ionic-liquid-functionalized graphite sheets with the assistance of an ionic liquid and water. This method produces graphene sheets in ionic liquid/water mixtures where hydroxyl and oxygen radicals produced by anodic oxidation of water start the oxidation of the edge planes of graphite, facilitating intercalation of anions from the ionic liquid. More recent studies have shown that graphene sheets with lateral size up to 30 μm can also be obtained by electrochemical exfoliation of highly oriented pyrolitic graphite (HOPG) in aqueous electrolyte solutions containing H_2SO_4 with high amount of defects.

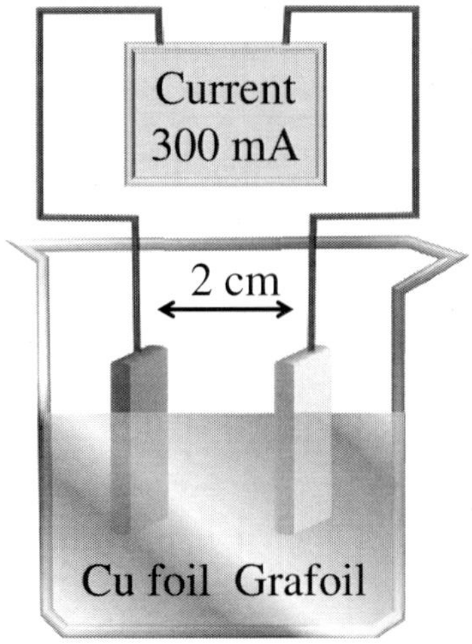

Figure 10. Schematic diagram of electrolytic exfoliation.

Lee et al. [64] reported the obtainment of a large area 2D graphene nanosheets on the surface of flexible graphite foils by electrochemical exfoliation of graphite in an effective electrolyte poly(sodium-4-styrenesulfonate) solution. Uniform expansion of the graphite foil was observed when a constant current of 150 mA cm^{-1} was applied to the vertically-aligned graphite (anode) and copper (cathode) sheet in the PSS electrolyte solution during a preset time for electrolytic surface exfoliation of the graphite sheet.

In ref [65] was reported a multiple electrochemical exfoliation approach to obtain graphene flakes by using graphite rod from spent zinc–carbon as graphene source. The graphite rod (anode) and platinum (cathode) were placed vertically at bottom and top of the electrochemical cell, with protonic acid (i.e, H_2SO_4, H_3PO_4 or $H_2C_2O_4$) aqueous solution as electrolyte; upon the application of a constant current of 0.1 A. The mechanism of electrochemical exfoliation is depicted in Eqs. ((1)–(5)) consisting of anodic oxidation of water as well as intercalation of anions into graphite rod (Cs-surface of graphite rod, Cb-bulk of graphite rod). This method produce graphene flakes with 2–8 nm thickness with lateral size about 1–5 μm (AFM image of exfoliated graphene flakes spin-coated onto a SiO_2 (285 nm)/Si).

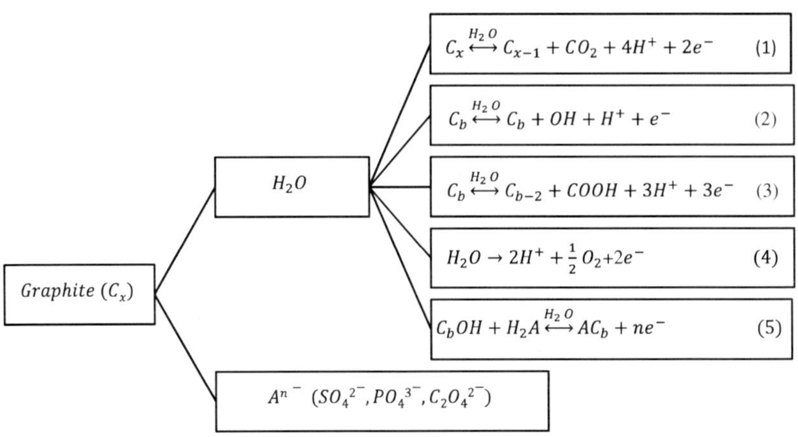

2.3. Electroanalytical Applications

Among all forms of carbon used as electrode material, carbon graphite constitutes one of the most basic and well known ways. This material is extremely soft and easy to handle is an excellent conductor of electricity [66, 67], so that electrode materials composite conductors embedded in insulating matrices were developed initially by Adams in 1958 [67] for his pioneering work on a new carbon paste electrode. Since then, various procedures have been developed for the manufacture of carbon paste electrodes. These composite materials consist of conductive microelectrodes a set of randomly dispersed on a surface, and whose "edge effects" grain carbon contributes significantly to the faradaic current. The conductive particles incorporated in the arrays have different sizes and shapes, and therefore do not conform to theoretical models used to describe the behavior of electrochemical electrodes being based conductive particles uniformly distributed on surfaces [68-70]. The electrodes made from carbon paste have been widely used in electrochemical measurements. The compositions vary over a wide range of materials, however, mostly using different types of carbon entrapped in a binder, the polymeric matrices being the most used. [71,72].

The electrodes of composite materials provide some advantages over traditional only one conductive phase such as glassy carbon, platinum or gold, because they can be manufactured with great flexibility in various shapes and sizes, making it easy to adapt them to the desired systems [73]. Are easily modifiable, allowing the incorporation of different filler components, e.g. enzymes, cofactors, mediators, etc. catalysts [74, 75]. Among the composite materials stand out as the graphite electrode conductive component which have good characteristics such as high mechanical strength, relative simplicity of preparation and renewal of the surface and low manufacturing cost [76]. Carbon particles incorporated in new materials are widely studied with the objective of developing electrodes polymeric matrices that are chemically, physically and electrochemically stable and good electrochemical properties for their applications in various situations [77-80].

The good electrical conductivity coupled with easy handling makes it a kind graphite carbon quite interesting for the electrochemical studies, and these features will be emphasized in this study their application as electrode material. For these reasons this material is very attractive for applications in electrochemistry specially because the electrodes are easily polarizable, however, its electrical conductivity is highly dependent on the thermal and chemical treatment used, depends on their composition, hybridization and different number of atoms present in its structure, in addition to its amphoteric nature, or their ability to behave as an acid or a base depending on the other reactant present in the medium, the carbon becomes both electron donor and receiver of this other reagent taking advantage of the electrochemical properties of this material [81].

2.4. EQCM Applications

Sensors like quartz crystal microbalance (QCM), has grown in many fields of science. Many researchers devoted special attention to the development of materials such as nanostructured carbon films with applications in mass sensors capable of performing real-time studies at the nanometer scale [82-87]. The high sensitivity of the QCM is the basis for their application in studies of thin films [88]. Making use of the correlation mass / frequency obtained one can study the processes of charge transfer and the mass interface electrode/electrolyte [89]. However, one factor which limits the use of the system is its operation with crystals coated with metallic films [90, 91] for replacement by another electrode material appears to be a need and opportunity for development. The principal requirements for a good electrode material are: long-term operational stability, low electrical current and large residual potential window. At this point the graphite carbon in its various forms emerged as a promising material as an alternative to metal electrodes for electrochemical sensors with applications, because it has good physical and mechanical properties such as hardness and toughness [92-94].

Electrode materials based on carbon have a wide potential window are not toxic and have good long-term stability and hence the development electrode based on carbon has been much interest in the academic [86]. The development of carbon films on substrates of quartz piezoelectric crystals, further studies should allow for conventional, applying EQCM to study the phenomena of adsorption, deposition and dissolution interfaces electrode/electrolyte. Recently, researchers has developed and characterized a new electrode materials based on carbon deposited on graphite Films 500 nm thicker and using sputtering techniques to produce a new sensor on piezoelectric crystals for electrochemical applications [86].

Pieces carbon graphite density 99.9995% ~ 1.9 g/cm^3 were sputtered on samples quartz crystal KVG, 0,28 cm^2 covered with 300 nm gold thin films. The deposition was carried out in a chamber with controlled internal pressure ~ 2.66 x10^{-4} Pa, $Ar_{(gas)}$ was injected until it reaches an operating pressure of ~ 0.39 Pa. The depositions of films were made by pulses of 60-120 μ s at a frequency range between 10-200 Hz [86].

The atomic force microscopy (AFM) was used in order to evaluate the morphological and structural characteristics of films formed on the gold QCM surfaces, Figure 11 shown 3D results obtained by Pinto et al. [86], carried out in AFM contact mode.

Reference [89] shows the gold substrate with a "quasi" uniformity with regard to the distribution of nanostructures across surface with structures with ~ 90 nm in diameter, maximum heights of ~ 12.83 nm, 100 nm in diameter. The average roughness (R_a), defined as the arithmetic mean of mean absolute deviation of the imperfections of a surface in relation to the length Sampling [85] has been calculated for these films with a sample area Geometric 9 μm^2. The gold films showed Ra ~ 1.1 nm.

The roughness average obtained for graphite films sputtered on gold substrate is superior in height, *R*a ~ 3.8 nm for an area of 9 μm^2, i.e. three times the one obtained for the gold substrate, these results are expected since like surfaces coated by carbon films constructed by pyrolysis of methane [27], composed by different types of carbon, with similar structures and whose composition is mostly comprised of carbon as graphite.

As observed in films coated by pyrolysis, graphite sputtered presents higher electrical resistance indicate that the composition and structure of films play an important role in the performance electrochemical electrodes. The films had structures partially species consisting of conductive carbon thus presenting little smaller conductivity obtained by electrochemical impedance spectroscopy and voltametric analyzes [86].

Structures with different sizes and diameters greater than 250 nm and dispersed heterogeneously over the surface presents cavities (holes) occurring mainly due to formation of species other than carbon, formed under the conditions Special inside the sputtering chamber, Figure 11 shows the results obtained for graphite sputtered films.

The Raman confocal spectroscopy was also used as a technique for characterizing the graphite films sputtered on quartz crystals elucidating the conductivity behavior of the deposited films. Taking in to account that the microstructures are performed by different carbon species including graphite and carbon like diamond, it was analyzed different regions on the surface, namely: (I) the surface microstructure and (II) of a flat section film. The spectra resulted from this analysis presents two Raman bands well defined, with one band of lesser intensity at ~ D 345 cm^{-1} and another with greater intensity, G at ~ 1594 cm^{-1}, both present in all samples.

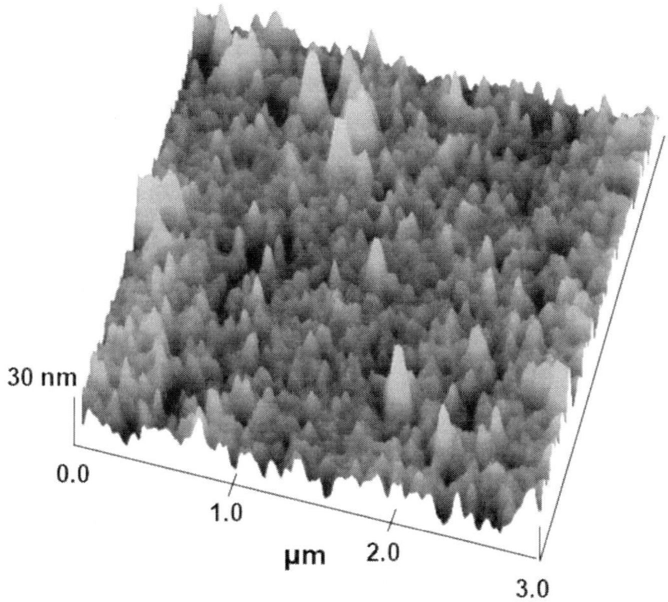

Figure 11. Micrograph obtained by AFM contact mode for, graphite sputtered film formed on gold substrate.

Carbon films are usually ordered by short-range connections sp^3 and sp^2 hybridizations, and diamond whit sp^3 hybridization. A single Raman mode to 1332 cm^{-1} and another at 1575 cm^{-1}, designated by *G-band* reflects the vibration mode of a kind of perfect graphite [86] Multi-crystalline graphite has an additional band at 1355 cm^{-1} a called *D-band* which appears due to the high degree of disorder of its structure [95].

In other words, the spectra presented in ref 86 shows the occurrence of two species of carbon, amorphous, carbon diamond-like and carbon graphite formed during the growth films process on these structures. This film forms have well-targeted carbon as they grow the diamond after a process orientation structures due to internal conditions of the deposition chamber. On the flat surfaces it is formed predominantly graphitic amorphous carbon structures, for which the spectrum has higher peak intensity in the *G- band*.

The electrochemical properties of the films, such as operational stability, potential window an electroative area were evaluated in by voltammetric studies; resistance and capacitance of the films were also evaluated by electrochemical impedance spectroscopy (EIS), all of measures carried out in 0.1 mol L^{-1} KCl solution at 20 mV s^{-1}.

The results obtained shown operating wide potential window between initial potential -0.9V and +0.9 V vs. SCE. The films presented good operational stability after five cycles maintaining the values of pick current after hundred several cycles, the electroactive area obtained was ~ 0.19 cm^2 lesser than the 0.28 cm^2 of the gold substrate, explained by the different carbon a forms previous detected by the Raman confocal spectroscopy.

The electrodes covered by graphite films named CG/Au presents double layer capacitance C_{dl} about three times lower than those obtained for the Au electrode, however the charge transfer resistance, R_{ct} were higher with maximum ~ 75.6 cm^2 at 0.40 V and minimum of 12.2 cm^2 to 0.30 cm^2 V. There is influence of the substrate and also, due to the different carbon species such as, HPOG, graphite, DLC among others, which interferes in their electrochemical properties. Decreases of the electrical conductivity of the electrodes CG/Au are attributed to the difficulties in the transfer of electrons at the interfaces gold/carbon and carbon/electrolyte. Other parameters such as the composition and structure of carbon films, roughness are dependent on the effective areas of the electrodes [95].

Electrodes performed by sputtered graphite films were used as substrate for electropolimerization of Poly Methylene blue (PMB), and compared with gold substrate films. Results shown the gravimetric response recorded, by the quartz crystal microbalance, during the cyclic voltammetry applied for electropolymerization of the PMB films on both substrates. The slops of the curves correspond to frequency variation recorded during the voltammetric cycles along the time. Decreasing on frequencies recorded by the electrochemical quartz crystal microbalance, EQCM, means increasing in mass adsorbed on substrate surface.

For both cases, the first profiles are directly related to the formation of the monomer that starts the process of electropolymerization. Thc PMB films forms strongly on CG/Au substrate, it is necessary only 30 cycles, while Au electrodes occur at lower rate of electropolymerization. The frequency variation recorded indicates the formation of a thicker film to the carbon electrodes since the total variation frequency is higher reaching ~1.32 kHz compared to 0.98 kHz obtained for the Au electrode [96].

The curve for the electrode PMB/CG/Au reveals three distinct profiles of frequency variation, first between the 1st and 5th cycles, ~ 55 Hz , however a new profile extends up to the 18th cycle presenting greater frequency variation. The total variation in frequency recorded for the electrode PMB/CG/Au was ~ 1.32 kHz corresponding to a mass of an electropolymerized film of ~ 16 mg cm^{-2}. This values is obtained by correlation by using of Sauebrey model [96]. For the Au electrode is also observed different profiles, the first starts already at 1st cycle extending up to 15th with changing ~ 980 Hz which corresponds to a total mass of ~ 12.08 mg cm^{-2}. Results shown that CG/Au, is more efficient for the process of polymerization of MB monomers , the electrodes with the same geometric area achieved films 33% more massive demonstrating the efficiency of the material as well as the promises of combined use of an electrochemical technique with the QCM technique [96].

ACKNOWLEDGMENTS

Authors are thankful to Coordenação de Aperfeiçoamento de Pessoal de Nível Superior (CAPES), Conselho Nacional de Desenvolvimento Científico e Tecnológico (CNPq),

Fundação Carlos Chagas de Amparo a Pesquisa do Estado do Rio de Janeiro (FAPERJ, E-26/102.971/2012, E-26/111.195/2011, E-26/110.092/2010, E-26/100.401/2010, E-26/101.719/2010, E-26/102.632/2010, E-26/102.791/2010), Fundação de Amparo a Pesquisa do Estado de São Paulo (FAPESP), and Pró-reitoria de Pesquisa, Pós-graduação e Inovação (Proppi-UFF) for constant financial support.

REFERENCES

[1] Henning, T.; Salama F. Carbon in the Universe; *Science*, 1998, 282, 2204-2210

[2] John Downard, A. J. Electrochemically Assisted Covalent Modification of Carbon Electrodes, *Electroanal*; 2000, 12, pp 1085-1096.

[3] Geim, A. K.; Novoselov, K. S. The rise of graphene, *Nature Materials*; 2007, 6, 183 – 191.

[4] Santos, R. M.; Lourenço, C. F.; Piedade, A. P.; Andrews, R.; Pomerleau, F.; Huettl, P.; Gerhardt, G. A.; Laranjinha, J.; Barbosa, R. M.; A comparative study of carbon fiber-based microelectrodes for the measurement of nitric oxide in brain tissue; *Biosensors and Bioelectronics.*, 2008, 24, 704 -709.

[5] Kang, Y. J.; Kim,Y. H.; Chang, J. Electrical transport properties of nanoscale devices based on carbon nanotubes; *Current Applied Physics*, 2009, 9, S7-S11.

[6] Zhao, Q. L.; Zhang, Z. L.; Bao, L.; Pang, D. W. Electrochem. Commun., Surface structure-related electrochemical behaviors of glassy carbon electrodes; 2008, 10, 181-185.

[7] Evstefeeva, Y. E.; Pleskov, Y. V.; Kutsay, A. M.; Bello, I. Electrode Properties of Tetrahedral Amorphous Carbon, *Russian Journal Electrochemistry*; 2005, 41, 772-777.

[8] Dekanski, A.; Stevanovic, J.; Stevanovic, R.; Nikolic, B. Z.; Jovanovic, V. M. Glassy Carbon Electrodes I - Characterization and Electrochemical Activation *Carbon*; 2001, 39, 1195-1205.

[9] Xu, Q.; Guo, R. X.; Wang, C. Y.; Hu X. Y. Application of activated glassy carbon electrode for the detection of nuciferine in lotus leaves; *Talanta*, 2007, 73, 262-268.

[10] Tokuda, K.; Morita, K.; Shimizu, Y. Cyclic Voltammetry at Microhole Array Electrodes; Analytical Chemistry; 1989, 61, 1763- 1768.

[11] Yao, Z.; Kane, C. L.; Dekker, C. High-Field Electrical Transport in Single-Wall Carbon Nanotubes; *Physical Review Letters*, 2000, 84, 2941-2944.

[12] Bachtold, A.; Fuhrer, M. S.; Plysunov, S.; Forero, M.; Anderson, E. H.; Zettl, A.; Mceuen, P.; Scanned probe microscopy of electronic transport in carbon nanotubes; *Physics Review Lett.*, 2000, 84, 6082-6085.

[13] Bertholdo, R. Síntese de novas formas de carbono obtidas a partir da pirólise de polímeros precursores impregnados em vidros porosos; Curitiba, 2001, 41 f. Dissertação (Mestrado em Química) – Departamento de Química, Universidade Federal do Paraná.

[14] Dresselhaus, M. S. Future directions in carbon science; Annual Review of Materials Science, Palo Alto: Annual Review Inc., 1997, vol. 27, pp 1-34.

[15] Jesus, C.; Ricardo N. Preparação e caracterização de carbono polimérico vítreo a partir da resina resol e modificações com íons metálicos.Ribeirão Preto, 2009, 80 f.

Dissertação (Mestrado em Ciências) – Departamento de Química, Universidade de são Paulo.

[16] Nossol, E. Novos eletrodos construídos a partir de diferentes nanoestruturas de carbono.Curitiba, 2009, 129 f, Dissertação (Mestrado em Química) – Departamento de Química, Universidade Federal do Paraná.

[17] Brett, A. M. O.; Brett, C. M. A. Experiências electroquímicas. In Electroquímica – Princípios, Métodos e Aplicações; Coimbra: Livraria Almedina, 1996, pp 147.

[18] Martins, R. N. Influência de recobrimentos de carbono no processo de deposição de parafinas.Rio de Janeiro, 2005, 69 f. Tese (Mestrado em Ciências em Engenharia metalúrgica e de materiais) – COPPE, Universidade Federal do Rio de Janeiro.

[19] Mahan, B. M.; Myers, R. J. Elementos representativos: Grupo I-IV. In: Química – Um curso universitário; 4 ed, Editora Blücher: São Paulo, 1995, pp 386 – 387.

[20] McCreery, R. L. Advanced carbon electrode materials for molecular electrochemistry; Chemical Reviews, Washington: ACS Publications, 2008, vol. 108(7), pp 2646-2687.

[21] Greenwood, N. N.; Earnshaw, A. Chemistry of the elements; 2 ed, 1998, pp 271-285

[22] Lee, J. D. Química inorgânica não tão concisa, 5th ed, 1999, Edgard Blucher, pp 202-210.

[23] Sengupta, R.; Bhattacharya, M.; Bandyopadhyay, S. Bhowmick, A. K. A review on the mechanical and electrical properties of graphite and modified graphite reinforced polymer composites; *Progress in Polymer Science* , 2011, 36, 638–670.

[24] Chung D. D. L.; Composite Materials Research Laboratory; State University of New York at Buffalo, Buffalo, NY, 14260-4400.

[25] Chung D. D. L. Review – Graphite; *Journal of Material Science,* 2002, 37, pp 1475-1489.

[26] Allen, M. J.; Tung, V. C.; Kaner, R. B. Honeycomb carbon: a review of grapheme. Chemical Reviews; Washington: ACS Publications, 2010, vol. 110 (1), 132-145.

[27] Stohr, J. NEXAFS Spectroscopy; Springer Series in Surface Sciences, 2003.

[28] Kroto, H. W.; Heath J. R.; Obrian, S. C.; Curl, R. F.; Smalley, R. E. C_{60}: Buckminsterfullerene; *Nature*, 1985, 318, 162-163.

[29] Atkins, P. W.; Jones, L. Princípios de Química: questionando a vida moderna e o meio ambiente; Porto Alegre: Bookman, 2012. pp 382-387.

[30] Taylor, R.; Hare, J. P.; Abdul-Sada, A. K.; Kroto, H. W. Journal of Chemical Society, *Chemical Communication*, 1990, 5, 1423.

[31] Rocha, F.; Romeu C. Os fulerenos e sua espantosa geometria molecular. Química nova na escola, Sociedade Brasileira de Química: São Paulo, 1996, 4, pp 7-11.

[32] Seed - Schlumberger Excellence in Educational Development. A descoberta dos fulerenos: Começando a pesquisa, 2000. <http://www.planetseed.com /pt-br/node/16748>.

[33] Santos, L. J.; Rocha, G. P.; Alves, R. B.; Freitas, R. P. Fulereno [C_{60}]: química e aplicações. Química nova, Sociedade Brasileira de Química: São Paulo, 2010, vol. 33 (3), pp 680-693.

[34] Morozov, S. V.; Novoselov, K. S.; Katsnelson, M, I,; Schedin, F.; Elias, D.C.; Jaszczak, J. A.; Geim, A. K. Giant intrinsiecarriemobilities in grapheme and its bilayer; *Physical Review Letters*, 2008, 100 (1), 016602.

[35] Balandin, A. A.; Ghosh, S.; Bao, Z.; Calizo, I; Teweldebrhan, D.; Miao, F.; Lau, C. N. Superior thermal conductivity of single-layer grapheme; *Nano Letters*, 2008, 8(3), 902-907.

[36] Lemme, M. C.; Echtermeyer, T. J.; Baus, M.; Kurz, H. A graphene field-effect device, IEEE Eletron Device Lett., 2007, 28(4), 282-284.

[37] Standley, B.; Bao, W.; Zhang, H. J.; Bruck, J.; Lau, C. N.; Bockrath, M. Graphen-based atomic-scale switches; *Nano Letters*, 2008.

[38] Józsa, C.; Popinciuc, M.; Tombros, N.; Jinkman, H. T.; Wees, B. J.; Eletronic spin drift in graphene field-effect transistors; *Physical Review Letters*, 2008, 100 (23), 236603.

[39] Li, D.; Kaner, R. B. Materials Science: Graphene-Based Materials, *Science*, 2008, 320 (5880), 1170-1171.

[40] Soldano, C.; Mahmood, A.; Dujardin, E. Production, properties and potencial of graphene. *Carbon*; Elsevier, 2010, 48, 2127-2150.

[41] Pereira, F. C.; Bergamo, E. P.; Zanoni, M. V. B.; UGO, L. M. M. P. Aplicações de nanoeletrodos como sensores na química analítica. Química Nova, Sociedade Brasileira de Química: São Paulo, 2006, vol.29 (5), pp 1054 -1060.

[42] Gong, K.; Yan, Y.; Zhang, M.; Su, L.; Xiong, S.; Mao, L. Electrochemistry and electroanalytical applications of carbon nanotubes: a review; *Analytical Sciences,* 2005, 21, 1383-1393.

[43] Ong, Y. T.; Ahmad, A. L.; Zein, S. H. S.; Tan, S. H. A review on carbon nanotubes in an environmental protection and green engineering perspective; *Brazilian Journal of Chemical Engineering*, 2010. 27 (2), 227-242.

[44] Potts, J. R.; Dreyer, D. R.; Bielawski, C. W.; Ruoff, R. S. Graphene-based polymer nanocomposites; *Polymer*, 2011, 52, 5-25.

[45] Hernandez, Y.; Nicolosi, V.; Lotya, M.; Blighe, F. M.; Sun, Z.; Sukanta D.; McGovern, I. T.; Holland, B.; Byrne, M.; Gun'Ko, Y. K.; Boland, J. J.; Niraj, P.; Duesberg, D.; Krishnamurthy, S.; Goodhue, R.; Hutchison, J.; Scardaci, V.; Ferrari, A. C.; Coleman, J. N. High-yield production of graphene by liquid-phase exfoliation of graphite; *Nature Nanotechnology*, 2008 (3) ,563 – 568.

[46] Alzari, V.; Nuvoli, D.; Scognamillo, S.; Piccinini, M.; Gioffredi, E.; Malucelli, G.; Marceddu, S.; Sechi, M.; Sanna, V.; Mariani, A. *J. Mater. Chem.,* 2011,21, 8727-8733

[47] Liu, Y. T.; Xie, X. M.; Ye, X. Y. High-concentration organic solutions of poly(styrene-co-butadiene-co-styrene)-modified graphene sheets exfoliated from graphite *Carbon*, 2011, 49 (11), 3529-3537.

[48] Puangbuppha, B.; Limsuwan, P.; Asanithi, P.Non-Chemically Functionalized Graphene Exfoliated from Graphite in Water using Ultrasonic Treatment; *Procedia Engineering*, 2012, 32, 1094-1099.

[49] Pu, N. W.; Wang, C. A.; Sung, Y.; Liu, Y. M.; Ger, M. D. Production of few-layer graphene by supercritical CO_2 exfoliation of graphite; *Materials Letters*, 2009, 63, 1987-1989.

[50] Rangappa, D.; Sone, K.; Wang, M.; Gautam, U. K.; Golberg, D.; Itoh, H.; Ichihara, M.; Honma, I. Rapid and direct conversion of graphite crystals into highyielding, good-quality graphene by supercritical fluid exfoliation; Chemistry - *A European Journal*, 2010, 16, 6488 6494.

[51] Sim, S. Y.; Kim, T. A.; Lee, Min, K. H. Park Preparation of graphene nanosheets through repeated supercritical carbon dioxide process; *Materials Letters*, 2012, 89, 343-346.
[52] Shin, H. J.; Kim, K. K.; Benayad, A.; Yoon, S. M.; Park, H. K.; Jung, I. S.; Jin, M. H.; Jeong, H. K.; Kim, J. M.; Choi, J. Y.; Lee, Y. H. Efficient reduction of graphite oxide by sodium borohydride and its effect on electrical conductance; *Advanced Functional Materials,* 2009, 19, 1987-1992;
[53] Si, Y.; Samulski, E. T. Synthesis of water soluble grapheme; *Nano Letters*, 2008, 8, 1679-1682.
[54] Stankovich, S.; Dikin, D. A.; Piner, R. D.; Kohlhaas, K. M.; Kleinhammes, A.; Jia, Y.; Wu, Y.; Nguyen, S. T.; Ruoff, R. S. Synthesis of graphene-based nanosheets via chemical reduction of exfoliated graphite oxide; *Carbon*, 2007, 45, 1558-1565.
[55] Wang, G. X.; Yang, J.; Park, J.; Gou, X. L.; Wang, B.; Liu, H. Chemical selfassembly of graphene sheets; *Journal of Physical Chemistry*, 2008, 112, 8192-8195.
[56] Stankovich, S; Piner, R. D.; Chen, X.; Wu, N.; Nguyen, S. T.; Ruoff, R. S. Stable aqueous dispersions of graphitic nanoplatelets via the reduction of exfoliated graphite oxide in the presence of poly(sodium 4-styrenesulfonate); *Journal of Materials Chemistry,* 2006, 16, 155-158.
[57] Schniepp, H. C.; Li, J. L.; Mcallister, M. J.; Sai, H.; Herrera, A. M.; Adamson, D. H.; Prud`homme, R. K.; Car, R.; Saville, D. A.; Aksay, I. A. Functionalized single graphene sheets derived from splitting graphite oxide; *Journal of Physical Chemistry*, 2006, 110, 8535-8539.
[58] Mcallister, M. J.; Li, J. L.; Adamson, D. H.; Schniepp, H. C.; Abdala, A. A.; Liu, J.; Herrera, A. M.; Milius, D. L.; Car, R.; Prud`homme, R. K.; Aksay, I. A. Single sheet functionalized graphene by oxidation and thermal expansion of graphite; *Chemistry of Materials*, 2007, 19, 4396-4404.
[59] Wu, Z. S.; Ren, W.; Gao, L.; Liu, B.; Jiang, C.; Cheng, H. M. Synthesis of high-quality graphene with a pre-determined number of layers; *Carbon*, 2009, 47, 493-499.
[60] Umar Khan, Arlene O'Neill, HarshitPorwal, Peter May, Khalid Nawaz, Jonathan N. Coleman Size selection of dispersed, exfoliated graphene flakes by controlled centrifugation *Carbon*, 2012, 50 (2), 470-475.
[61] Stankovich, S.; Dikin, A. D.; Piner, R. D.; Kohlhaas, K. A.; Kleinhammes, A.; Jia, Y.; Wu, Y.; Nguyen, S. T.; Ruoff, R. S. Synthesis of graphene-based nanosheets via chemical reduction
of exfoliated graphite oxide; *Carbon*, 2007, 45 (7), 1558-1565.
[62] Zhang, T.; Zhang, D.; Shen, M. A low-cost method for preliminary separation of reduced graphene oxide nanosheets *Materials Letters*, 2009, 63 (23), 2051-2054.
[63] Liu, N.; Luo, F.; Wu, H. Y. L.; Zhang, C.; Chen, *J. Advanced Functional Materials*, 2008, 18, 1518.
[64] Lee, S. H.; Seo, S. D.; Jin, Y. H.; Shim, H. W.; Kim, D. W. A graphite foil electrode covered with electrochemically exfoliated graphene nanosheets; *Electrochemistry Communications*, 2010, 12, 1419-1422.
[65] Liu, J.; Poh, C. K.; Zhan, D.; Lai, L.; Lim, S. H.; Wang, L.; Liu, X.; Sahoo, N. G.; Li, C.; Lin, Z. S. J. Improved synthesis of graphene flakes from the multiple electrochemical exfoliation of graphite rod; Nano Energy, In Press, Corrected Proof, Available online 15 November 2012

[66] Edström, K. A.; Andersson, A. M.; Bishop, A.; Fransson, L.; Lindgren, J.; Hussénius, A. Carbon electrode morphology and thermal stability of the passivation layer, *Journal Power Sources*, 2001, 97-98, 87-91.

[67] Adams, R. N. Carbon Paste Electrodes, *Analytical Chemical*, 1958, 30, 1576-1576.

[68] Ramírez-García, S.; Alegret, S.; Céspedes, F.; Forster, R. J. Carbon composite electrodes: surface and electrochemical properties, *Analyst*, 2002, 127, 1512-1519.

[69] O'Hare, D.; Macpherson, J. V.; Willows, A. On the microelectrode behaviour of graphite–epoxy composite electrodes; *Electrochemical Communication*, 2002, 4, 245-250.

[70] Kalcher, K.; Kauffmann, J. M.; Wang, J.; Svankara, I.; Vytras, K.; Neuhold, C.; Yang, Z. Sensors based on carbon-paste in electrochemical analysis – a review with particular emphasis on the period 1990–1993; *Journal of Electroanalytical Chemistry*, 1995,7, 5-22.

[71] Céspedes, F.; Martínes-Fábregas, E.; Alegret, S. New materials for electrochemical sensing I. Rigid conducting composites; *Trends in Analytical Chemistry*, 1996, 15, 296-304.

[72] Tallman, D. E.; Petersen, S. L. Composite electrodes for electroanalysis: Principles and applications; *Journal of Electroanalytical Chemistry*, 1990, 2, 499-510.

[73] Ramírez, G. S.; Céspedes, F.; Alegret, S. Development of Conducting Composite Materials for Electrochemical Sensing in Organic Media; *Journal of Electroanalytical Chemistry*, 2001, 13, 529-535.

[74] Morales, A.; Céspedes, F.; Alegret, S. M. Graphite–methacrylate biocomposite material with renewable sensing surface for reagentlessamperometric biosensors based on glucose dehydrogenase; *Journal of Materials Science and Engineering*, 2000, 7, 99-104.

[75] Stulík, K. Activation of solid electrodes, *Journal of Electroanalytical Chemistry*, 4, 829-834.

[76] Luque, M. C.; Ríos, A.; Valcárcel, M. A poly (vinyl choloride) graphite composite electrode for flow-injection amperometric determination of antioxidants; *AnalyticaChimicaActa*, 1999, 395, 217-223.

[77] Dávila, M. M.; Elizalde, M. P.; Mattusch, J.; Wennrich, R. Capability of a carbon–polyvinylchloride composite electrode for the detection of dopamine, ascorbic acid and uric acid, *Electrochimica Acta*, 2004, 49, 851-859.

[78] Barsan, M. M.; Pinto, E.M.; Florescu, M.; Brett, C. M. A. Development and characterization of a new conducting carbon composite electrode; *Analytica Chimica Acta*, 2009, 635, 71-78.

[79] Cataldi, T. R. I.; Centonze, D. Nickel oxide dispersed in a graphite/poly(vinyl chloride) composite matrix for an electrocatalyticamperometric sensor of alditols in flow-injection analysis; *Analytica Chimica Acta*, 1995, 307, 43-48.

[80] Frackowiak, E.; Béguin, F. Carbon materials for the electrochemical storage of energy in capacitors; *Carbon*, 2001, 39, 937-950.

[81] Barsan, M. M.; Pinto, E. M.; Brett, C. M. A. Methylene blue and neutral red electropolymerisation on AuQCM and on modified AuQCM electrodes: an electrochemical and gravimetric study; *The Journal of Physical Chemistry*, 2011, 13, 5462-5471.

[82] Nomura, T.; Ijima, M. Behaviour of piezoelectric quartz crystals in solutions with application to the determination of iodide; *Analytica Chimica Acta*, 1985, 175, 107-116.
[83] Soares, D. M.; Tenan, M. A.; Wasle S. Determination of the electromechanical parameters of the electrochemical quartz crystal *microbalance*, 1998, 44, 263-268.
[84] Bruckenstein, S.; Shay, M. Experimental aspects of use of the quartz crystal microbalance in solution; *Electrochimica Acta*, 1985, 30, 1295-1300.
[85] Soares, D. M. A quartz microbalance with the capability of viscoelasticity measurements for in situ electrochemical investigations; *Measurement Science and Technology*, 1993, 4, 549-553.
[86] Pinto, E. M.; Gouveia, C. C.; Soares, D. M.; Brett, C. M. A. Electrochemical and surface characterization of carbon-film-coated piezoelectric quartz crystals; *Applied Surface Science*, 2009, 255, 8084-8090.
[87] Watling, K.; Hope, G. A.; Woods, R. SERS Investigation of Gold Dissolution in Chloride and Cyanide Media; *Journal of Electrochemical Society*, 2005, 152, D103-D108.
[88] Sauerbrey, G. Use of quartz vibration for weighing thin films on a microbalance; *Z. Phys*., 1959, 155, 206-212.
[89] Buron, C. C.; Filiâtre, C.; Membrey, F.; Perrot, H.; Foissy, A. Mass and charge balance in self-assembled multilayer films on gold. Measurements with optical reflectometry and quartz crystal microbalance; *Journal of Colloid and Interface Science*, 2006, 296,c409-418.
[90] Uslu, B.; Ozkan, S. A. Electroanalytical Application of Carbon Based Electrodes to the Pharmaceuticals; *Analytical Letters*, 2007, 40, 817-853.
[91] Stozhko, N. Y.; Malakhova, N. A.; Fyodorov, M. V.; Brainina, K. Z. Modified carbon-containing electrodes in stripping voltammetry of metals Part I. Glassy carbon and carbon paste electrodes; *Journal of Solid State Electrochemistry*, 2008, 12, 1185-1204.
[92] Stozhko, N. Y.; Malakhova, N. A.; Fyodorov, M. V.; Brainina, K. Z. Modified carbon-containing electrodes in stripping voltammetry of metals. Part II. Composite and microelectrodes; *Journal of Solid State Electrochemistry*, 2008, 12, 1219-1230.
[93] Gouveia-Caridade, C.; Brett, C. M. A. Electrochemical impedance characterization of nafion-coated carbon film resistor electrodes for electroanalysis; *Electroanalytical*, 2005, 17, 549-555.
[94] Binnig, G.; Rohner, H.; Gerber, C.; Weibel, E. *Physical Review Letters*, 1982, 49, 57.
[95] Gouveia, C. C.; Soares, D. M.; Liess, H. D.; Brett, C. M. A. Electrochemical, morphological and microstructural characterization of carbon film resistor electrodes for application in electrochemical sensors; *Journal of Applied Surface Science*, 2008, 254, 6380-6389.
[96] Barsan, M. M.; Pinto, E. M.; Brett, C. M. A. Methylene blue and neutral red electropolymerisation on AuQCM and on modified AuQCM electrodes: an electrochemical and gravimetric study. *PCCP. Physical Chemistry Chemical Physics*, 2011, 13, 5462-5471.

In: Graphite
Editor: Quinton C. Campbell

ISBN: 978-1-62618-576-0

Chapter 6

MICROSTRUCTURAL FACTORS WHICH GOVERN THE OXIDATIVE PROPERTIES OF GRAPHITE

Heinrich Badenhorst*, Brian Rand and Walter W. Focke
SARChI Chair in Carbon Materials and Technology,
Department of Chemical Engineering, University of Pretoria, South Africa

ABSTRACT

Graphite is widely utilized in high temperature applications under oxidative conditions. As such an understanding of the oxidative behaviour of graphite and hence the factors which govern these properties is essential. The exact kinetic mechanism remains a complex and disputed topic. However the effect of the microstructure on the oxidation is often neglected. The primary microstructural factors which influence the oxidation and graphite properties which lead to these are discussed. This includes the complex and varied microstructures found in real graphite materials from different origins. As well as the influence of both catalytically active and inert impurities on the development of the microstructure. This affords a comprehensive view of the influence of the microstructure on the oxidative properties of graphite. It is possible to represent the simplest of these behaviours qualitatively through simulation of the microstructural development. This indicates that the use of intricate and arbitrary kinetic reaction schemes is not necessary to accurately describe and sensibly compare the oxidative properties of different graphite materials.

1. INTRODUCTION

Graphite is a very important industrial material used in a very wide variety of applications. By far the largest application of bulk graphite is the steel and aluminium production industries. In the steel industry graphite is mostly used as electrodes in electric-arc furnaces but also as refractory materials in blast furnaces. The electrodes must have good electrical conductivity and low thermal expansion due to the extreme working conditions.

* Corresponding author: carbon@up.ac.za.; Lynwood Road, Pretoria, 0002, South Africa.

However, low cost is a priority since the electrodes gradually erode in use and at times fracture completely. Aluminium is processed electrolytically using similar electrodes as arc-furnaces for anodes. The aluminium collects at the cathodes which are large blocks lining the electrolytic cell. The cathodes used to be made from lower grade cokes but the use of moulded graphite has increased. Moulded graphite also has a wide variety of applications in other liquid metal processing industries, including casting moulds, dies, guides, plungers, sintering containers, susceptors and furnace linings.

In recent years there has been an increasing use of graphite in speciality applications, for example as crucibles in the semi-conductor industry for crystal-pulling. This also includes the ever increasing list of expanded natural graphite uses in electronic and industrial heat management, intumescent polymer additives, insulation panels and pressed graphite sheets for automotive and fluid sealing applications. Due to its electrical conductivity, chemical inertness and ability to withstand arcing, graphite also has several smaller uses in electrical applications like brushes, railroad shoes, heating elements, fuel-cell components and electrodes for batteries. Similarly, a wide range of secondary uses exist in mechanical applications like seal rings, o-rings, bearings and rocket nozzles. Finally, moulded graphite is also used in nuclear reactors, both as a neutron moderator and as a structural component.

This is not even to mention the numerous applications of carbon-carbon composites and graphite fibres. Most of these applications directly or indirectly involve the generation of high temperature operating conditions, usually in the presence of air. Thus knowledge of the high temperature oxidative behaviour of graphite is critical to understanding and possibly preventing any structural changes which may occur during normal operation or in some cases during abnormal situations such as accidental air-ingress. Over the past century, the gasification of different carbon materials has received widespread attention owing to their industrial relevance in terms of energy and electricity generation. Gulbransen, one of the early investigators of graphite oxidation, remarked in a paper written in 1952 that, "The reaction of carbon with oxygen has been the subject of many studies over the past 100 years. A review and evaluation of this literature would be an extremely long and arduous task" (Gulbransen, 1952). In the sixty years that have passed since then, the number of studies conducted has risen exponentially. However, a comprehensive review is still not forthcoming. At this point, it is probably an impossible task. Despite all the attention that gasification has received, much still remains unclear regarding the fundamental mechanisms that govern the "simple" reaction of carbon with oxygen. Of all the carbon allotropes, graphite has arguably one of the seemingly simplest, two-dimensional structures available at relatively low cost. This makes it an ideal starting point for oxidative studies and also gaining insights into the oxidation of carbon materials as a whole. Five factors which contribute to the uncertainty in the intrinsic oxidation rate for graphite are:

- Ambiguity surrounding the exact kinetic reaction mechanism
- The very complex microstructures found in real materials classified as graphite
- Discrepancies between the perceived crystalline structure and true levels of crystalline perfection found in different graphitic materials
- The impact of extremely low levels of catalytic impurities on the microstructural evolution
- The contributions of inhibiting impurities to the microstructure

Each of these factors will be discussed in the following sections to afford a comprehensive view of the primary microstructural factors which influence the oxidation of graphite across the entire conversion range. But first a general discussion on the theoretical structure of ideal carbon materials is provided as background, before functional industrial materials are examined.

2. Theoretical Background

Unlike most elements carbon has a virtually limitless number of allotropes or polymorphs. These materials are comprised entirely of carbon but have vastly different physical structures and properties. This is in part due to the ability of carbon to form three different structural hybrids, namely sp^1, sp^2 and sp^3.

The sp^3 hybrid occurs when all four orbitals (s and p) of the L shell are combined to form four sp^3-orbitals. To maintain the lowest energy configuration, these four orbitals are arranged in a tetrahedral formation with the maximum equal angle (109.5°) between each, as shown in Figure 1.

When such a hybrid orbital overlaps with an orbital from another atom, the nuclei share a pair of electrons and a strong covalent bond is formed. Due to the small size of the carbon nucleus and the fact that this hybrid structure allows four of carbon's six electrons to form bonds, the resulting covalent bonds are quite strong. By convention these strongly directional (or stereospecific) bonds are called sigma (σ) bonds.

This hybridization is found in all aliphatic carbon compounds. Due to the heavily lopsided nature of these orbitals, a substantial overlap is possible when the orbitals combine with other sp^3-orbitals from other carbon atoms. This further enhances the strength of the bonds and allows a three dimensional, entirely covalent, tetrahedral lattice structure. This forms the well known three dimensional arrangement which is the basic structural unit of the diamond crystal, as shown in Figure 2.

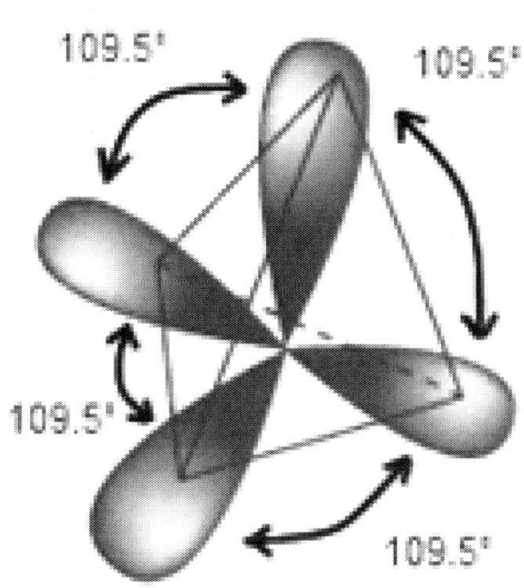

Figure 1. Tetrahedral formation of sp^3-hybrid orbitals (negative lobes omitted for clarity).

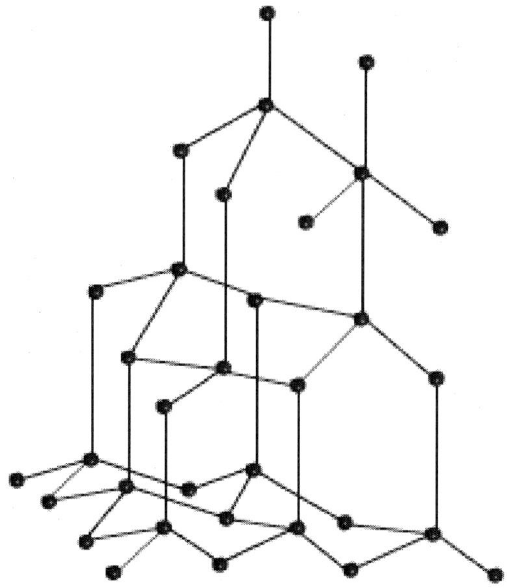

Figure 2. Tetrahedral structure of diamond.

The next hybrid is sp^2 in which case the s-orbital combines with two p-orbitals to form three sp^2-orbitals. The remaining p-orbital remains in its unhybridized ground state. Three of the four electrons occupy the three hybrid orbitals, while the final electron exists in a "free" or delocalised state within the remaining p-orbital. This arrangement forms the basis of all graphitic structures and aromatic compounds. To maintain the lowest energy configuration, these orbitals are arranged in a planar, trigonal fashion with the maximum equal angle (120°) between them. When bonded in plane to other sp^2 hybridized carbon atoms, the 120° bond angle leads to the formation of a planar hexagonal structure. Furthermore, when several of these planes are arranged on top of each other, a layered structure results.

This arrangement forms the basic structural unit of the graphite crystal, as shown in Figure 3.

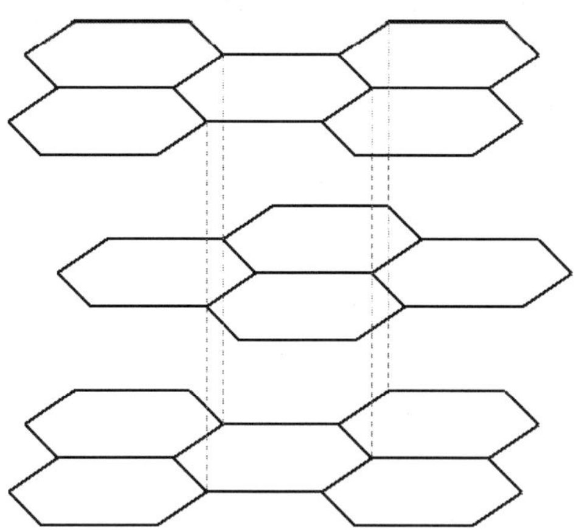

Figure 3. Hexagonal planar structure of graphite.

The orbitals are similar to the sp^3-orbitals but in this case even more pronounced overlap is possible with other sp^2 hybridized carbon atoms. This leads to a shorter bond length and higher bond energy in the orbital plane, which implies that in-plane graphite has a higher strength than diamond. The fourth, unhybridized orbital lies perpendicular to the plane of the three sp^2-orbitals. The electron in this orbital is capable of forming a significantly weaker bond with adjacent atoms, pi (π) bond. In the graphitic structure, the delocalised electron has a weak van der Waals interaction with the adjacent layer. This leads to the extremely low in-plane shear resistance of graphite. The final hybridization state is sp^1, whereby the s-orbital and one p-orbital combine to form two sp^1-orbitals and two remaining unhybridized p-orbitals. A common example of this allotrope is ethyne. In chain form however, these carbynes are usually unstable (Korshak, 1987).

These three hybridized configurations (or combinations thereof) account for the molecular structure of all the stable carbon allotropes. This includes the so-called "nanostructured" carbons such as fullerenes, nanotubes and graphene sheets. Graphene is considered to be a single sheet of pristine graphite. Fullerenes and nanotubes may be thought of as three-dimensional derivatives of single graphene sheets that have been folded into specific conformations, as can be seen in Figure 4. These structures have many unique properties and interesting applications, but they are difficult and expensive to manufacture in bulk quantities.

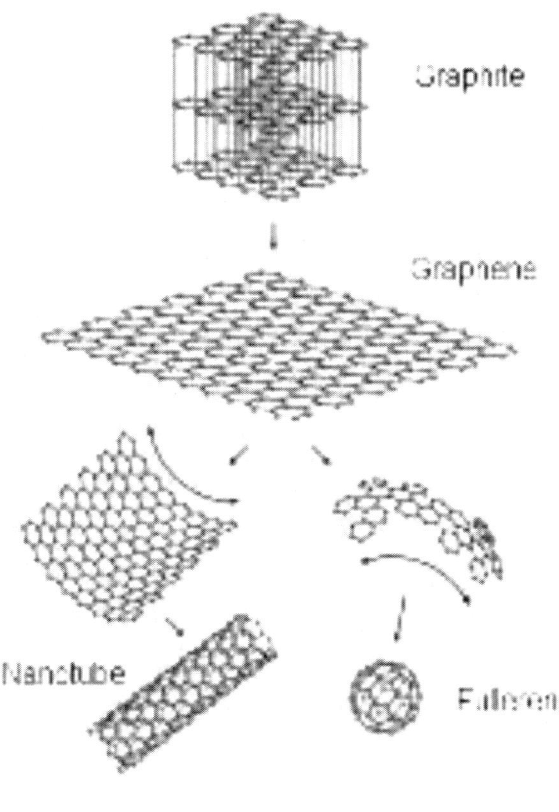

Figure 4. Configurations of graphene, fullerenes, nanotubes and graphite.

In addition to these ideal allotropes, a general group of carbon materials exists which may be broadly labelled as amorphous carbons. This group includes char, coal, anthracite, carbon black, activated carbon, vitreous carbon, etc.

Superficially these carbons may be classified as amorphous due to their lack of long range order and having a molecular structure related to either diamond or graphite or some combination of the two. Their actual structure is still poorly understood despite numerous investigations into a wide variety of these materials (Lu et al., 2001; Harris, 2005).

Recent work (Kumar et al., 2005) suggests that these materials may be entirely sp^2 hybridized, consisting of complex, three dimensional arrangements, shown in Figure 5, of varying sub-varieties and combinations of the structures shown in Figure 4.

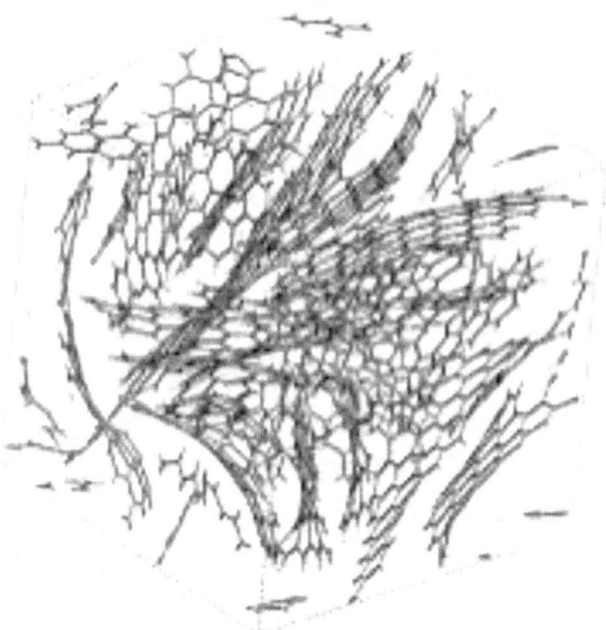

Figure 5. Theoretical structure of amorphous carbon (Reproduced with permission from Kumar et al., 2005).

The oxidation kinetics of this group of carbon materials is by far the most exhaustively investigated. However, due to the indeterminate atomic arrangement and the resulting complex microstructures of these materials, it remains difficult to truly decouple the kinetic mechanism from structural effects.

3. Kinetic Background

The random pore model has been widely applied to a variety of amorphous carbons, mostly chars (Su and Perlmutter, 1985; Bhatia and Vartak, 1996). It models the growth and coalescence of pores within the material and hence is a representation of how the surface area of the char particle develops. This implies that the reaction rate is heavily dependent on the surface area.

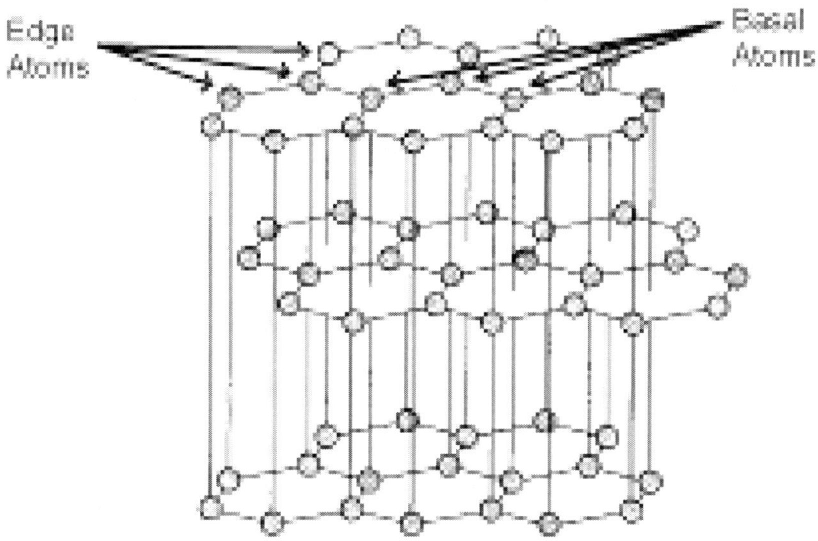

Figure 6. Reaction anisotropy of the graphite crystal structure.

In general, this is found to be true for amorphous carbons, but as the degree of graphitization, or rather the crystalline order, of a carbon increases, this assertion becomes less accurate.

The cause lies in the atomic structure of the graphite crystal. From a schematic representation of the graphite crystal, shown again in Figure 6, the reason for the reactive anisotropy becomes clear.

The crystals' basal atoms are linked in three directions with adjacent carbon atoms. These form very strong covalent σ bonds with each other and hence are comparatively difficult to remove from the lattice. This is further accentuated by the fact that there are additional π bonds between these atoms and thus the carbons' electronic requirements are fulfilled. These atoms share four electrons with other atoms, filling their outer shell completely. Hence the basal plane of graphite may be considered inert up to very high temperatures.

It should be borne in mind that this is the basal plane of the ideal, perfect graphite crystal. Crystal and lattice defects will dramatically influence the reactivity of the basal atoms around it (Thomas, 1969). Furthermore, the planar, layered structure of graphite shields atoms located in successive basal planes below the outer layer, making it very difficult for reactive gases to gain access to these atoms to attack and remove them.

The edge atoms, however, are quite different. They are exposed at all layers in the structure to attack from gaseous reactants. In addition, these atoms are less tightly bound to the crystal structure since there are only two covalent σ bonds to the adjacent carbon atoms.

The electronic requirements of these atoms are not fulfilled and they can be easily bound by gaseous reactants. The atomic structure shown in Figure 6 is potentially incomplete as it represents a dangling carbon atom. These dangling atoms have an unpaired electron, potentially two depending on the nature of the π bond structure of the adjacent atoms. Such an atomic configuration is possibly stable under inert conditions (Radovic and Bockrath, 2005), but would be highly reactive in the presence of oxygen. Thus, in general, these atoms are presumed to be bound to a variety of oxygen-based surface complexes.

Thus it is also clear why the surface area of a graphitic particle is no longer representative of its reactivity. The ratio of edge atoms to basal atoms is a strong function of the crystallite size. Therefore these exposed, reactive edges are referred to as the active surface area (ASA). For a example ideal, disc shaped, perfectly crystalline graphite flakes, with a 250 μm diameter, have an ASA of approximately 0.008 m^2/g. This very low value is very difficult to measure accurately and cheaply in practise at present, especially *in situ* during oxidation. The technique of using thermal desorption was first applied by Laine et al. (1963) to measure the development of the active surface area as a function of conversion for Graphon, which is a graphitized carbon black, Spherocarb. The most simplistic reaction of these active sites with oxygen would be represented by:

$$C(s)+O_2 \rightarrow CO_2^*(s) \rightarrow CO_2(g) \tag{1}$$

where $CO_2^*(s)$ represents the adsorbed surface complex. However it soon became apparent that the reaction was not this simple. Studies conducted on a wide variety of carbon materials was summarised by Walker et al. (1991) and implies the existence of at least two, one "fleeting" and one "stable" surface complex.

Since the original work by Laine et al. numerous studies focussed on these complexes (Smith et al., 1966; Starsinic et al., 1983; Marchon et al., 1988a; Marchon et al., 1988b; Brown and Haynes, 1992; Henschke et al., 1994; Zhuang et al., 1994a; Zhuang et al., 1994b; Zhuang et al., 1995; Figueiredo et al., 1999; Sibraa et al., 2000; Szymanski et al., 2002; Zhou et al., 2007), using mainly temperature programmed desorption (TPD) and combinations of Fourier transform infrared spectroscopy (FTIR) and X-ray photoelectron spectroscopy (XPS) techniques. Examples of some of the most commonly proposed surface groups for zigzag sites are shown in Figure 7.

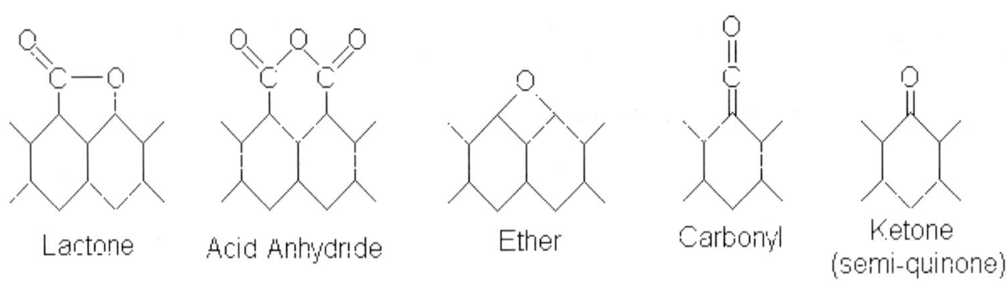

Figure 7. Possible surface complexes.

The wide variety of possible surface intermediates has led to the development of countless kinetic schemes postulating a variety of mechanisms. Some are fairly simple (Ahmed and Back, 1985; Haynes 2001), similar to the three step intrinsic mechanism proposed by Hurt and Calo (2001). Others are extremely complex (Backreedy et al., 2001; Zhu et al., 2002; Sendt et al., 2005), some with pathways for surface migration of and interconversion between complexes.

In most cases, these studies were conducted on carbon materials with much larger active surface areas than graphite, in which the surface complexes are present in measurable quantities. As such these materials cannot be considered crystalline, having the complex microstructures mentioned previously for amorphous carbons.

As was pointed out by Hurt and Calo (2001) this raises the additional question of surface site heterogeneity. The three dimensional, possibly sp^2 and sp^3 hybridized, interlinked structure of these materials gives rise to the potential for far more intricate surface complex structures than those shown in Figure 6. Thus despite identical composition, each may have differences in the required desorption energy.

Thus it is impossible to determine whether the observed behaviour is caused by differences in the active site configurations of the same surface complex or by different surface complexes on the same type of site. Connecting these complexes to reaction kinetics is a difficult task despite the direct link to experimentally observable TPD spectra. It was stated in the review by Walker et al. (1991) and later by Zhuang et al. (1995) based on detailed isotopic studies, that some surface complexes may act as intermediaries for the desorption of other complexes. Thus these complexes may effectively act as catalysts for others due to the lowering of the desorption energy as a function of surface coverage. For all of these reasons it appears to be impossible at this point to determine *in situ* during oxidation, the temperature dependent behaviour of these intermediaries.

In response to this general ambiguity and extremely complex reaction behaviour, molecular simulations have been used to validate reaction pathways (Sendt et al., 2005; Chen and Yang, 1998; Montoya et al., 2001; Espinal et al., 2004; Sendt and Haynes, 2005; Allouche and Ferro, 2006; Jones and Jones, 2007). However, these approaches provide very little motivation for proposed transitions between two, completely arbitrarily proposed surface structures. In most cases, no tangible link with measurable kinetic data is possible. This approach has led to a proliferation of arbitrary reaction schemes with no *a priori* way of selecting between them.

A recent reaction scheme has been proposed by (Campbell and Mitchell, 2008) which contains fifteen reaction pathways, five surface structures and roughly thirty adjustable parameters. With models of such high complexity and heavy dependence on adjustable parameters, extrapolation to areas outside the test region and applicability to samples from other origins becomes limited and very risky.

The solid state kinetic approach has been widely applied to the thermal decomposition of solids (Galwey and Brown, 1999). It has also been applied to other solid transitions like crystallization or crystallographic transitions and autocatalytic reactions. These reactions are subject to diverse rate controlling factors like nucleation and growth, nucleus coalescence, ingestion and generation, or in some cases, where a product barrier layer is formed, diffusion through this layer. For most solid state decompositions other complex phenomena are also occurring such as melting, sintering, etc.

These reactions are fundamentally different from classical homogeneous reactions. For example, the concept of concentration is usually not applicable to the solid reactant.

Crystalline solids contain defective or boundary regions which are more susceptible to chemical change than identical material elsewhere within the solid. This leads to heterogeneous reaction behaviour where the kinetics are usually not determined by the amount of the reactant available.

The reactions frequently occur only within a thin contact zone, the reaction interface, which advances into the remaining reactant.

The zone undergoing reaction may be the total surface area, as in the case of a sphere, or a specific reaction interface, as in the case of a disc, or even in the case of a porous solid, the pore surface area.

Thus the reaction rate expressions derived for these cases are fundamentally determined by geometry, under the assumption that the surface is homogeneous and the material is isotropic.

This is analogous to the case of graphite where reaction only occurs at the ASA. The classical solid state kinetic expression for the reaction rate (R_{ss}) is given by:

$$R_{SS} = \frac{d\alpha}{dt} = k(T)f(\alpha)P^{m} \tag{2}$$

where α is the dimensionless conversion which may be defined as:

$$\alpha = \frac{m_0 - m_t}{m_0 - m_\infty} \tag{3}$$

where m is the mass of the sample and the subscripts $0,t$ and ∞ denote the initial mass, the mass at any time during the experiment and the mass at the end of the experiment respectively.

For materials with very low or zero ash content the last term may be neglected. The partial pressure of the reactant gas may be incorporated via the term, P^m, with a simple power law dependence of order m.

However for investigations conducted in, for example, pure oxygen this term reduces to one. The term $k(T)$ in eqn. (2) is the temperature dependent reaction rate constant and is usually parameterized through the Arrhenius equation:

$$k(T) = k_0 \exp\left(\frac{-E_A}{RT}\right) \tag{4}$$

The final term $f(\alpha)$ is the conversion function. This term incorporates all physical and chemical transformations which occur as the conversion progresses. For the simple case where the reaction interface alone controls the reaction rate, this expression may be analytically derived, as shown below:

$$f(\alpha) = (1-\alpha)^{\frac{n-1}{n}} \tag{5}$$

where n is equal to three for a sphere and two for a disc reacting only at the edges. It has recently been shown (Badenhorst et al., 2013) that this expression may be generalized to include any geometric shape where the active surface area governs the reaction as:

$$R_{SS} = \frac{d\alpha}{dt} = \frac{ASA_0}{m_0} k^{ASA}(T) f^{ASA}(\alpha) \tag{6}$$

Here $k^{ASA}(T)$ is the reaction rate constant based on active surface area which has units of:

$$\left[\frac{mass\ of\ carbon\ reacted\ per\ \mathrm{second}\ at\ time\ t}{m^2\ ASA\ at\ time\ t} \right]$$

and $f^{ASA}(\alpha)$ is defined as:

$$f^{ASA}(\alpha) = \frac{ASA}{ASA_0} \tag{7}$$

where the subscript *0* denotes the initial ASA. This function conveniently starts at an initial value of one and allows qualitative comparison of the ASA progression as a function of conversion.

This expression is equivalent to the specific reaction rate expression for carbon (R_{sp}) proposed by Radovich and co-workers (Lizzio et al., 1990), based on the expression originally suggested by Walker and co-workers (Laine et al., 1963). Which is given by:

$$R_{\mathrm{sp}} = (1 - X_C)^{-1} \frac{dX_C}{dt} = k\, C_{CS} \tag{8}$$

where X_C is the conversion. The only distinction is a slightly different definition of the reaction rate. Substituting eqn. (3) into eqn. (2) and assuming the residual mass is zero the units of the reaction rate (R_{SS}) is:

$$\left[\frac{mass\ of\ carbon\ reacted\ per\ \mathrm{second}}{mass\ of\ carbon\ at\ the\ initial\ time\ t_0} \right]$$

where as in the case of eqn. (8) the specific reaction rate (R_{sp}) has units of:

$$\left[\frac{mass\ of\ carbon\ reacted\ per\ \mathrm{second}}{mass\ of\ carbon\ at\ any\ time\ t} \right]$$

Hence it can be shown that:

$$R_{SS} * \frac{m_0}{m} = \frac{d\alpha}{dt} * \frac{m_0}{m} = \frac{d\alpha}{dt} * \frac{m_0}{m_0(1-\alpha)} = (1-\alpha)^{-1} \frac{d\alpha}{dt} = R_{sp} \tag{9}$$

The reactive surface area (C_{CS}) in eqn. (8) is defined as the atomic concentration of carbon atoms on which the carbon-oxygen surface complex is formed and actively decomposes to give gaseous products.

If numerator and denominator are multiplied by the molecular mass of carbon, this term has units of:

$$\left[\frac{mass\ of\ active carbon}{mass\ of\ carbon\ at\ any\ time\ t}\right]$$

Then the reaction rate constant (k) has units of:

$$\left[\frac{mass\ of\ carbon\ reacted\ per\ \mathrm{second}}{mass\ of\ active\ carbon}\right]$$

The mass of active carbon can be converted to the ASA by multiplying with the area occupied by a single carbon atom, $A_{C\,ATOM} = 0.083\ \mathrm{nm}^2$ (Laine et al., 1963). Thus:

$$f^{ASA}(\alpha) = C_{CS}\frac{A_{C\,ATOM}\ m_t}{ASA} \tag{10}$$

It should however be noted that the term reactive surface area (RSA) used by Lizzio et al. (1990) is equivalent to the term ASA as used here. The distinction being that the ASA used by Lizzio et al. (1990) is partially occupied by "stable" surface complexes, which do not participate in the reaction. This is relevant if the mass of active carbon is to be measured by TPD and/or if secondary reactions occur between the "stable" and "fleeting" surface complexes. Thus eqn. (6) effectively assumes that all active sites are covered by "fleeting" surface complexes. The expression given by eqn. (6) provides a more convenient basis for the comparison of the oxidative reactivity of graphite samples from different origins. This is because the conversion function $f^{ASA}(\alpha)$, will always start at a value of one and represent the ensuing ASA development. This allows the ASA development of materials to be qualitatively compared and interpreted. In addition, the International Confederation for Thermal Analysis and Calorimetry (ICTAC) has a clear and structured method for performing kinetic computations on thermal analysis data based on the solid state kinetic approach (Vyazovkin et al., 2011).

Specifically iso-conversional methods have been developed and standardized (Vyazovkin et al., 2011) to easily calculate the activation energy for eqn. (4) without any prior knowledge of the conversion function $f^{ASA}(\alpha)$.

4. MICROSTRUCTURE OF GRAPHITE

The idealized microstructure of graphite is a highly crystalline, lamellar flake, such as the natural graphite flake shown in Figure 8.

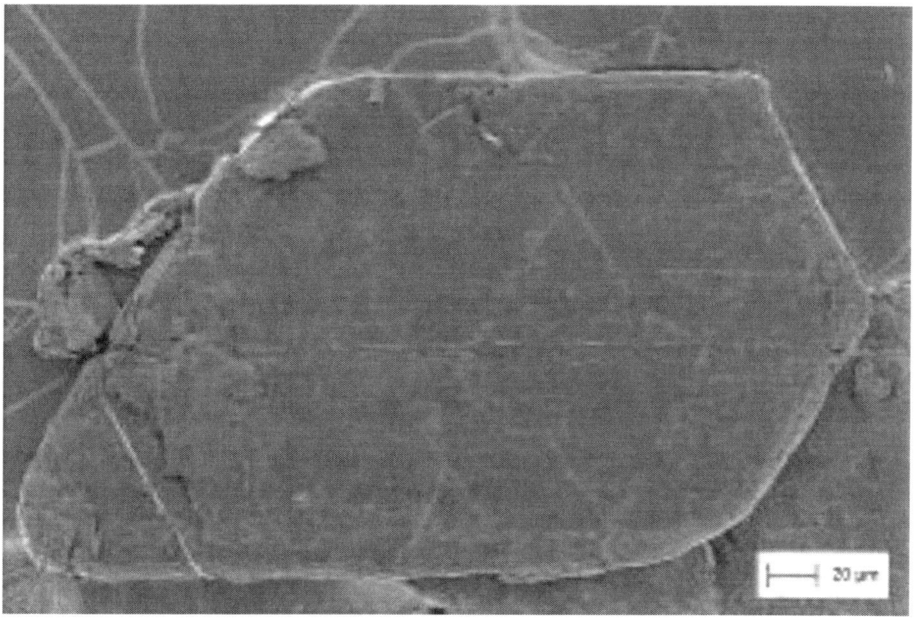

Figure 8. Graphite flake.

Figure 9. Graphite flake edge.

When examined edge on, the layered structure of the graphite is immediately evident, as can be seen from Figure 9.

The closest approximation to this ideal structure is expected for mined, natural graphite flakes (Wissler, 2006). Despite this, the microstructures found in these materials often differ significantly from this ideal shape, for example the potato shaped graphite particle shown in Figure 10.

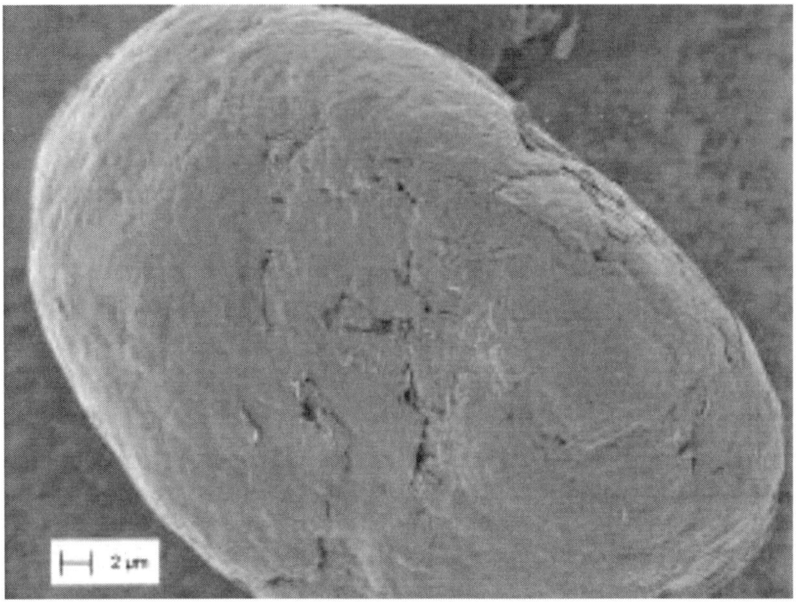

Figure 10. Potato shaped graphite particle.

Figure 11. Sonicated graphite particle.

These particles are a consequence of the milling step during beneficiation. Graphite is highly malleable and even a light sonication can damage the flake structure. This can be seen in Figure 11 where even after only a few minutes of sonication, curling is evident along the flake edge.

However, in most cases some level of pre-treatment, such as sieving, is required for graphite powders since these inevitably have a distribution of particles which tend to agglomerate and clump together as shown in Figure 12.

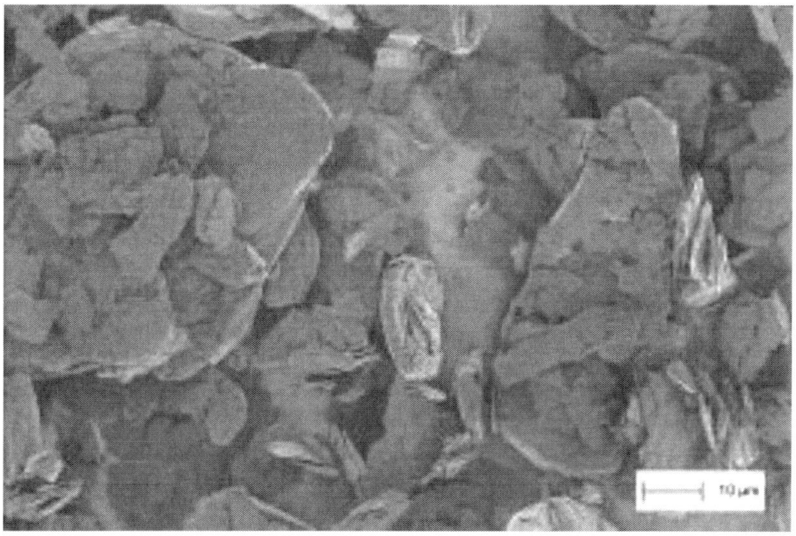

Figure 12. Natural graphite particle distribution.

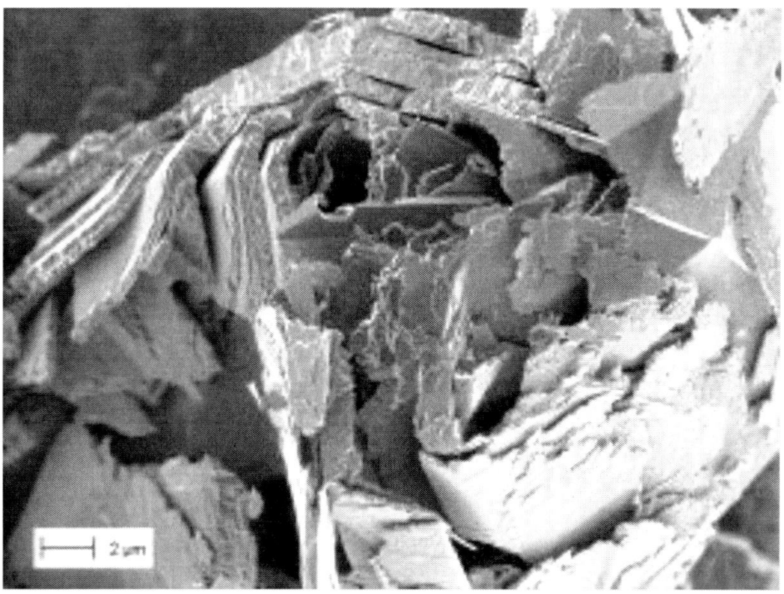

Figure 13. Oxidized potato shaped graphite particle.

Such a particle distribution would result in a skewing of the obtained conversion function (Badenhorst et al., 2012). When oxidized the complex underlying microstructure of the potato shaped or milled graphite particles become evident, as seen in Figure 13.

It is clear that the ASA development of this particle would be vastly different from the ideal graphite flake shown in Figure 8. The situation becomes even more complex when synthetic graphite is examined. Synthetic graphite is produced via a multi-step, re-impregnation process resulting in very complex microstructures and porosity (Pierson, 1993). During the production of most of the high quality graphite materials mentioned earlier, a premium grade filler material known as needle coke is utilized. The needle coke has distinctive, elongated, needle-shaped particles as a direct consequence of the delayed coking process and the use of selected feed stocks with low levels of insoluble impurities. This filler is then mixed with a binder which can be either a coal tar or petroleum derived pitch. The pitch is in a molten state when added and the mixture is then either extruded or moulded. The resulting artefact can then be re-impregnated with pitch after carbonization if a high density material is required. In the case of synthetic graphite powders, it is possible to discern between particles derived from these two materials, for example the needle coke derived particle shown in Figure 14.

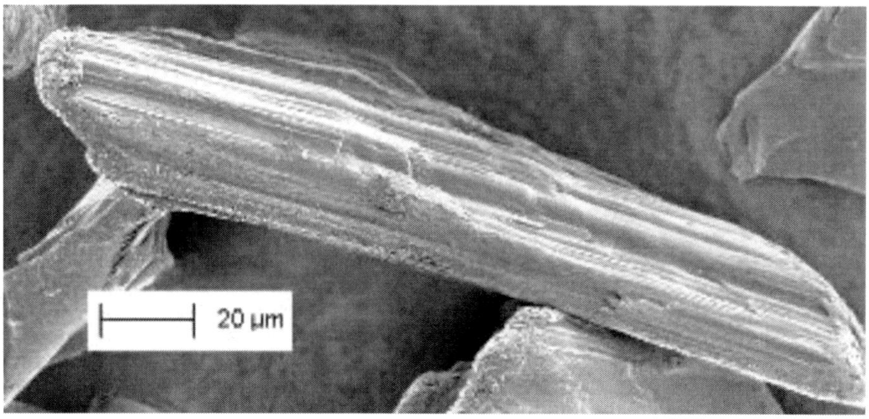

Figure 14. Needle coke graphite particle.

When examined edge on, the layered structure of the graphite is still evident. Although in this case the layers are severely distorted and resembles a stack of paper which has been repeatedly folded lengthwise, as can be seen in Figure 15.

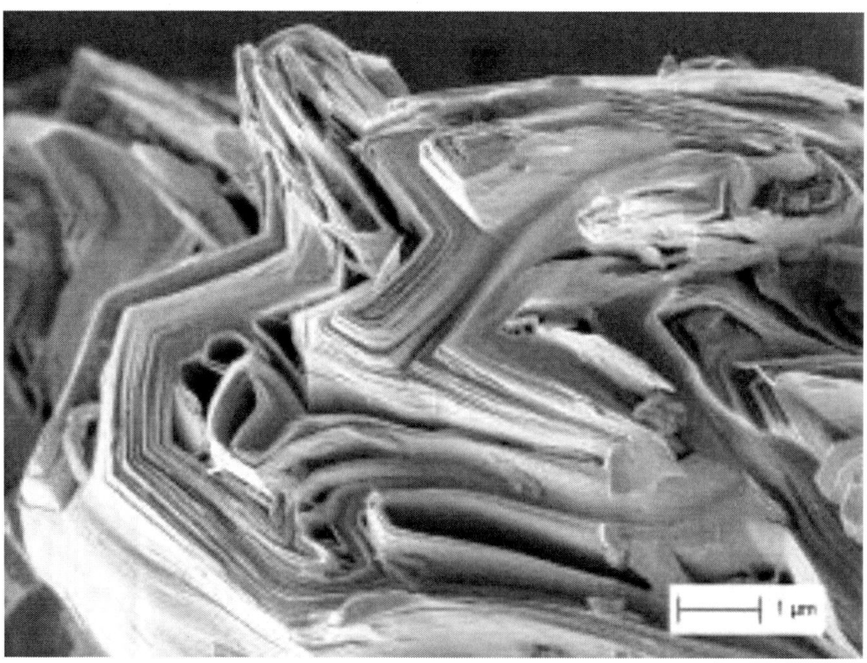

Figure 15. Needle coke graphite particle edge on.

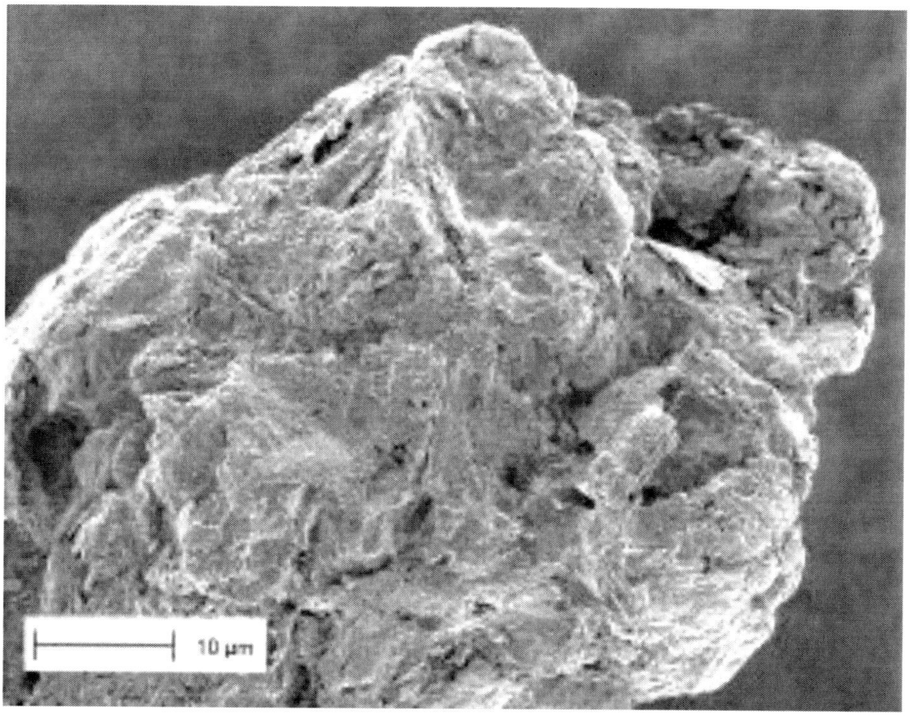

Figure 16. Pitch graphite particle.

The pitch was in the liquid phase during impregnation and hence is expected to result in a complex behaviour. A particle derived from the pitch is shown in Figure 16.

When examined more closely in Figure 17, the highly disordered, characteristic mosaic textures of these particles are evident, as a result of the flow phenomena during impregnation.

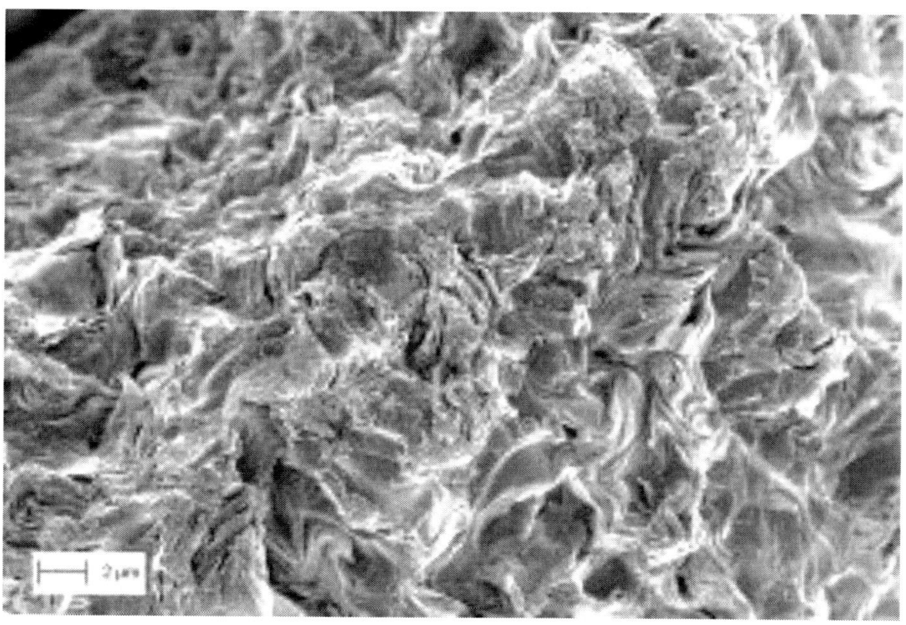

Figure 17. Close-up of pitch graphite particle.

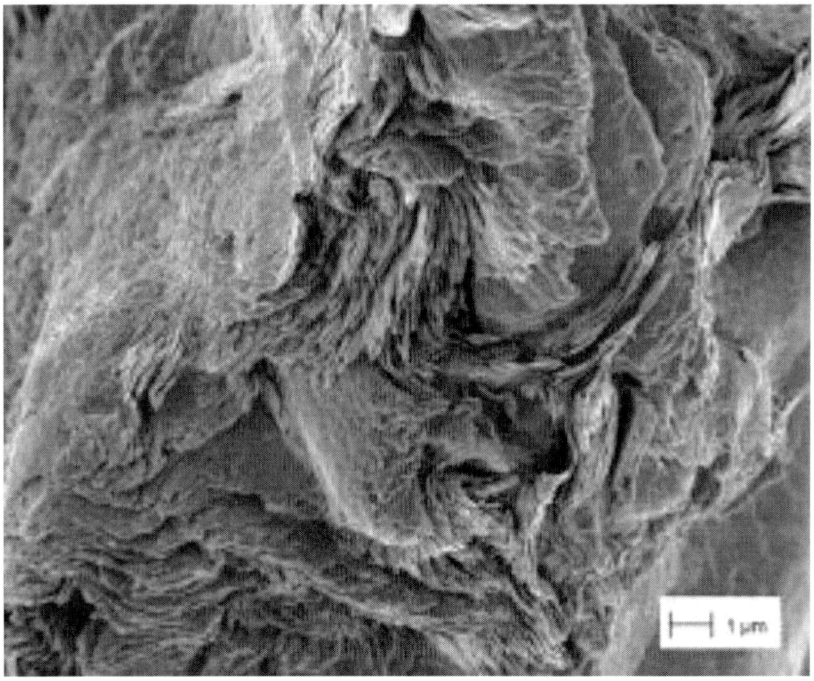

Figure 18. Complex microstructural development in synthetic graphite.

It is clear that when oxidized, the ASA development of these particles will again be vastly different from the ideal flake shown in Figure 8. In fact in most cases, very complex microstructural development takes place as can be seen from Figure 18.

When the XRD spectra for a natural graphite material (NG) and a synthetic graphite sample (SG) are compared (both purified and approved as nuclear grade), they appear very similar as shown in Figure 19.

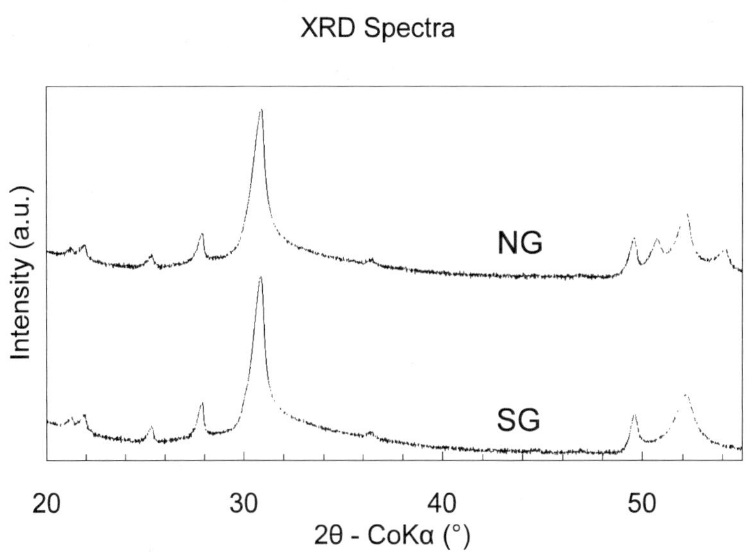

Figure 19. XRD spectra of natural and synthetic graphite.

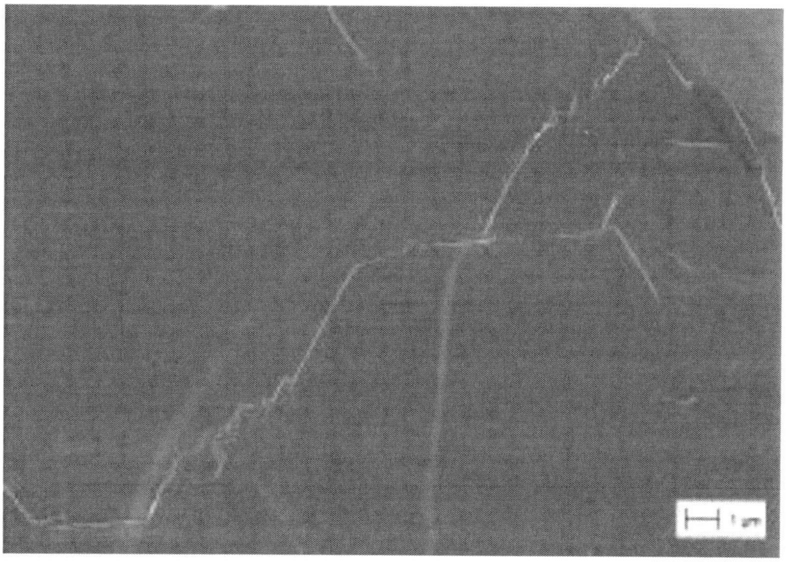

Figure 20. Oxidized natural graphite basal plane.

The only noticeable difference being the presence of the rhombohedral phase in the natural graphite spectra (Parthasarathy, 2006). Using the empirical Scherrer equation (Warren, 1941) it is possible to calculate the crystalline domain sizes for the two samples from their XRD spectra, in order to get a quantitative idea of the differences between their respective levels of crystallinity. It is found to be 38 nm for the natural graphite and 26 nm for the synthetic sample. As expected the result is higher for the natural material, but the difference would appear to be marginal. When the oxidized basal plane of the natural graphite is examined, the results are in accordance with those expected for a highly crystalline graphite, as can be seen from Figure 20.

The basal plane is intact with only some slight layering effects. When examined edge on in Figure 21, the results are again as expected.

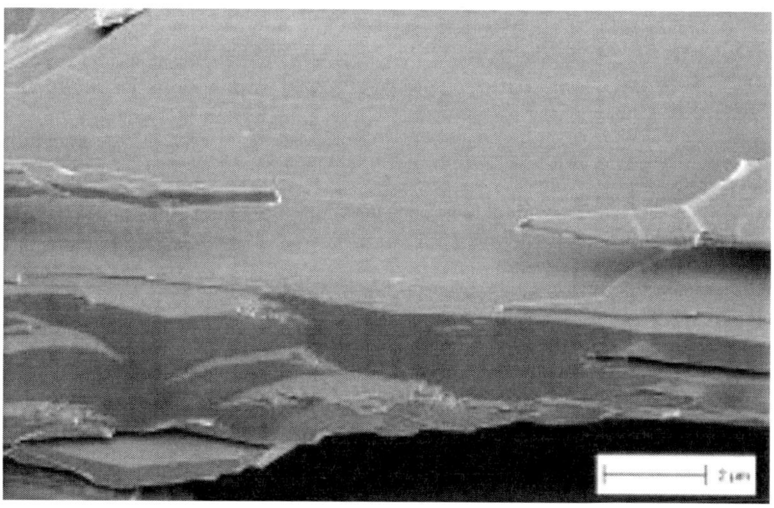

Figure 21. Oxidized natural graphite edge.

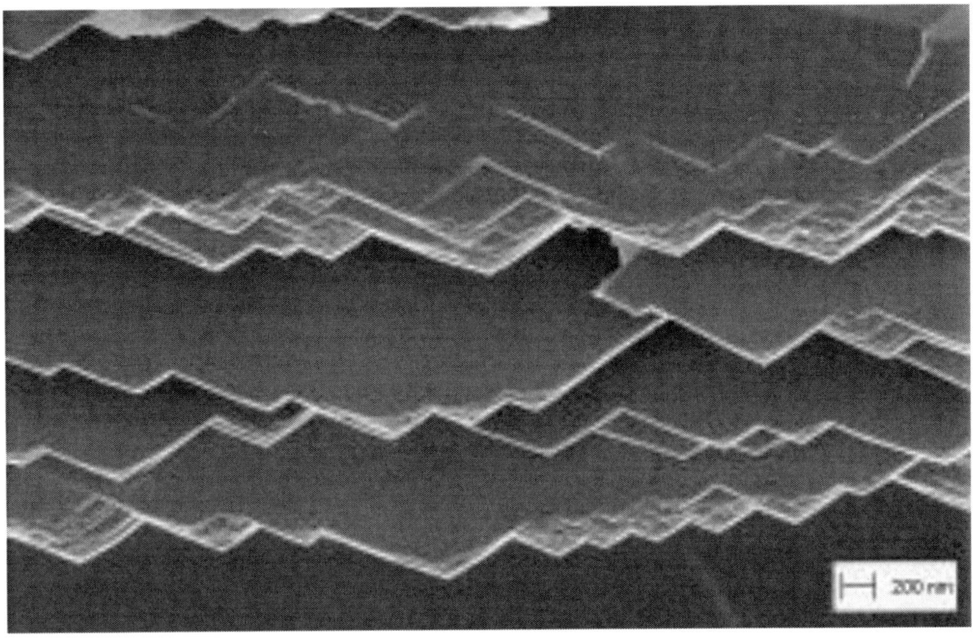

Figure 22. Oxidized natural graphite edge viewed from above.

The edge is oxidized fairly uniformly and remains intact across a couple of micron width. Also when examined from above the underlying crystallinity is evident as shown in Figure 22.

The edges form perfect 120° angles, consistent with the internal angles of a hexagon, as expected from the graphite crystal structure. However, this is in stark contrast to the synthetic graphite edges, seen in Figure 23.

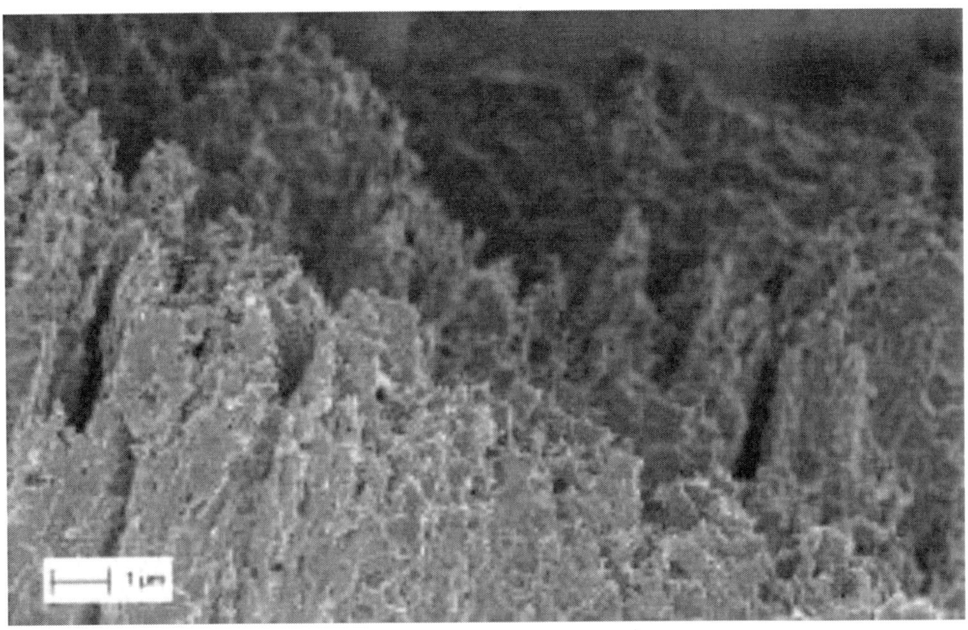

Figure 23. Oxidized synthetic graphite edge viewed from above.

Figure 24. Oxidized synthetic graphite edge.

The direction of the basal plane of this needle coke derived graphite particle is still discernable, but the smooth, layered edge has been disintegrated. When examined edge on the surface is irregular and regions where the edge face remains intact are very small, no more than a couple of hundred nanometers, as confirmed by Figure 24.

When examined more closely, the basal plane of the synthetic material is highly degraded. As can be seen in Figure 25 the surface is heavily roughened, indicating that attack by oxygen was possible virtually anywhere on the basal surface.

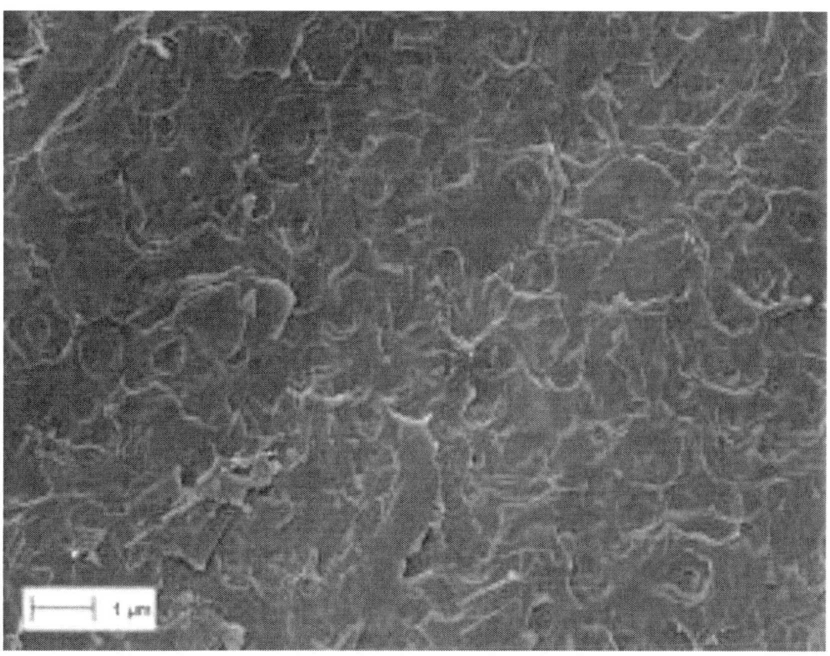

Figure 25. Oxidized synthetic graphite basal plane.

The behaviour of the synthetic graphite is very different from the natural material. This is not expected when considering the similarity of the XRD spectra of both materials. The behaviour of the natural material is entirely consistent with that expected for highly crystalline graphite. The synthetic graphite on the other hand exhibits very low levels of crystalline perfection and is expected to have an entirely different conversion behaviour despite the apparent similarity in measured crystalline parameters. Thus the structure of synthetic graphite may be more accurately represented by the amorphous structure shown in Figure 5, with comparatively small regions of high crystalline order arranged throughout the structure. Such a molecular structure clearly has a vastly different oxidative reactivity compared to the ideal crystalline structure. Hence it remains quite necessary to examine and understand the microstructural evolution taking place in a specific graphite sample in order to sensibly compare the reactivity.

5. Influence of Catalytic Impurities

This section will focus mainly on the impact of catalytic impurities on the microstructural evolution, rather than the direct impact on the reaction rate. The exact mechanism at the carbon–metal interface is still disputed (Ci et al., 2008) for a variety of gases. The theories associated with the catalytic mechanisms may be broadly classified into two categories (Tomita, 2001). The gas spill–over mechanism whereby the gas species adsorbs and dissociates on a metal catalyst. These atoms migrate through or on the particle to nearby carbon atoms and gasification occurs. In the alternative carbon dissolution mechanism, carbon dissolves in the catalyst and migrates through or over the particle to the surface where reaction with the gas takes place. A myriad of catalytic behaviours may be observed in a natural graphite sample, despite extensive purification. The variety of catalytic behaviours shown in Figure 26 were observed in a natural graphite concentrated, to a purity of >99.9 %wt, using both an acid wash and a high temperature soda ash burn purification. Regardless of the high sample purity the catalytically active material still has a clear and pronounced impact on the observed microstructure. In general the observed catalyst behaviours can be classified into four categories based on their impact on the graphite microstructure. The first and most easily recognizable is the random, erratic edge roughening visible in Figure 27.

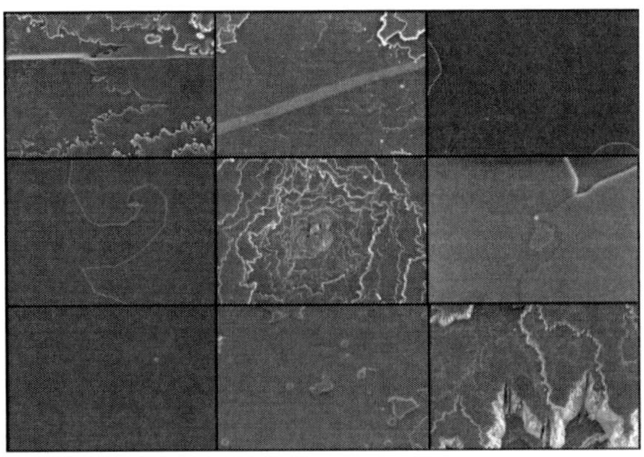

Figure 26. Catalytic presence in purified natural graphite.

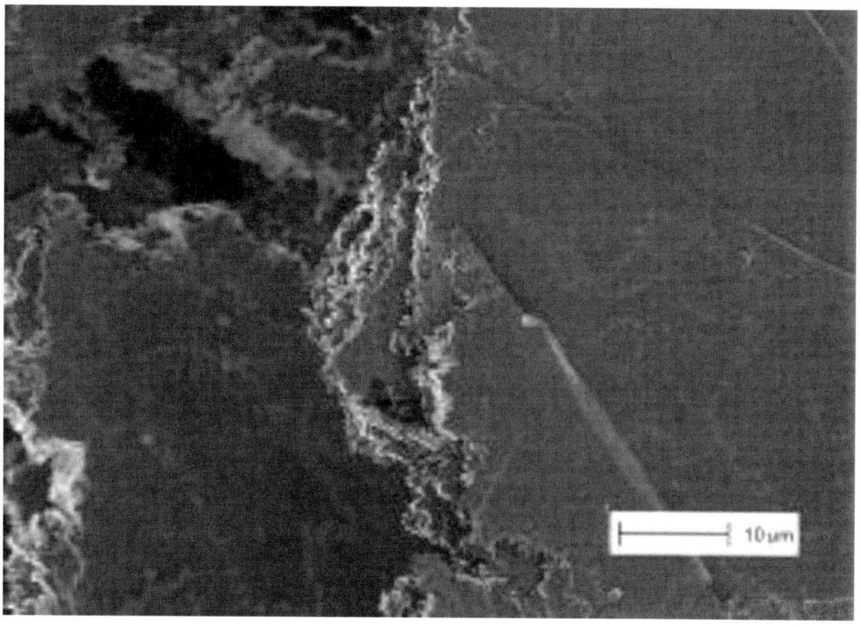

Figure 27. Random, erratic edge roughening.

Upon closer inspection it may be seen that these edges are comprised of multiple, jagged channels traced in the graphite by numerous small particles, which coalesce and overlap, as seen in Figure 28.

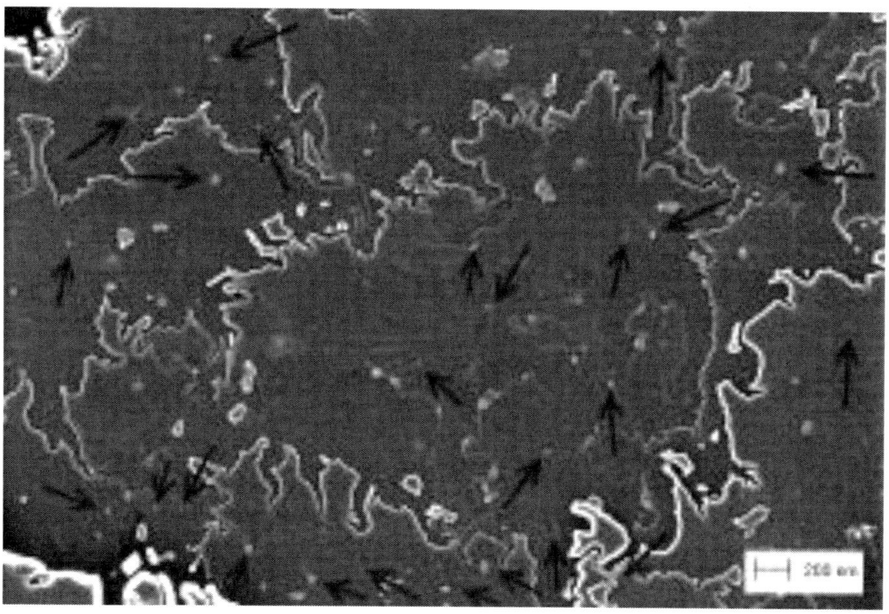

Figure 28. Close-up of roughened edge.

When individual particles are examined as in Figure 29, they can be seen to have a rough, irregular shape and move in channels through the graphite with curved and unpredictable shapes.

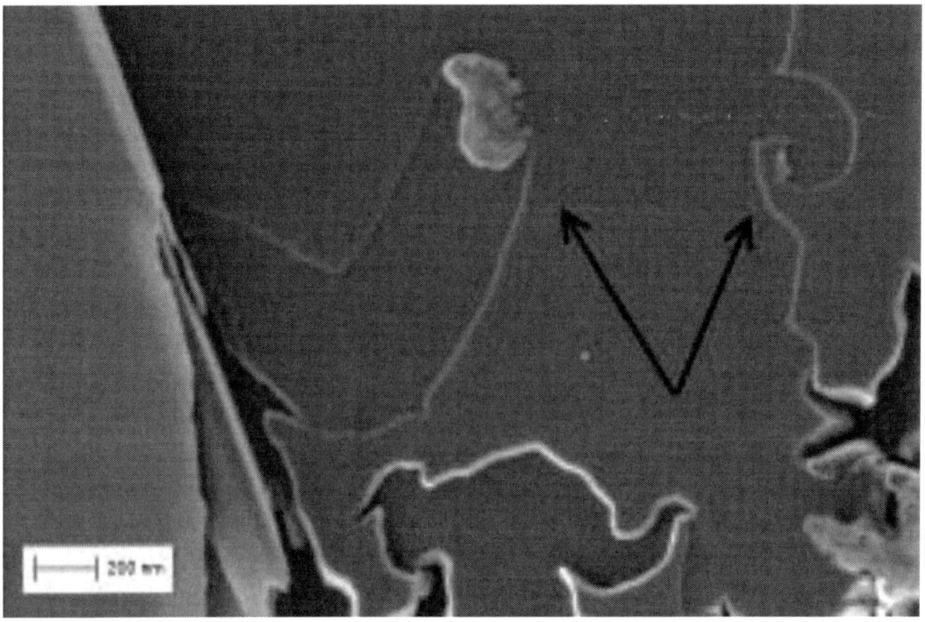

Figure 29. Randomly channelling catalyst particles.

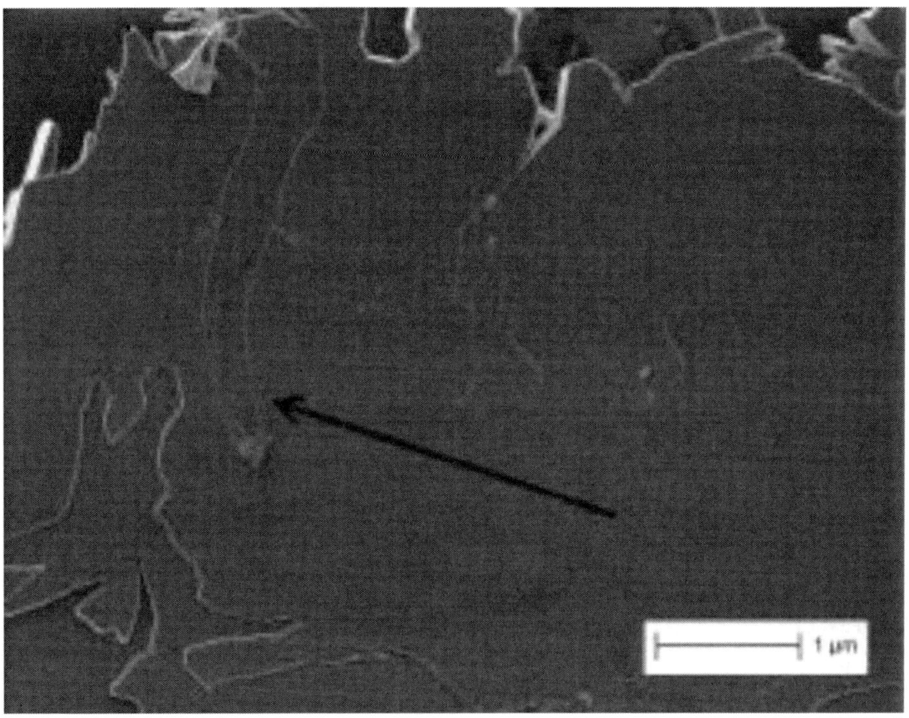

Figure 30. Catalyst particle channelling across two distinct levels.

In some cases the random shape of these particles even allow the tracing of a single channel with two distinct levels, such as the one shown in Figure 30.

Alternatively catalyst particles are found which consistently channel along preferred directions, presumably associated with the underlying crystal structure, as seen in Figure 31.

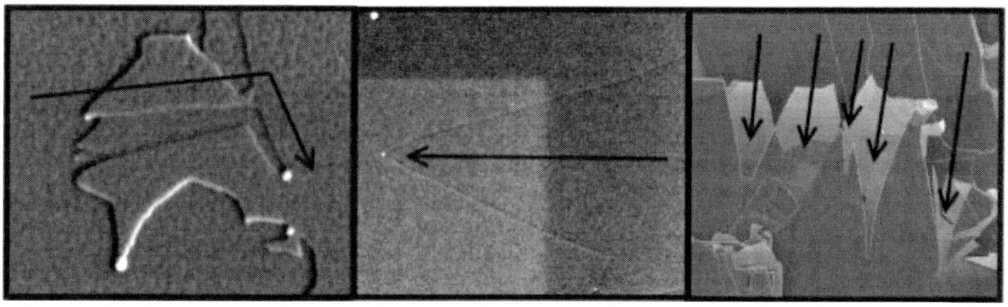

Figure 31. Catalytic channelling along preferred directions.

Often these particles can be seen to execute predictable 120° angle turns, as seen in Figure 32. It is thought that these turns occur at lattice vacancies or other defects (Ci et al., 2008).

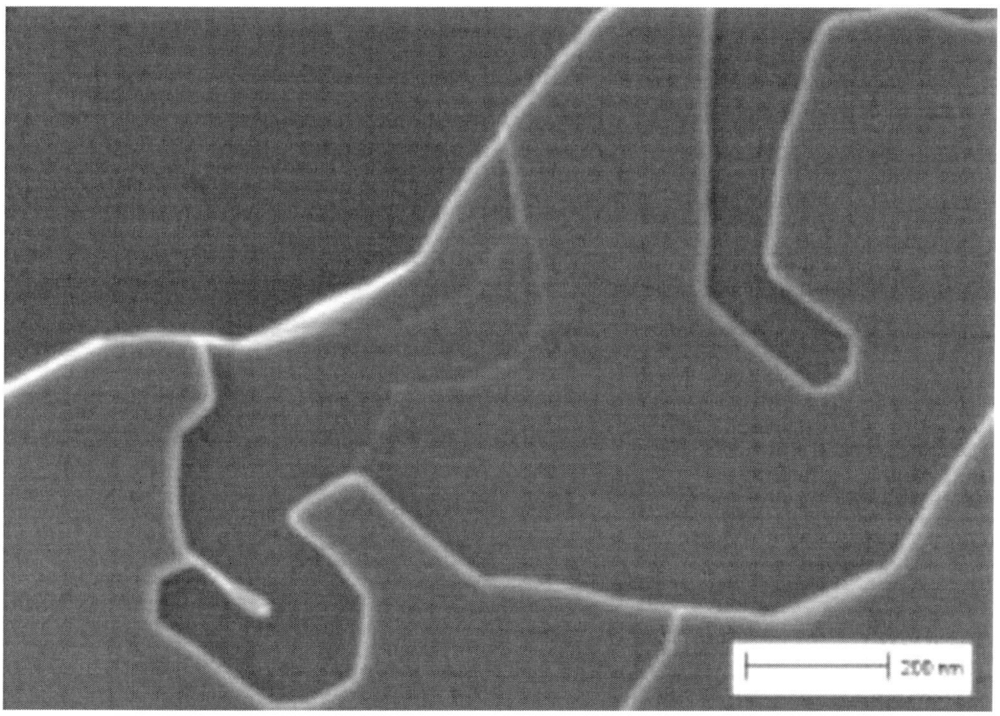

Figure 32. Catalyst particles turning along crystal directions.

The main impact of these particles is the creation of large, triangular shaped channels along the graphite edge, such as those seen in Figure 33.

The insert in Figure 33 shows that the catalyst particle responsible for the channel has a diameter of less than 20 nanometers.

The figure serves to demonstrate the immense impact, extremely low catalyst concentrations can have on graphite reactivity. A single nanometer sized catalyst particle has created several micron of additional surface area where uncatalyzed oxidation can proceed, greatly increasing the overall reaction rate. The third type of catalyst behaviour generally observed is pitting, as shown in Figure 34.

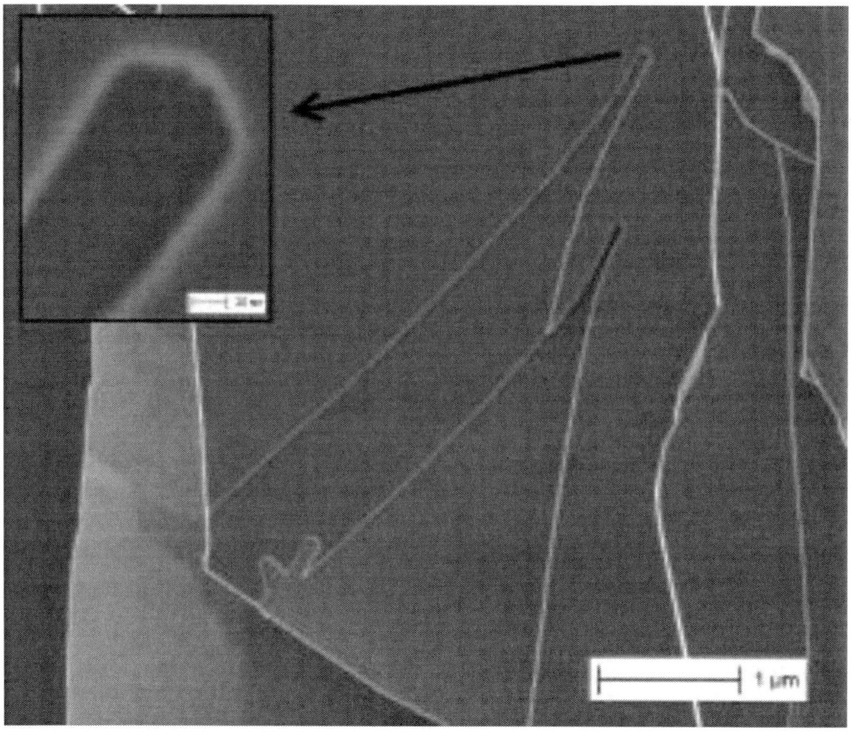

Figure 33. Triangular channels along graphite edge.

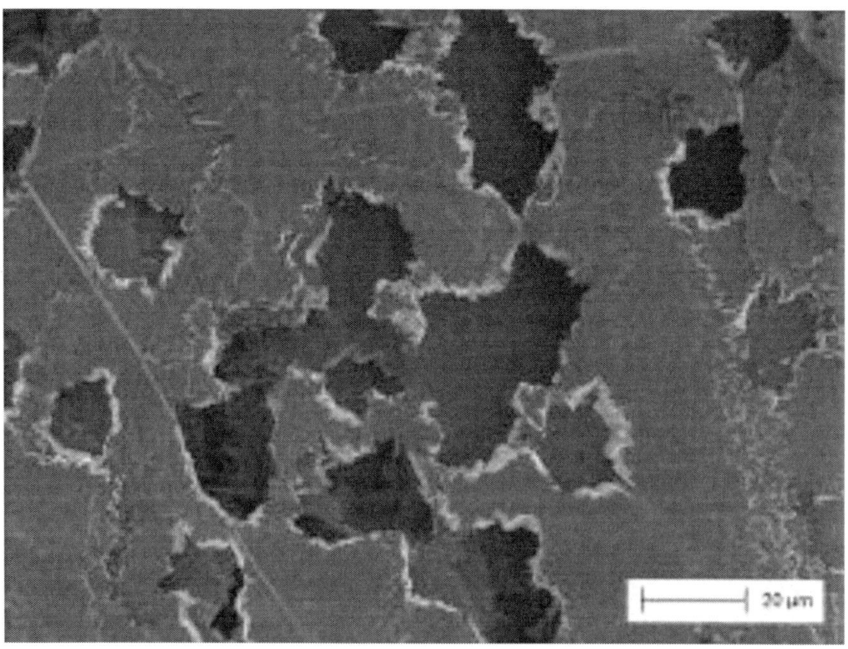

Figure 34. Catalytic pitting.

In some cases, randomly shaped catalyst particles may be found as the origin of irregular, roughly circular pits, as seen in Figure 35.

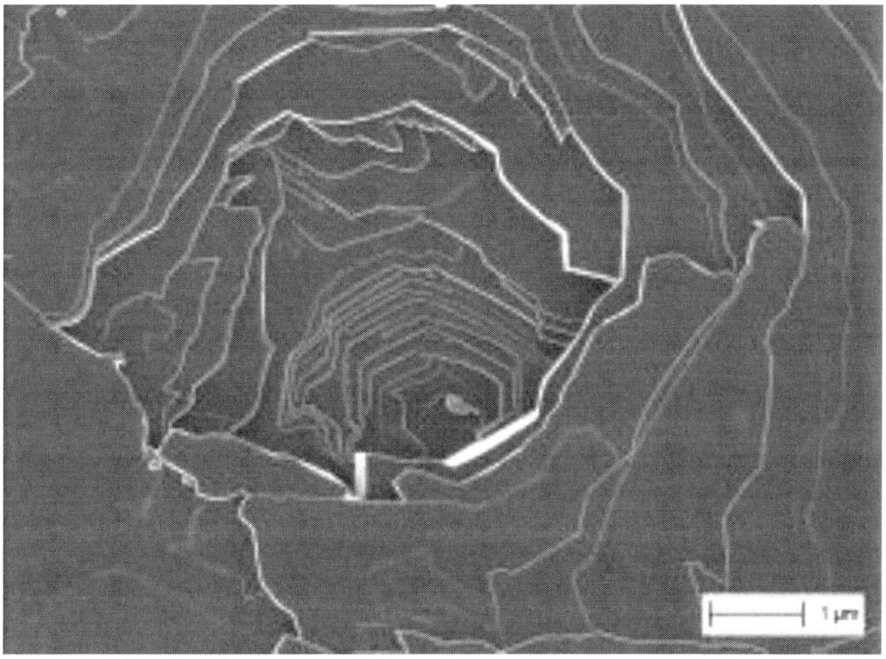

Figure 35. Irregular, roughly circular pits.

However on other occasions, catalyst particles may be found which have resulted in regular, hexagonally shaped pits, such as the one shown in Figure 36.

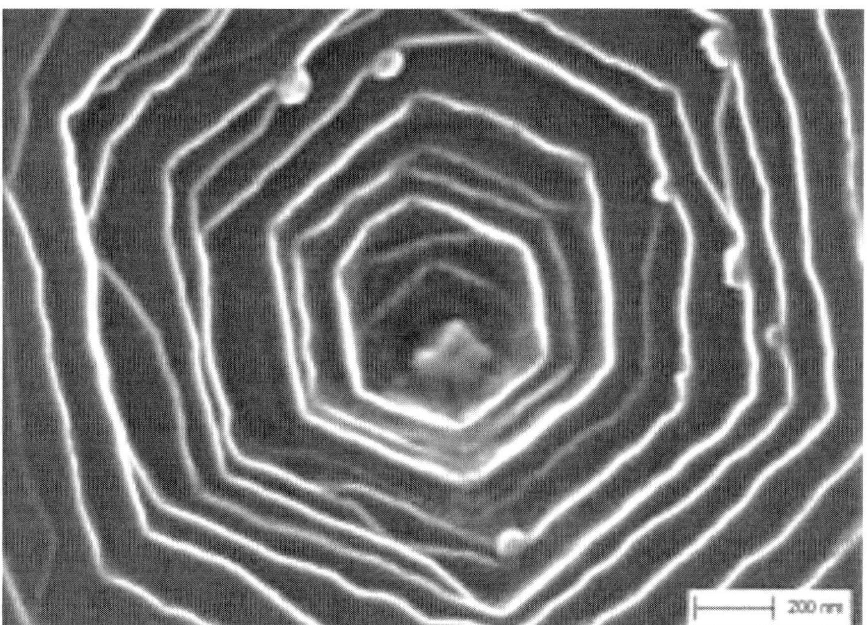

Figure 36. Regular, hexagonal pits.

The latter pits may also on occasion be found with no associated catalyst particle but the characteristic corkscrew shape of a screw dislocation, as can be seen in Figure 37.

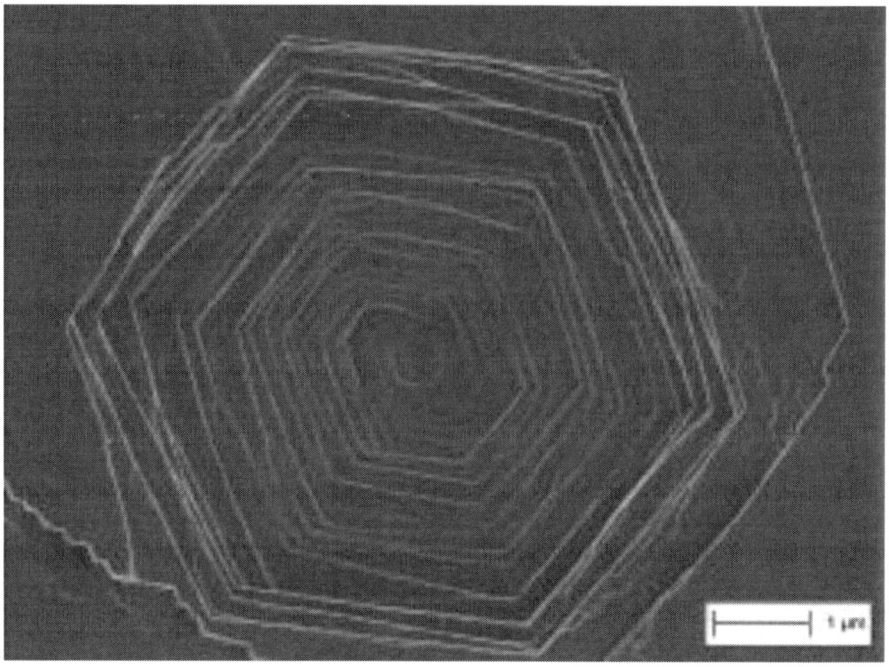

Figure 37. Hexagonal pit caused by a screw dislocation.

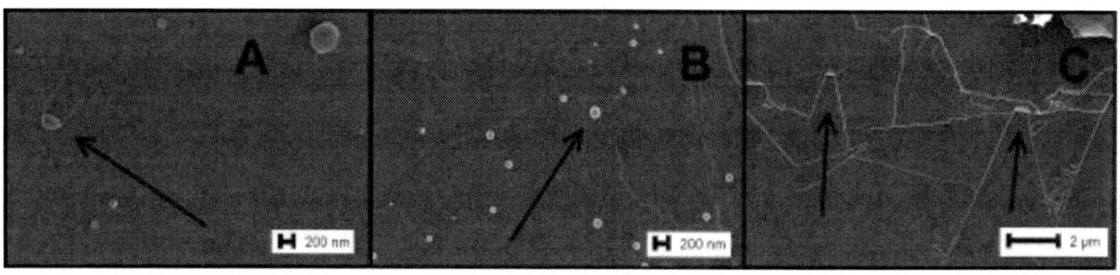

Figure 38. Unclassified catalyst behaviours.

The final group includes behaviours which cannot be readily included in the previous categories, for example those shown in Figure 38.

The catalyst particle in Figure 38 A is undergoing spreading as channelling proceeds, depositing catalyst material all along the channel wall, presumably eventually depleting itself. Whilst in Figure 38 B a catalyst particle appears to have stalled, with uncatalyzed oxidation proceeding from its final position in a circular shape, this may be the result of particle agglomeration leading to de-activation. In Figure 38 C the peculiar behaviour is observed where catalyst particles encounter a different graphite edge and proceed to channel under the graphite surface. Evidently the layer height must be of a specific dimension since the particle on the left encountered a layer prior to its final position where tunnelling did not occur. Furthermore, judging from the channel shading in the image, the channel depth from the surface appears to increase towards the channel tip (as indicated by the progressively darker shading as less electrons penetrate the thin film). This may indicate that the particle is on a downward trajectory or, more likely, the channel is in fact wedge (or pyramid) shaped. All of the discussed catalytic behaviours have a distinct effect on the microstructural development and hence also on the reaction rate.

6. Microstructural Impact of Inhibiting Impurities

A final influence on the microstructural development during oxidation is the impact of inhibiting impurities. This section does not include the effect of oxidation inhibition on the chemical level by reactants which bind to the carbon or scavenge the oxygen. Instead focus is placed on impurities which are unaffected by the oxidation and impact the microstructural development by blocking oxidative attack. The most noticeable macroscopic feature of this effect are the saw-tooth edge formations seen in Figure 39.

Figure 39. Saw-tooth edge formations due to inhibiting impurities.

Upon closer inspection of these formations it is found that the tip of each formation is covered by a single particle, as is visible in Figure 40.

Figure 40. Close-up of inhibiting impurities.

Thus these formations are caused by inhibiting particles which shield the underlying graphite from attack. These layers shield subsequent layers leading to the formation of pyramid like structures topped by a single particle. In some cases smaller particles will accumulate at the graphite edges. Then as oxidation proceeds they are forced into straight lines, forming inhibition ridges as seen on the left and right hand side of Figure 41 respectively.

Figure 41. Formation of inhibition ridges.

In some cases these ridges will accumulate across an entire exposed edge face, as shown in Figure 42.

Figure 42. Edge face covered in inhibition ridges.

The circular catalytic impurities observable in images in the previous section, most likely liquid at the reaction temperature, shown in Figure 43, are also capable of inhibiting oxidation.

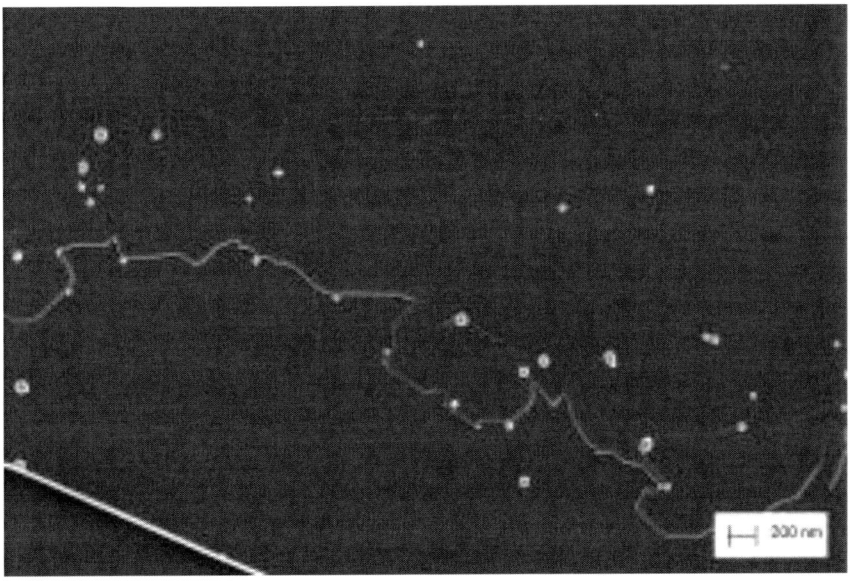

Figure 43. Circular catalytic impurities.

As mentioned earlier and demonstrated in Figure 38B, these impurities appear to agglomerate and deactivate. Inevitably as the conversion proceeds and the flake shrinks, these particles end up at the edges, as observable in Figure 44.

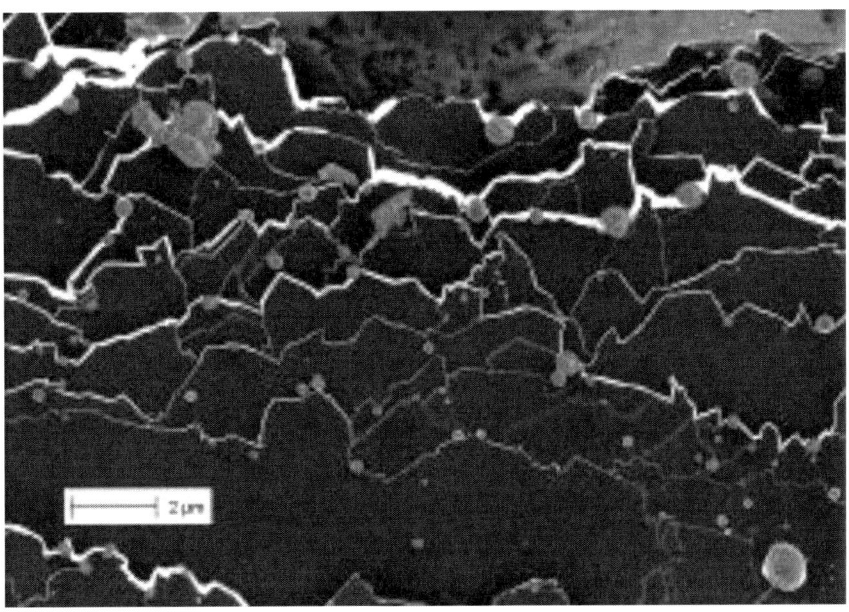

Figure 44. Edge accumulation of deactivated catalytic impurities.

Finally as more of these particles accumulate at the edges, they agglomerate further until their sheer size is capable of shielding an entire edge, as seen in Figure 45. They do not however create the same saw-tooth edge structures (also visible in the left hand side of Figure 41), possibly due to their liquid state during oxidation.

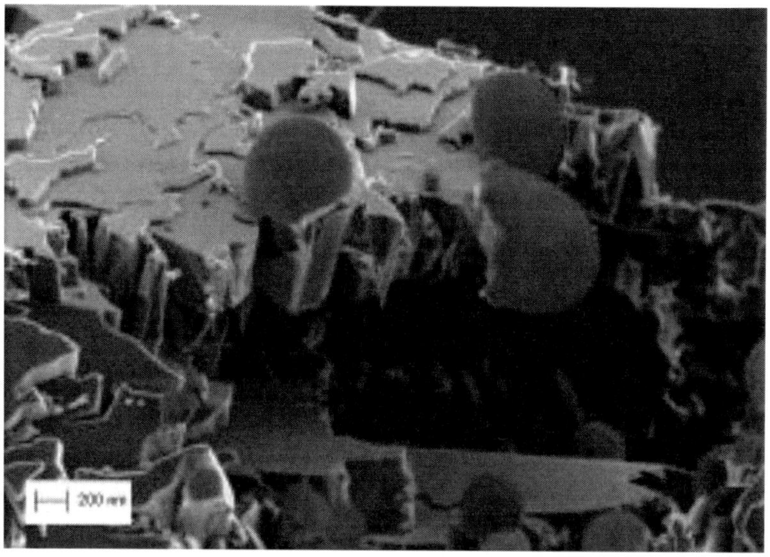

Figure 45. Edge coverage of liquid impurities.

Due to the small size of these impurities and initial homogenous distribution, they are not expected to impact the reaction significantly during the early stages of oxidation. However, at very high conversions these impurities will significantly impact the reaction rate, by reducing the available ASA.

7. Simulating Microstructural Development in Graphite

The simplest representation for the oxidation of highly crystalline graphite flakes is that of a shrinking disc, given by $n=2$ in eqn. (5). This behaviour is plotted in Figure 46.

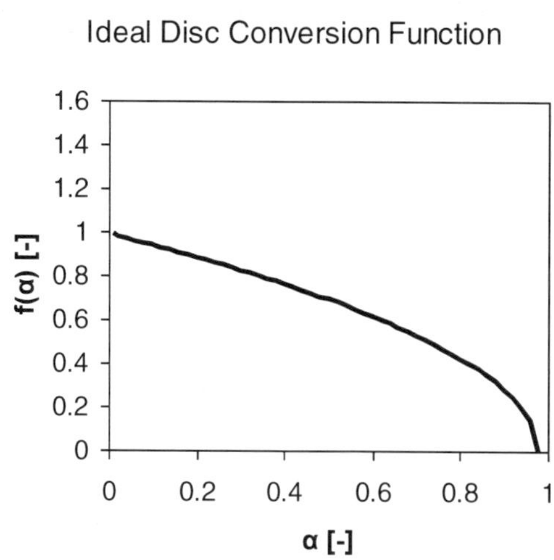

Figure 46. Ideal disc conversion function.

However as can be seen from Figure 47, where the conversion functions for a variety of powdered graphite samples are plotted, this behaviour is not representative of real graphite behaviour.

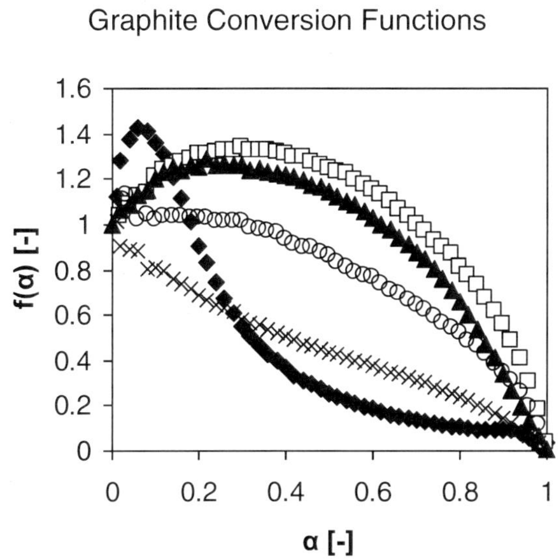

Figure 47. Ideal disc conversion function.

This is not unexpected, given the complex microstructural and catalytic effects discussed in the previous sections. Understanding the relative influence of these factors is difficult, especially when attempting to ascertain the rate controlling effects. In addition, the convoluted reaction schemes mentioned in section 3 further complicate the situation. These provide multiple complex pathways for reaction with no *a priori* method of choosing between them. Instead consider the pitting behaviour mentioned earlier, shown again in Figure 48.

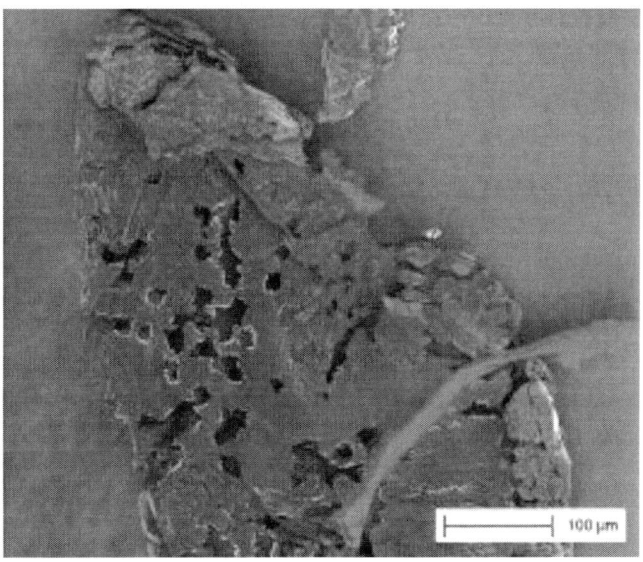

Figure 48. Pit development in graphite flakes.

As these pits grow they create additional active surface area whilst the circumference of the flake recedes. Thus it is expected that the ASA will initially increase until the holes overlap and coalesce, at which point the overall ASA will decline rather rapidly.

This is confirmed by the conversion behaviour for this flake natural graphite which is plotted in Figure 49.

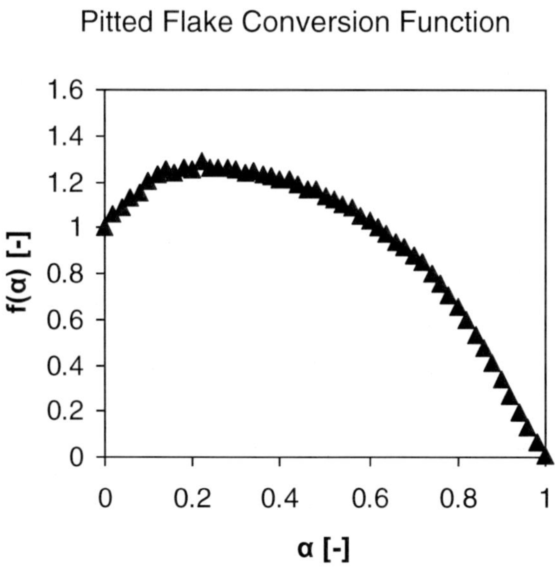

Figure 49. Conversion function for pitted graphite flakes.

The foremost obstacle in representing this behaviour is the random nature of these phenomena. The holes created by the initial pitting, are randomly distributed across the flake. It is impossible to analytically represent such a structure. Instead a block cellular automaton (Badenhorst, 2013) may be used to qualitatively simulate the ASA progression of such a structure. To do this the simulated structure is divided into discrete sub-elements, such as the circular disc shown in Figure 50.

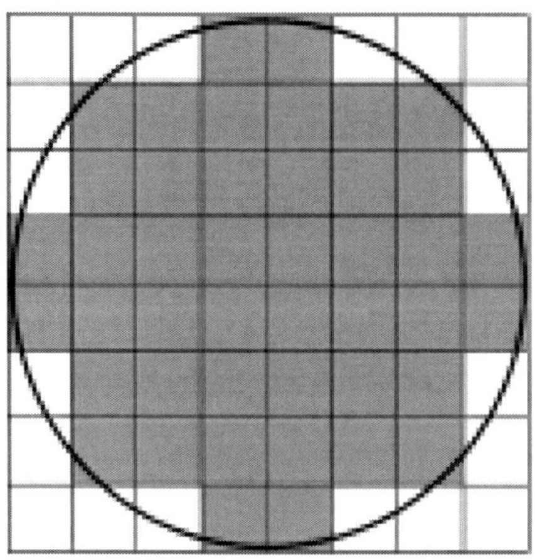

Figure 50. Discretized representation of a disc.

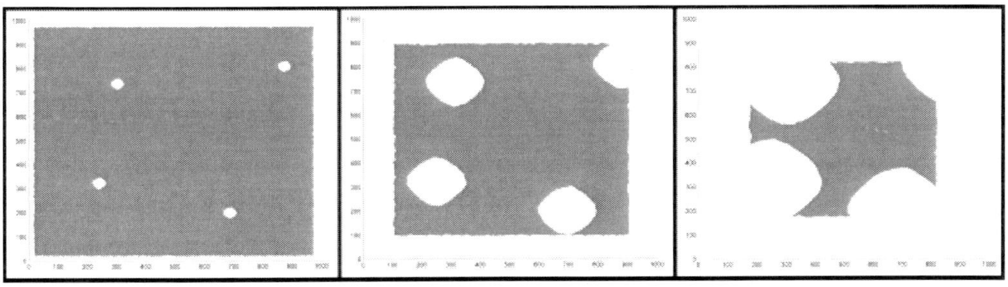

Figure 51. Simulation of pit formation in square flake.

By choosing a given reactivity and using a random number generator the progression of any arbitrary shape can be followed over time, such as the square flake with four random pits shown in Figure 51.

The random placement of the holes and the large number of flakes present in a given sample, make it necessary to repeat and average the simulations to get a comparable representation. The average for ten repeats of the simulated structure shown in Figure 51 is given in Figure 52, compared to the conversion function for the real graphite sample.

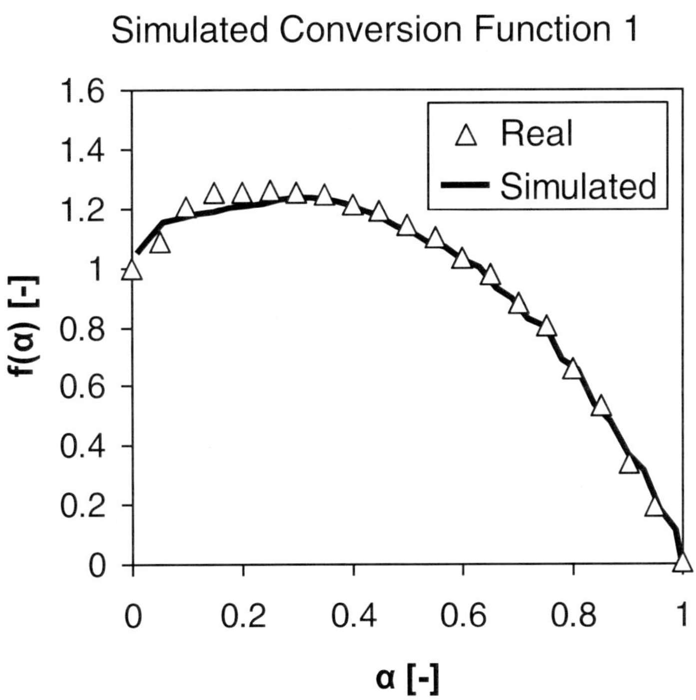

Figure 52. Simulated conversion function for pitted structure.

The simulation is in good agreement with the observed conversion behaviour. Thus it is possible to interpret the real conversion behaviour and represent it qualitatively using a simple simulation without the need for complex reaction schemes. If this graphite is purified using a high temperature heat-treatment, the presence of pitting is removed. Instead a fine edge structure is created seen in Figure 53, presumably where impurities present initially were removed during the treatment.

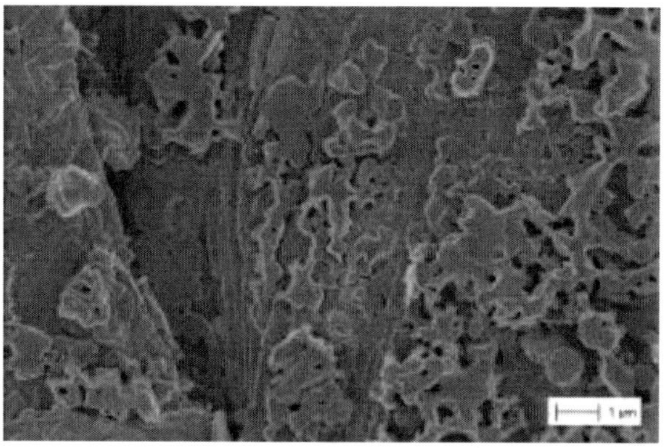

Figure 53. Fine edge structures in purified graphite.

The conversion function for this material is shown in Figure 54.

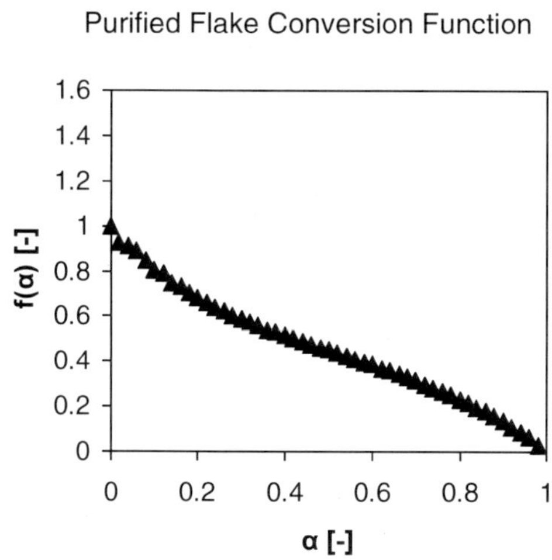

Figure 54. Conversion function for purified graphite flakes.

The fine edge structures represent small, localized regions of high active surface area which are rapidly removed, leaving behind the macro flake body. This was approximated as a flake structure with a variety of small, randomly shaped notches at the edge, as shown in Figure 55.

Figure 55. Simulation of fine edge structures.

This simulated conversion function is compared to the real, experimentally observed conversion function for this purified graphite sample in Figure 56.

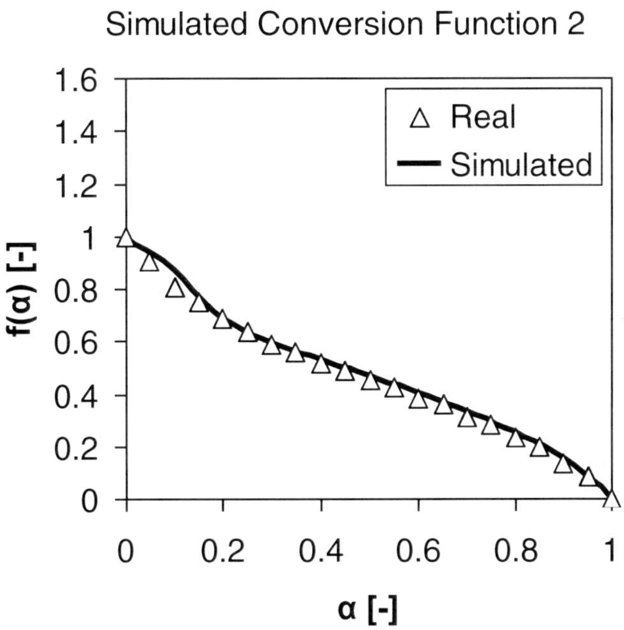

Figure 56. Simulated conversion function for purified graphite flakes.

Again the agreement between the simulated and real conversion functions is good. Thus this approach allows sensible comparison of graphite samples with different conversion behaviours, from different origins. However, finding accurate and representative simulations for all the possible microstructures found in graphite remains a complex and arduous task.

Conclusion

Graphite is considered one of the structurally simplest allotropes of carbon, available at a reasonable cost. As such it provides an ideal starting point for understanding the reaction of carbon with oxygen. This allows a clear insight into the oxidative properties of different graphite materials and ultimately the rate controlling factors. Amongst these factors the active surface area is known to be of vital importance. However, as demonstrated, the microstructural development during the oxidation of materials classified as graphite is a very complex subject with a wide variety of possible behaviours. This is further complicated by the differences visible in materials with apparently similar levels of crystallinity. The microstructural development is influenced by a diverse assortment of causal factors such as the material fabrication process, sample origin, catalytically active impurities and impurities which inhibit the oxidation. A wide variety of complex kinetic schemes have been proposed to explain the oxidation reaction. But these are derived from intricate surface processes which are difficult to characterize *in situ*. Recent schemes are largely based on arbitrary molecular simulation with very little justification for the selection of the chosen reaction pathway.

Furthermore the idealized behaviour of a shrinking disc does not accurately represent the behaviour of even high purity, highly crystalline natural graphite. Qualitative simulation of the more simplistic microstructures found in real graphite indicates that the microstructure is responsible for the observed behaviour without the need for arbitrary kinetic reaction schemes. However, finding accurate and representative simulations for all the possible microstructures found in graphite remains a complex and arduous task.

REFERENCES

Ahmed, S.; Back, M.H. *Carbon* 1985, 23, 513-524.

Allouche, A.; Ferro, Y. *Carbon* 2006, 44, 3320-3327.

Backreedy, R.; Jones, J.M.; Pourkashanian, M.; Williams, A. Faraday *Disc.* 2001, 119, 385-394.

Badenhorst, H, Comput. *Chem. Eng.* submitted Nov 2012.

Badenhorst, H.; Focke, W.W. *J. Therm. Anal. Calorim.* 2012, 108, 1141-1150.

Badenhorst, H.; Rand, B.; Focke, W.W. *Thermochim. Acta* submitted Nov 2012.

Bhatia, S.K.; Vartak, B.J. *Carbon* 1996, 34, 1383-1391.

Brown, T.C.; Haynes, B.S. *Energy Fuels* 1992, 6, 154-159.

Campbell, P.A.; Mitchell, R.E. *Combust. Flame* 2008, 1, 47-66.

Chen, N.; Yang, R.T. *J. Phys. Chem.* A 1998, 102, 6348-3656.

Ci, L.; Xu, Z.; Wang, L.; Gao, W.; Ding, F.; Kelly, K.F. *Nano Res.* 2008, 1, 116-122.

Espinal, J.F.; Montoya, A.; Mondragón, F.; Truong, T.N. *J. Phys. Chem. B* 2004, 108, 1003-1008.

Figueiredo, J.L.; Pereira, M.F.R.; Freitas, M.M.A.; Orfao, J.J.M. *Carbon* 1999, 37, 1379-1389.

Galwey, A.K.; Brown, M.E. "Thermal decomposition of ionic solids", *Elsevier*, Amsterdam,1999.

Gulbransen, E.A.; Andrew, K.F. *Ind. Eng. Chem.* 1952, 44, 1039-1044.

Harris, P.J.F. Crit. Rev. *Solid State Mater. Sci.* 2005, 30, 235-253.

Haynes, B.S. Combust. *Flame* 2001, 126, 1421-1432.

Henschke, B.; Schubert, H.; Blöcker, J.; Atamny, F.; Schlögl, R. *Thermochim. Acta* 1994, 234, 53-83.

Hurt, R.H.; Calo, J.M. Combust. *Flame* 2001, 125, 1138-1149.

Jones, J.M.; Jones, D.H. *Carbon* 2007, 45, 668-689.

Korshak, V.V. *Carbon* 1987, 25, 735-738.

Kumar, A.; Lobo, R.F.; Wagner N.J. *Carbon* 2005, 43, 3099-3111.

Laine, N.R.; Vastola, F.J.; Walker Jr, P.L. *J. Phys. Chem.* 1963, 67, 2030-2034.

Lizzio, A.A.; Jiang, H.; Radovic, L.R. *Carbon* 1990, 28, 7-19.

Lu, L.; Sahajwalla, V.; Kong C.; Harris D. *Carbon* 2001, 39, 1821-1833.

Marchon, B.; Carrazza, J.; Heinemann, H.; Somorjai, G.A. *Carbon* 1988b; 26, 507-514.

Marchon, B.; Tysoe, W.T.; Carrazza, J.; Heinemann, H.; Somorjai, G.A. *J. Phys. Chem.* 1988a, 92, 5744-5749.

Montoya, A.; Truong, T.; Mondragón, F.; Truong, T.N. *J. Phys. Chem.* A 2001, 105, 6757-6764.
Parthasarathy, G.; Sreedhar, B.; Chetty, T.R.K. *Curr. Sci.* 2006, 90, 995-1000.
Pierson, H.O. "Handbook of carbon, graphite, diamond and fullerenes. Properties, processing and applications" Noyes Publications, *NJ*, 1993, 90-95.
Radovic, L.R.; Bockrath, B. *J. Am. Chem. Soc.* 2005, 127, 5917-5927.
Sendt, K.; Haynes, B.S. *J. Phys. Chem.* A 2005, 109, 3438-3447.
Sendt, K.; Haynes, B.S. *Proc. Combust. Inst.* 2005, 30, 2141-2149.
Sibraa. A.; Newbury, T.; Haynes, B.S. Combust. *Flame* 2000, 120, 515-525.
Smith, I.W. 19th Symposium on Combustion 1982, 1045-1065.
Smith, R.N.; Young, D.A.; Smith, R.A. *Trans. Faraday Soc.* 1966, 62, 2280-2286.
Starsinic, M.; Taylor, R.L.; Walker Jr, P.L.; Painter P.C. *Carbon* 1983, 21, 69-74.
Su, J.-L.; Perlmutter, D.D. *AIChE J.* 1985, 31, 973-981.
Szymanski, G.S.; Karpinski, Z.; Biniak, S.; Swiatkowski, A. *Carbon* 2002, 20, 2627-2639.
Thomas, J.M. *Carbon* 1969, 7, 350-364.
Tomita, A. *Catal. Surv.* Japan 2001; 5:17–24.
Vyazovkin, S.; Burnham, A.K.; Criado, J.M; Pérez-Maqueda, L.A.; Popescu, C.; Sbirrazzouli, N. *Thermochim. Acta* 2011, 520, 1-19.
Walker, Jr, P.L.; Taylor, R.L.; Ranish, J.M. *Carbon* 1991, 29, 411-421.
Warren, B.E. *Phys. Rev.* 1941, 59, 693-698.
Wissler, M. *J. Power Sources* 2006, 156, 142-150.
Zhou, J.–H.; Sui, Z.–J.; Zhu, J.; Li, P.; Chen, D.; Dai, Y.–C. *Carbon* 2007, 45, 785-796.
Zhu, Z.H.; Finnerty, J.; Lu, G.Q.; Yang, R.T. *Energy Fuels* 2002, 16, 1359-1368.
Zhuang, Q.–L.; Kyotani, T.; Tomita, A. *Carbon* 1994b, 32, 539-540.
Zhuang, Q.–L.; Kyotani, T.; Tomita, A. *Energy Fuels* 1994a, 8, 714-718.
Zhuang, Q.–L.; Kyotani, T.; Tomita, A. *Energy Fuels* 1995, 6, 494-497.

In: Graphite
Editor: Quinton C. Campbell

ISBN: 978-1-62618-576-0

Chapter 7

APPLICATION OF GRAPHITE NANOPARTICLES IN ULTRAFAST FIBER LASERS

Yung-Hsiang Lin and Gong-Ru Lin*
Graduate Institute of Photonics and Optoelectronics,
Department of Electrical Engineering, National Taiwan University

ABSTRACT

Ultrafast optical pulse sources have impacted scientific researche and industrial applications for many years, especially in the fields of optical communication, optical measurements, semiconductor fabrication and diagonostics, and medicine. Furthermore, ultrashort laser pulses with extremely wide band spectra also show potential associated with the development of future device technologies. In view of previous reports, the majority of ultr fast lasers are generated by mode-locking techniques. Most ultrafast lasers with sub-pspulsewidths can be easily attainedby using a passive mode-locking technique, due to the net gain window induced by the transient transparency in a saturable absorber with thefast carrier relaxation time.The passive mode-locking technique with a fast saturable absorber is more suitable to generate an ultrafast laser than the active mode-locking technique because the reaction time of the fast saturable absorber is faster than most optoelectronic devices.

Graphenesaturable absorbers have demonstrated to passively mode-lock versatile lasers in 2009. Graphene with a broadband nonlinear transmission feature makes it a superior saturable absorber for all kinds of lasers operating at a wide wavelength range, which replaces the typical saturable absorbers such as semiconductor saturable absorber mirrrors (SESAMs) and carbon nanotubes (CNTs). Since 2011, the graphite nano-particle has been found to possess similar optical properties asgraphene, including an ultrafast carrier relaxation time, a large nonlinear modulation depth, a wideband wavelength tuning range and a high optical damage threshold. The graphite nano-particle has been used to mode-lock the erbium-doped fiber laser. It is noted that fabrications of graphite nano-particles are convenient such as with the direct polishing process. A complex chemical vapor deposition (CVD) process or chemical solution assistanceisnot needed.

*Corresponding author's email: grlin@ntu.edu.tw.

Moreover, the generated passively mode-locked pulse width with a graphite nano-particle is comparable to that which uses a graphene saturable absorber.

Except for its unique optical properties, the graphite nano-particle is electrically and thermally conductive due to the distribution of theπ orbital over the graphene sheet, which can freely move the electron on the plane. To date, graphite nano-particles also have important roles in a variety of fields including with batteries, conductive coatings, metallic alloys and nuclear reactors.

In this chapter, the fabrications of a graphite nano-particle are discussed. Subsequently, the structural and optical properties of the graphite nano-particle are studied. The structural properties of the graphite nano-particle are investigated by using X-ray diffraction (XRD) and Raman scattering spectroscopy. The optical properties of the graphite nano-particle include linear absorption and nonlinear absorption. At last, the applications of graphite nano-particles as being the saturable absorber for the passively mode-locked fiber lasers are introduced.

1. INTRODUCTION

Graphite is a three dimensional (3D) carbon allotrope, which is made out of layered graphene that consists of carbon atoms in a two-dimensional honeycomb structure with two carbon atoms in one unit cell, and the invariant angle of each atom is 120°according to van der Waals forces, After the isolation of single-layer graphene from graphite foil in 2004 [1], the unique electronic and optical properties of graphene made it a promising material in versatile applications [2-6]. The distance between each planar graphene sheet in graphite is 0.142 nm. Electrons can move inside the graphite, due to the electron delocalization within the graphene layers, to make it conductive. In addition to the high electrical conductivity, graphite also possesses several unique properties such as high thermal resistance, low friction and self-lubrication, etc. From an early stage, graphite has been widely used in batteries, steelmaking, lubricants and pencils. Recently, nano-scale graphite has also demonstrated to be the key role in ultrafast photonics.

Since 2009, single and few-layer graphene has been reported to show saturable absorption with ultrafast carrier relaxation time [7-10]. From then on, graphene becomes a good candidate as a saturable absorber for passively mode-locked lasers [11-43], which gradually replaces the need of semiconductor saturable absorber mirrors (SESAMs) [44-48] and carbon nanotubes (CNT) [49-53], typically used for initiating the mode-locking procedure in lasers. Besides its ultrafast carrier relaxation time, graphene also reveals several distinctive optical properties such as wideband absorption, high optical damage thresholds, and large optical nonlinearity. These outstanding characteristics of graphene help to deliver passively mode-locked lasers with excellent *performances*. For example: (1) the mode-locked laser pulse width can be compressed to thefs-degree due to the ultrafast carrier relaxation time of graphene exfoliated from graphite [20, 24]; (2) the zero-bandgap property of graphene leads to a wavelength-insensitive saturable absorption, which facilitates the graphene to be a "full-band" mode locker. Up to now, the exfoliated graphene-based materials have been employed in various laser systems such as the erbium-doped fiber laser (EDFL) [8-18], the $Nd:YVO_4$ laser [27], the Ti:sapphire laser [41], the Yb:KLuW laser [35], the $Tm:YAlO_3$ laser [42] and the $Nd:GdVO_4$ laser [43]. The wavelength tunability of the passively mode-locked lasers starting from exfoliated graphene are extended from visible to infrared regions; (3)

Graphene based materials can sustain high optical intensity to avoid its damage inside the laser cavity, which benefits from the fabrication of the passively mode-locked lasers with high output power [27].

The first pace of passively mode-locked EDFL with agrapheme-saturable absorber was demonstrated by Baoet al. At that time, the chemical vapor deposition (CVD) was employed to grow graphene for obtaining an ultra short soliton pulse with a 756-fs pulse width from EDFL [9]. Sun et al. also fabricated a graphene-polymer composite via a wet-chemistry technique, which was able to mode-lock the EDFL with a 460-fs pulse width [12]. Various graphene-saturable absorbers have emerged to mode-lock the fiber lasers such as a single-layer grapheme [9,11,16,21,22,26,31], multilayer graphene [13,14,23,29,32], graphene polymer [12,19,20], graphene oxide [27,28,39,42] and graphene solution [15]. Several synthesized methods have also been shown to fabricate the graphene-based materials including the CVD [9,11,14,21,22,26,31], wet-chemistry method [12,28,29,39], electrospinning [19], electrochemical reduction [18], mechanical exfoliation [13,23], ultrasonication [20] and optical assisted deposition method [15,32].

In 2011, Lin's group preliminarily demonstrated the important application of graphite nano-particles obtained by mechanically polishing the graphite foil as the saturable absorber [54]. They directly brushed the graphite nano-particles on the end-face of a single-mode fiber (SMF) patchcord to the passively mode-locking EDFLs [55]. The direct brushing method greatly simplifies the fabrication of a graphene-based saturable absorber as compared to inconvenient chemical methods such as CVD, wet-chemistry and electrochemical reduction. Simultaneously, the fs-degree, passively mode-locked EDFL can also be obtained by using graphite nano-paricles; the generated EDFL pulse width is comparable with the single-layer graphene saturable absorber [56]. The graphite nano-particle is found to possess similar optical properties as graphene, especially its wavelength-insensitive saturable absorption. Structural properties of the graphite nano-particle have also been investigated. Ever since the demonstration of passively mode-locked EDFL, the graphite nano-particle is not only important in batteries, conductive coatings, metallic alloys and nuclear reactors, etc., but it also has a great impact in ultrafast photonics. In this chapter, we firstly demonstrate versatile fabrication methods of graphite nano-particles. Subsequently, the structural and optical properties of graphite nano-particles are investigated. Later on, the mechanism of passive mode-locking with the graphite nano-particle-based saturable absorber is depicted. At last, the performance of the thegraphite nano-particle inducingpassive mode-locking in EDFL is discussed.

2. Fabrication of the Graphite Nano-Particle Saturable Absorber

The two-dimensional (2D) graphene sheet is bound by one s orbital electron and two in-plane π orbital electrons (so-called sp^2 orbit). The stacks of graphene layers construct a 3D graphite structure connected by the van der Waals forces between each plane. The weak coupling between each graphene layer makes the graphite mechanically unstable and easily cleaved.

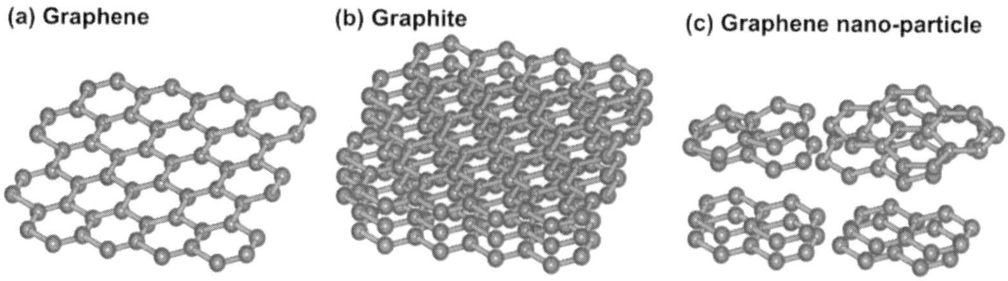

Figure 1. The schematic diagrams of (a) graphene, (b) graphite and (c) graphite nano-particle structures.

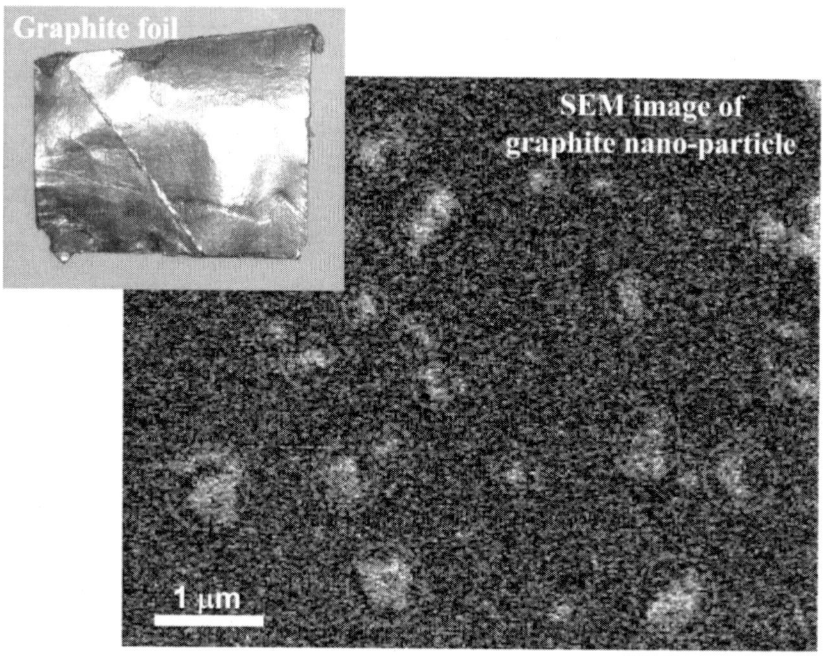

Figure 2.The photograph of the graphite foil and the SEM image of the graphite nano-particle [56].

Therefore, direct trituration via mechanical polishing is a convenient way to obtain the graphite nano-particles; however, some dispersants may be considered to be added; Figure 1 illustrates the schematics of graphene, graphite and graphite nano-particle structures. In 2011, Lin et al. reported mechanically polishing the commercial graphite foil (Alfa AesarCo. thickness: 250 mm, density: 1.12 g/cm^3, purity: 99.8%) with a commercial ceramic triturator for obtaining graphite nano-particles [54, 55]. The triturated graphite nano-particle is dispersed in a chemical solvent to prevent aggregation of the graphite nano-particle. Afterwards, the aqueous solution with graphite nano-particles can be dipped onto any desired substrate, which is heated in an oven to evaporate the chemical solvent in an atmosphere for obtaining the pure-graphite nano-particles. Figure 2 shows the photograph of the graphite foil and the scanning electron microscope (SEM) image of graphite nano-particles. According to Figure 2, the average size of the graphite nano-particle is about 500±200 nm.

To function as the saturable absorber, graphite nano-particles should be inserted into the laser cavity. In the first step, the graphite nano-particleis directly brushed on the fiber patchcord (SMF with FC/APC type connectors) with the assistance of a polymer polyvinyl alcohol (PVA) film [54]. Figure 3(a) illustrates the process of the direct brushing method. At the beginning, the PVA aqueous solution was prepared by dissolving the PVA powder in DI water with or without dispersants. The PVA aqueous solution was dripped on the end-face of the fiber patchcord, and it was flattened by using a plastic knife edge. The thickness of the scribed PVA film was about < 5 μm. Subsequently, the graphite nano-particle was directly imprinted on the fiber end-face with the ultra-thin PVA film. Afterward, the PVA-coated fiber patchcord with graphite nano-particles was baked at 75°C for 2 hours to stabilize the thin-film saturable absorber. Figure 3(b) shows the schematic of the passively mode-locked erbium-doped fiber laser (EDFL) system. After preparing the PVA-coated fiber patchcord with graphite nano-particles, the latter was connected to the other fiber patchcord with an FC/APC adapter. These two connected fiber patchcords with the confined graphite nano-particles were inserted into the EDFL system. As a result, a passively mode-locked EDFL can be successfully performed. It is the first demonstration of a graphite nano-particle serving as the saturable absorber for passively mode-locked EDFL.

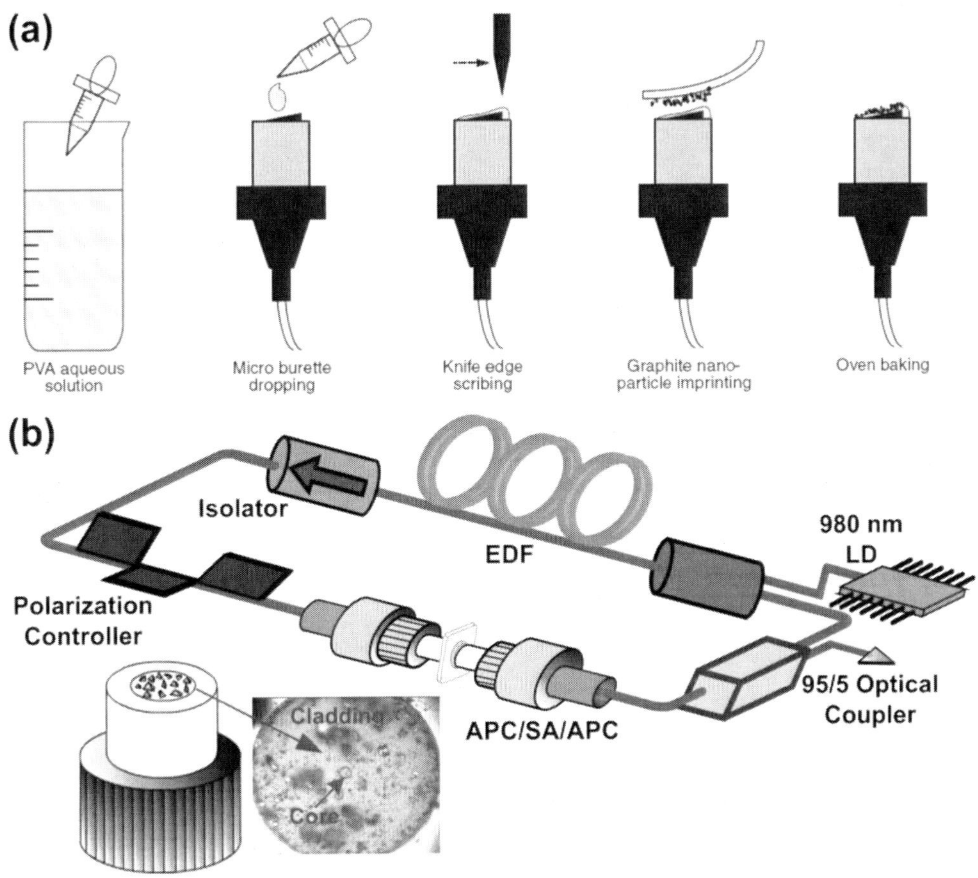

Figure 3. (a) The direct brushing method of a graphite nano-particle on the fiber end-face with the assistance of a polymer polyvinyl alcohol (PVA) film. (b) The schematic diagram of a graphite nano-particle-based passively mode-locked EDFL system. Inset is the optical microscope image of the fiber end-face with graphite nano-particle [54, 55].

During the development of the graphite nano-particle-based, passively mode-locked laser, the free-standing graphite nano-particles were never without the assistance of PVA film to successfully generate the passively mode-locked EDFL [55]. However, the self aggregation of the free-standing graphite nano-particle increases the thickness (or the layer number as well as the linear absorption coefficient) of the saturable absorber, which induces large cavity losses to shorten the pulsewidth. After that, a novel imprinting-exfoliation-wiping method was developed to solve this problem [56]. As mentioned, graphite is made out of the layered graphene with weak coupling by the van der Waals force. Each graphene layer can be separated easily by pressing and tearing the graphite. The first exfoliation of graphene is demonstrated by using sticky tape [1].

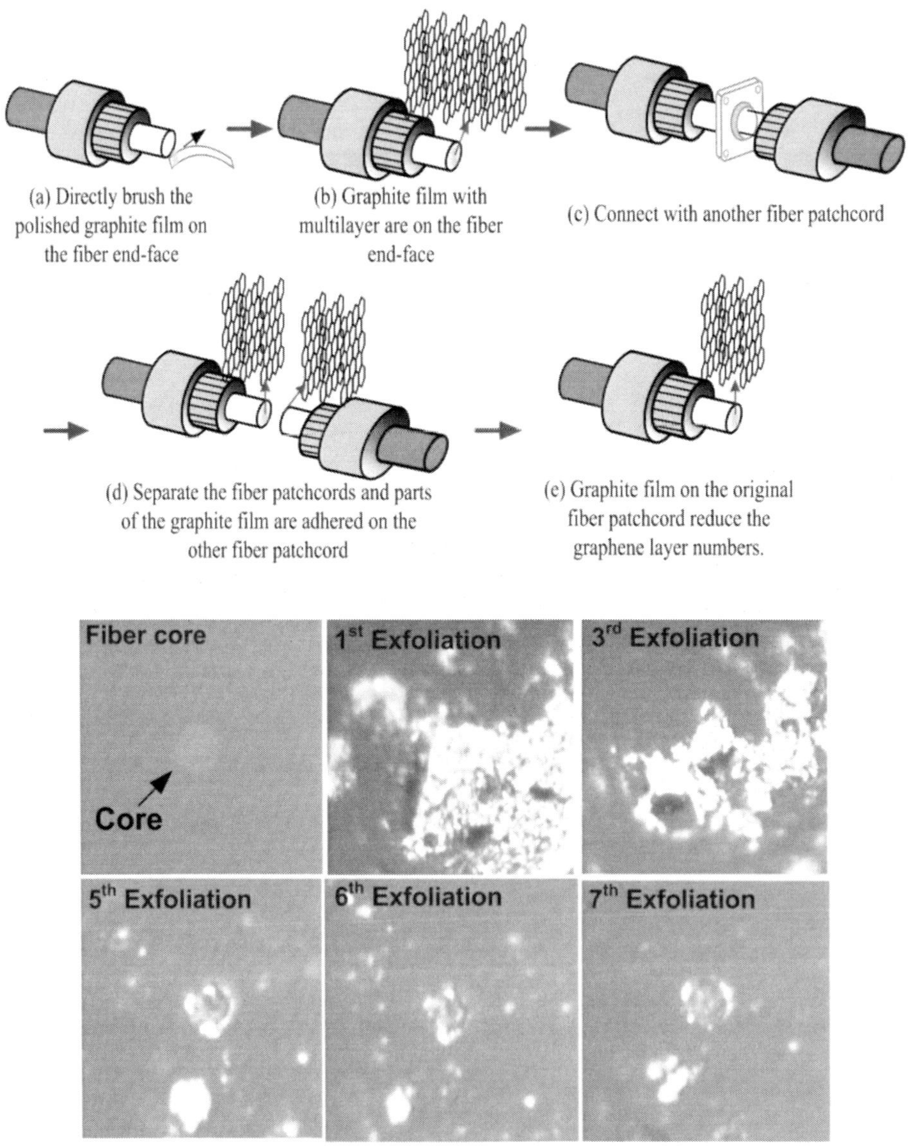

Figure 4. Upper: the flow chart of the imprinting-exfoliation-wiping method. Lower: the optical microscope images of the graphite nano-particle on the fiber end-face with different exfoliations [56].

The single-layer or multilayer graphene saturable absorber can also be produced by a sticky tape and be transferred to the fiber end-face [13]. This physical method inspired researchers to invent an imprinting-exfoliation-wiping method to reduce the number of layers of the graphite nano-particle [56]. After brushing the graphite nano-paticle onto the fiber end-face, the fiber patchcord connectors were contacted and disconnected continually several times to replace the sticky tape-based exfoliation for directly thinning the layers of the graphite nano-particle.

Figure 4 illustrates the flow chart of the proposed imprinting-exfoliation-wiping process. After the adhesion of the graphite nano-particle on the fiber patchcord connector end-face, ahard contact with another patchcord connector would imprint the graphite nano-particles on both sides.

Separating the connected patchcords would transfer part of the graphite nano-particle to another connector end-face of the patchcord, and the remaining graphite nano-particle on the original patchcord would reduce its layer number. Subsequently, one end-face of a patchcordis wiped to remove the graphite nano-particle. Afterwards, two patchcords are re-connected again to further continually exfoliate the graphite nano-particle. After the first imprinting-exfoliation-wiping process, the layer number of the graphite nano-particles can be efficiently decreased, and the transmittance of the graphite nano-particle is about 0.44. By repeating the process seven times, the layer number of the graphite nano-particle can be further reduced to increase a transmittance of 0.9. The imprinting-exfoliation-wiping process is a convenient method, which can also avoid the damage of a saturable absorber transferring from the tape to the fiber end-face. This method has been used to successfully generate a passively mode-locked EDFL with better performance [56].

3. STRUCTURAL PROPERTIESOF THE GRAPHITE NANO-PARTICLE

X-ray diffraction (XRD) and Raman scattering spectroscopy are two useful tools to provide structural in formation associated with the graphite nano-particle. Both are nondestructive and sensitive to the crystallinity and defect of the detected material. Figure 5 demonstrates the XRD spectrum of the graphite nano-particle. Two signals located at 26.22^{o} and 54.37^{o} are the {002} and {004} diffraction peaks [55, 57-60],which represent the perpendicular crystallinity of the hexagonal graphite plane. A highly ordered pyrolytic graphite (HOPG) exhibits a {002} diffraction peak at 26.54^{o}. In comparison with the HOPG, the diffraction peak of the graphite nano-particle shifts to 26.22^{o} due to the defects occurring inside the graphite nano-particle. The defects induced during the polishing process dominate the azimuth angle shifting, such as the fluctuation and curvature of graphene layer, as well as the dislocation and vacancy of carbon atom. In a theoretical calculation, the defects inside the graphite nano-particle also cause a broadened {002} peak with a full-width at half maximum (FWHM) $> 1^{o}$. However, the grain size of the graphite nano-particle limits the FWHM of the{002} peak [61].

In general, the Raman spectrum is the necessary evidence and the unique fingerprint for recognizing graphene both in industry and academic research. Not only identifying information associated with crystallinity and defect, Raman scattering spectroscopy can also identify the graphene layer number with prominent features. Figure 6 demonstrates the Raman spectra of graphene, graphite, and the graphite nano-particle. For a single-layer graphene, two prominent peaks, located at 1,580 and 2,685 cm^{-1}, represent the G band and the G' band, respectively [62-66]. The G band is the main feature of carbon materials with sp^2 bonding, and it is induced by the Raman active E_{2g} phonon mode at the Brillouin zone center. The other prominent G' band at 2,685 cm^{-1} is the second-order, two-phonon mode, which is sensitive to the graphene layer number. The G' band degrades its intensity and broadens its linewidth by increasing the layer number of graphene. The G' band intensity of a single-layer graphene is almost two times larger than the G band, but the intensity would degrade to the half maximum for a bilayer graphene.

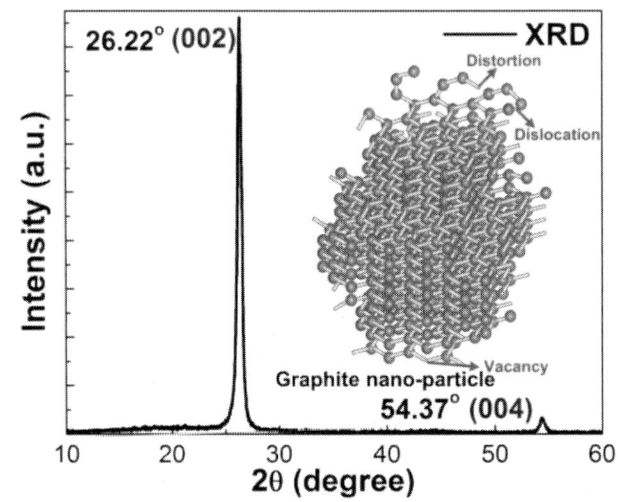

Figure 5. XRD spectrum of the graphite nano-particle. Inset: schematic diagram of the graphite nano-particle [55].

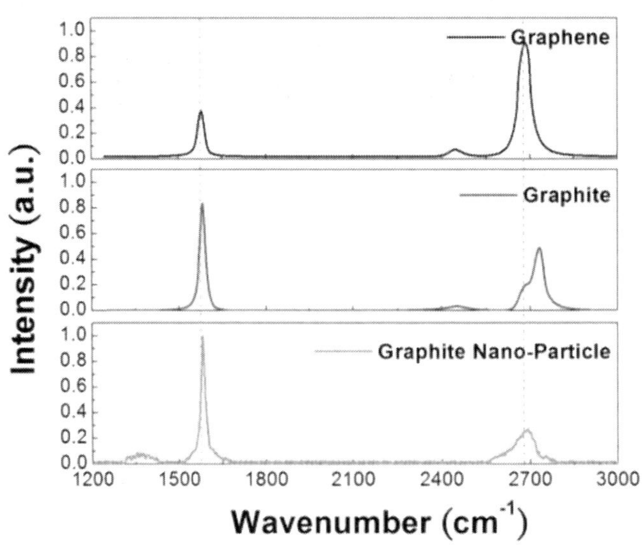

Figure 6. Raman spectra of graphene, graphite and graphite nano-particle [56].

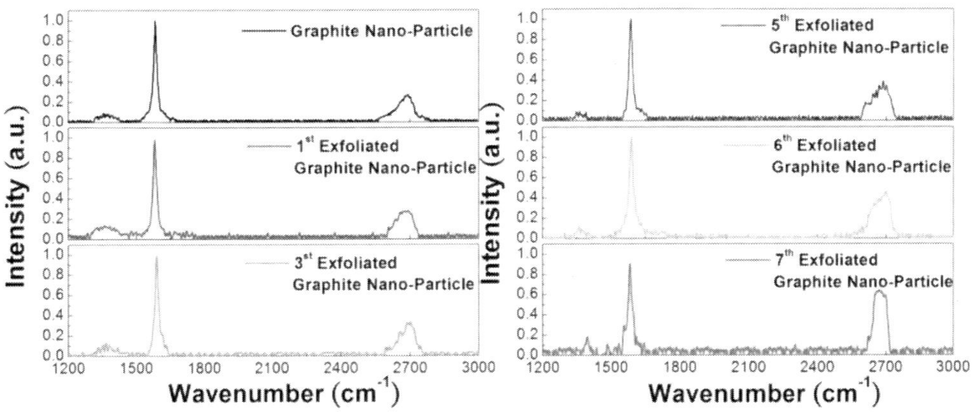

Figure 7. Raman spectra of graphite nano-particles with different exfoliations [56].

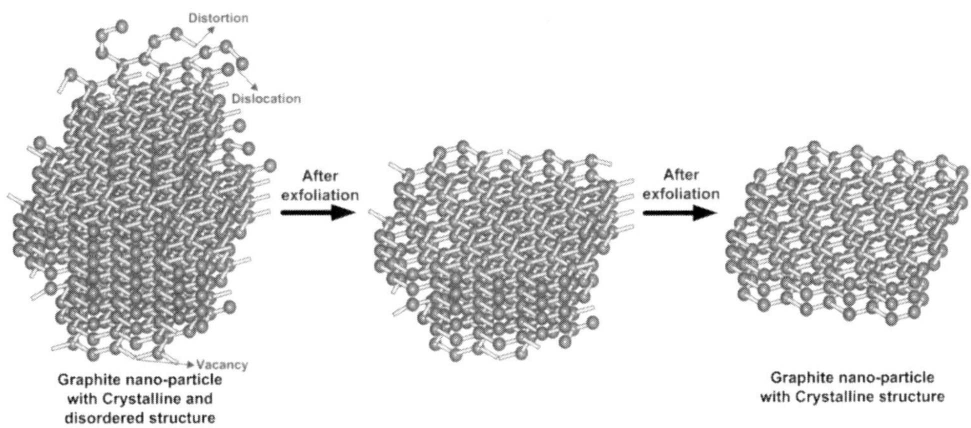

Figure 8. Schematic diagram of a graphite nano-particle with an improved crystalline quality after the imprinting-exfoliation-wiping process.

Therefore, the G' band becomes the Raman fingerprint to determine the graphene layer number. For graphite, the G band locates at the same position of 1,580 cm^{-1} with graphene. In contrast, the G'band of graphite blue shifts the location to 2,730 cm^{-1} with decreasing intensity. In addition, the asymmetric G' band separates into two modes due to the splitting of the electronic bands [62, 63].

According to Figure 6, the G band of the graphite nano-particle appears at the same position with graphite, but the slightly redG' band shifts its position to 2,690 cm^{-1}. A distinct D band at 1,360 cm^{-1}occurs,indicating the existence of defects in the graphite nano-particle. Although the direct polish is a rapid process to the fabricate graphite nano-particle, it also causes defects in the latter.

Figure 7demonstrates the Raman spectra of a graphite nano-particle with different exfoliations. With increasing the imprinting-exfoliation-wiping process from 1 to 7 times, the G' band of the graphite nano-particle gradually increases in intensity with the linewidth shortening. This phenomenon indicates the decreasing graphene layer number of the graphite nano-particle, and the property of graphite nano-particle approaches to the single layer graphene, especially for the 7th-exfoliated graphite nano-particle. The G' band of the 7th-exfoliated graphite nano-particle becomes sharp and symmetric to show the existence of few-layer graphene inside the graphite nano-particle (the layer number is almost < 5). Moreover, the D band slightly shrinks in intensity with increases in exfoliation, which means that the defects are removed during the imprinting-exfoliation-wiping process. During the polishing process, the mechanical stress presses the graphite nano-particle to produce defects.

Typically, most of the defects exist outside the graphite nano-particle. However, the defects outside the graphite nano-particle are expected to be stuck on the fiber end-face and be removed by using the imprinting-exfoliation-wiping method, as shown in Figure 8. Therefore, the 7th-exfoliated graphite nano-particle shows an improved crystalline quality with a decreasing D band intensity. A high lycrystalline graphite nano-particle with less graphene layers is expected to be a good saturable absorber. A discussion on the latter is given in the next section.

4. Optical Properties of the Graphite Nano-Particle

The optical property of graphene is unique due to the linear dispersion of massless Dirac fermions. A single-layer graphene demonstrates a huge optical linear absorption with a theoretical value of 2.3%, and the linear absorption is investigated to be frequency independent because of the zero-bandgap property, which makes graphene a wideband tunable saturable absorber [9, 12]. A saturable absorber is a material with power-dependent optical absorption. Figure 9 illustrates the schematic diagram of electron transition induced by photon absorption in a graphene saturable absorber.

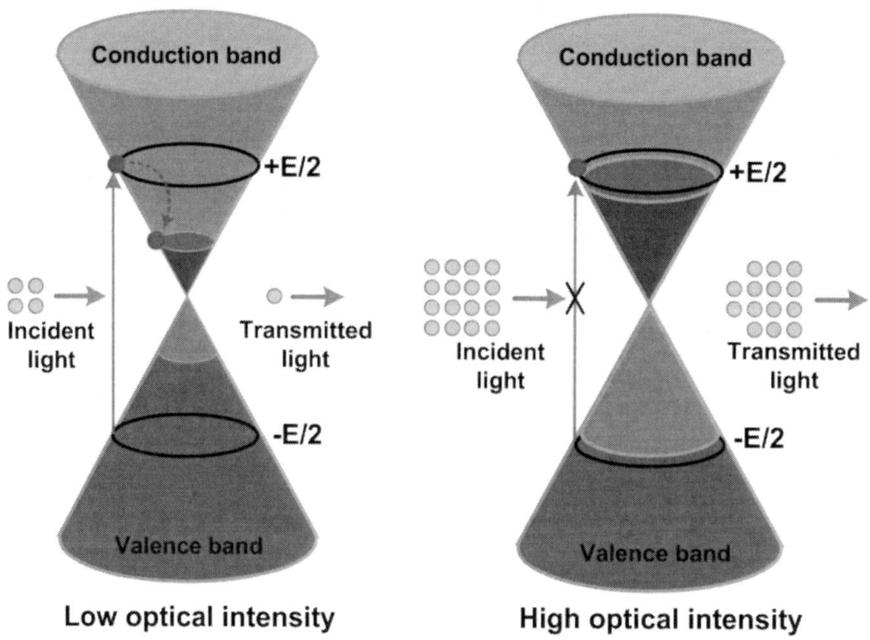

Figure 9. Schematic diagram of electron transition induced by photon absorption.

Typically, when light with low optical power incidents on the saturable absorber, part of the photons are absorbed to induce the photonelectron transiting from the valence band to the conduction band, thereby attenuating the transmitted light. Increasing the optical power would enlarge the probability of photonelectron transition. In this case, the conduction band is almost occupied by the photonelectron to deplete the electron number in the valence band, which would forbid the further electron transition. As a result, the optical absorption of the saturable absorber is decreased and the transmission is increased. The phenomenon of this nonlinear absorption can be described by the following equation [55, 56]:

$$\alpha_{non} = \frac{\alpha_0}{1 + P_{in} / P_{sat}} \tag{1}$$

α_0 represents the saturable absorption component, P_{in} and P_{sat} are the input power and the saturable power of the saturable absorber. Eq. (1) can be modified as:

$$\alpha_{non} = \alpha_0 (1 - \frac{P_{in}}{P_{sat}}) \tag{2}$$

The optical absorption of a saturable absorber includes the linear absorption α_{lin} and the nonlinear absorption α_{non}. The absorbance of a saturable absorber is expressed as the summation of α_{lin} and α_{non}:

$$\begin{aligned} \alpha_{SA} &= \frac{\alpha_{non} L_s}{1 + P_{in}/P_{sat}} + \alpha_{lin} L_s \\ &\equiv \frac{q_{non}}{1 + P_{in}/P_{sat}} + q_{lin} \\ &\cong (q_{non} + q_{lin}) - \frac{q_{non} P_{in}}{P_{sat}} \\ &= (q_{non} + q_{lin}) \left(1 - \frac{\gamma P_{in}}{(q_{non} + q_{lin})} \right) \end{aligned} \tag{3}$$

where $q_{non}= \alpha_{non}L_s$ (or $q_{lin}= \alpha_{lin}L_s$) is the nonlinear absorbance (or linear absorbance), andL_s is the thickness of the saturable absorber. $\gamma = q_{non}/P_{sat}$ denotes the modulation coefficient and the normalized term $\gamma/(q_{non}+q_{lin})$ represents the modulation depth.

The graphite nano-particle has been found to possess similar optical properties as graphene [55,56]. The linear absorption of the graphite nano-particle is found to be frequency independent from infrared-to-visible regions. The saturable absorption of the graphite nano-particle can be induced under an intense pulsed laser [54]. Moreover, the quality of saturable absorption is dependent with the graphene layer number.

Figures 10(a) and (b) show the saturable transmittance and absorbance of a graphite nano-particle with the 1st-exfoliated graphite nano-particle (measured by a pulsed laser with a wavelength of 1,560 nm and a pulsewidth of 700 fs). The linear transmittance under low optical intensity is 0.44, which is controlled by the size of the graphite nano-particle and the coverage ratio on the fiber end-face. By enhancing the optical power to 30 mW, the transmittance increases to 0.53 with aΔT of 0.09, and the corresponding normalized absorbance decreases to 0.78 with a modulation depth of 22%. The transmittance of a saturable absorber T_{SA} can be depicted as the following equation:

$$\begin{aligned} T_{SA} &= \exp(-\alpha_{SA}) \\ &= \exp\left(-\frac{q_{non}}{1+P_{in}/P_{sat}} - q_{lin}\right) \\ &= \exp\left[-(q_{non}+q_{lin})\left(1-\frac{\gamma\left|A_{peak}\right|^2}{(q_{non}+q_{lin})}\right)\right], \end{aligned} \tag{4}$$

where the simulated parameters are given: q_{non} = 0.2, q_{lin} = 0.61 and P_{sat} = 3.1 mW. The simulated results are also shown in Figures 10(a) and (b), which correlate well with the experimental results.

By increasing the imprinting-exfoliation-wiping process from 1 to 7 times, the graphene layer number of the graphite nano-particle can be reduced to enlarge its transmittance from 0.44 to 0.9 (Figure 11(a)). In particular, the normalized absorbance of the exfoliated graphite nano-particle decreases from 0.78 to 0.43 with the modulation depth increasing from 22% to 57%, as shown in Figure 11(b). Table 1 summarizes the saturation power, linear and non-linear losses, as well as the modulation depth of the graphite nano-particles with different exfoliations. The authors' observations correlate with Bao et al.'s statement that the large modulation depth of a graphite nano-particle with fewer layers is attributed to the less nonsaturable loss [9]. In contrast, the multilayer graphite nano-particle has a large nonsaturable loss caused by the enhanced scattering in the defects, which can induce a low modulation depth. More importantly, the graphite nano-particle with fewer layers of graphene is more suitable to be the sturable absorber than the one with multilayer graphene.

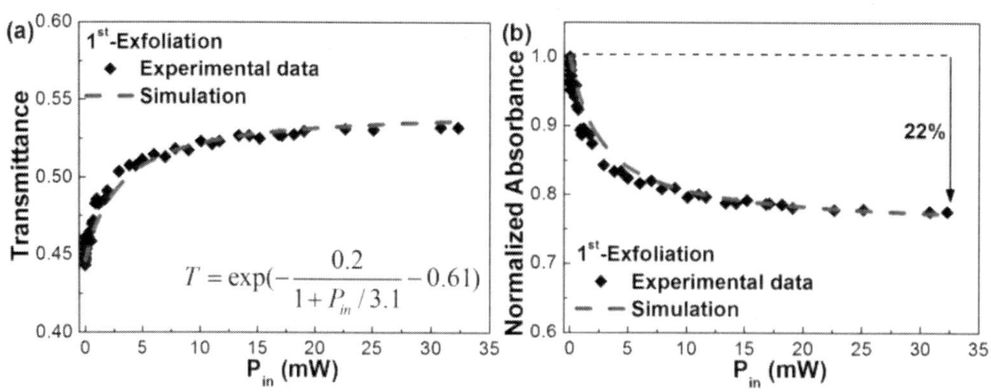

Figure 10. (a) Saturable transmittance and (b) absorbance of the graphite nano-particle with the1st-exfoliated graphite nano-particle.

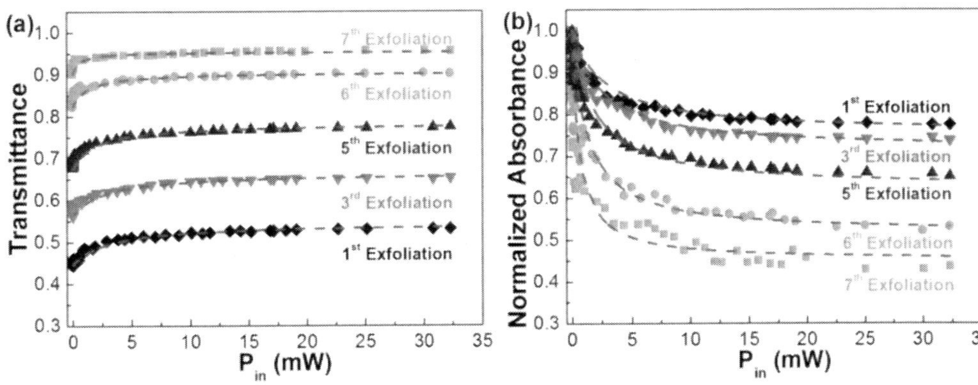

Figure 11. (a) Saturable transmittance and (b) absorbance of the graphite nano-particle with different exfoliations [56].

Table I. Saturation power, linear and non-linear losses, as well as the modulation depth of the graphite nano-particles with different exfoliations

Process #	P_{sat}(mW)	q_{non}	q_{lin}	*Modulation depth*
1st Exfoliation	3.1	0.2	0.61	22%
3rd Exfoliation	2.4	0.165	0.41	26%
5th Exfoliation	2	0.15	0.245	35%
6th Exfoliation	1.2	0.095	0.1	47%
7th Exfoliation	0.55	0.057	0.047	57%

Because the property of low linear loss can reduce the mode-locking threshold, and the property of large modulation depth is expected to form a sharp pulse, these points of view can be interpreted by the mechanism of passive mode-locking.

5. Mechanism of Passive Mode-Locking with a Saturable Absorber

Currently an ultrafast light source is an important tool in several technologies. It can be utilized to carry data in telecommunications [67, 68]; it can be also used to analyze material characteristics, including carrier lifetime, photoluminescence, nonlinearities, and saturable absorption [69-74]. To achieve a pulsed laser with ultrashort pulsewidth (ps or sub-ps degree), the mechanism of passive mode-locking by inserting a saturable absorber into the resonant cavity is a general method. The illustration of an experimental setup and the passive mode-locking mechanism is shown in Figure 12. It is a convenient method with only a gain medium and a saturable absorber inside the resonant cavity, and can generate a pulsed laser with excellent performances.

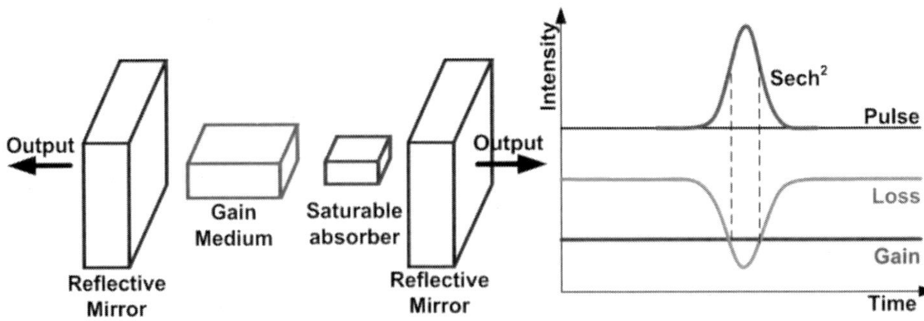

Figure 12. Schematic of a passively mode-locked laser with a saturable absorber, and the passive mode-locking mechanism in the time domain.

After the steady state formed in the resonant cavity, the cavity gain is considered to be saturated, equal to the cavity loss. Inserting a saturable absorber inevitably induces an additional loss which makes the cavity loss greater than the cavity gain. The locking of multiple modes already forms a short pulse circulated in the cavity. When the pulse transmits the saturable absorber once, the absorption is reduced to decrease the cavity loss, which can enhance the circulated pulse due to the sufficient gain in the cavity. In contrast, the noise in the resonant cavity has insufficient gain to grow up. By increasing the roundtrip time of the circulated pulse, the loss modulation in the cavity can build up an intensive and short pulse [75, 76]. An ultrashort pulse can be formed due to the locking of abundant longitudinal modes. In this condition, the spectrum of the ultrashort pulse is expected to be broadened, but limited by the finite bandwidth of the gain medium. Passive mode-locking can be used to generate an ultrashort pulse. Because the pulse is shaped by the net gain window opened by the saturable absorber, the response time is usually faster than ps. In contrast, active mode-locking is not an appropriate method to produce an ultrashort pulse due to the limitation of the modulation frequency of electronic products. But active mode-locking is a general method to produce a pulsed laser with a high repetition rate [77-80]. Haus established a differential equation to simulate the mode-locking pulse dynamic in the time domain [75] after demonstrating the CW mode-locking of dye lasers [81].

Typically, a mode-locked pulse formed in the cavity experiences the net gain window opened by the saturable absorber. During the pulse formation, the pulse amplitude is affected by the gain g and loss l. When the pulse propagates in the cavity, the phase of each frequency component is altered by group velocity dispersion (GVD) and self-phase modulation (SPM), to cause the chirp and the spectral broadening (or shortening). The master equation of a passively mode-locked laser can be expressed as [75, 82-83]:

$$T_R \frac{\partial}{\partial T} A(T,t) = \left(g - l + D_g \frac{\partial^2}{\partial t^2} - iD \frac{\partial^2}{\partial t^2} \right) A(T,t) + \left(-\alpha_{non}(T,t) + i\delta |A|^2 \right) A(T,t) \tag{5}$$

where T_R and $A(T, t)$ are the round-trip time and the pulse amplitude. D_g is the gain dispersion. D and δ are the GDD coefficient and the SPM coefficient, respectively. α_{non} denotes the saturable absorption. The solution of Eq. (5) is a hyperbolic secant function. The author utilizes the master equation to simulate the pulse shape of the passively mode-locked laser. The comparison of the simulated and the experimental results are given in the next section.

6. Passively Mode-Locked EDFL with the Graphite Nano-Particle

The graphite nano-particle has been used to generate the ring-type passively mode-locked EDFL in the infrared region; the configuration is shown in Figure 3(b). An erbium-doped fiber (EDF) is served as the gain medium and it is pumped by a laser diode with a wavelength of 980 nm. An optical coupler is utilized to provide 95% feedback and 5% output power. A polarization plays the role to optimize the circulated polarization, which is set in front of the saturable absorber. In general, an isolator is inserted to determine a one-way propagation for preventing the damage caused by the reflected photon.

Figure 13 demonstrates the passively mode-locked EDFL performances with the 1st-exfoliated graphite nano-particle, including the autocorrelation trace and the optical spectrum. According to the Haus master equation, a hyperbolic secant function is the solution of the equation [75]; therefore, the autocorrelation trace is well fit by a hyperbolic secant squared (sech^2) function with a linewidth of 1.16 ps.

For a sechfunction in an intensity autocorrelation, the linewidth of the autocorrelation trace is 1.54 times larger than the real pulseiwidth. Hence, the real pulsewidth of 755 fs is obtained by multiplying the linewidth of autocorrelation trace with a deconvolution factor of ~0.65. The central wavelength of the passively mode-locked EDFL is at 1,592 nm and the spectral full-width at half maximum (FWHM) is 3.55 nm. The time-bandwidth product (TBP) is 0.317, which represents a transform-limited pulse.

In an ultrafast laser, the pulsewidth is limited by its optical spectrum, and the minimum TBP of a sech^2-shaped pulse is ~0.315. If the TBP of a pulsed laser approaches the transform limit, this means that the cavity chirp induced by the chromatic dispersion and the nonlinearities is close to zero.

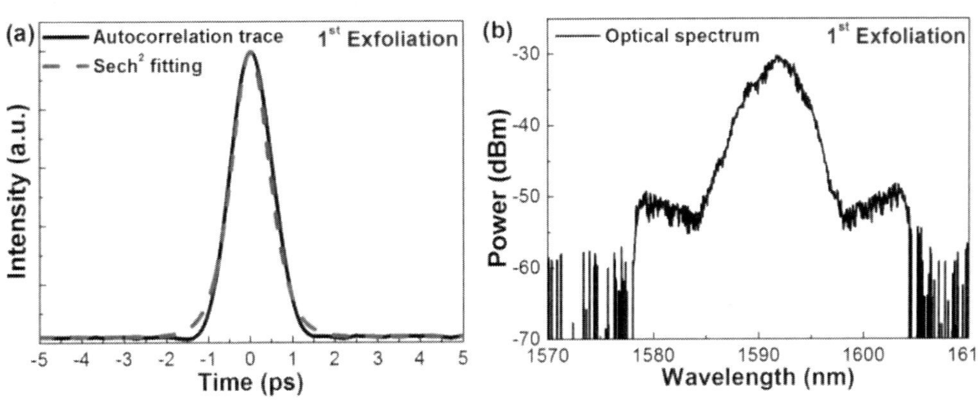

Figure 13. The passively mode-locked EDFL performances with a 1st-time exfoliated graphite nano-particle, including (a) the autocorrelation trace and the sech2 fitted curve, as well as (b) the optical spectrum [56].

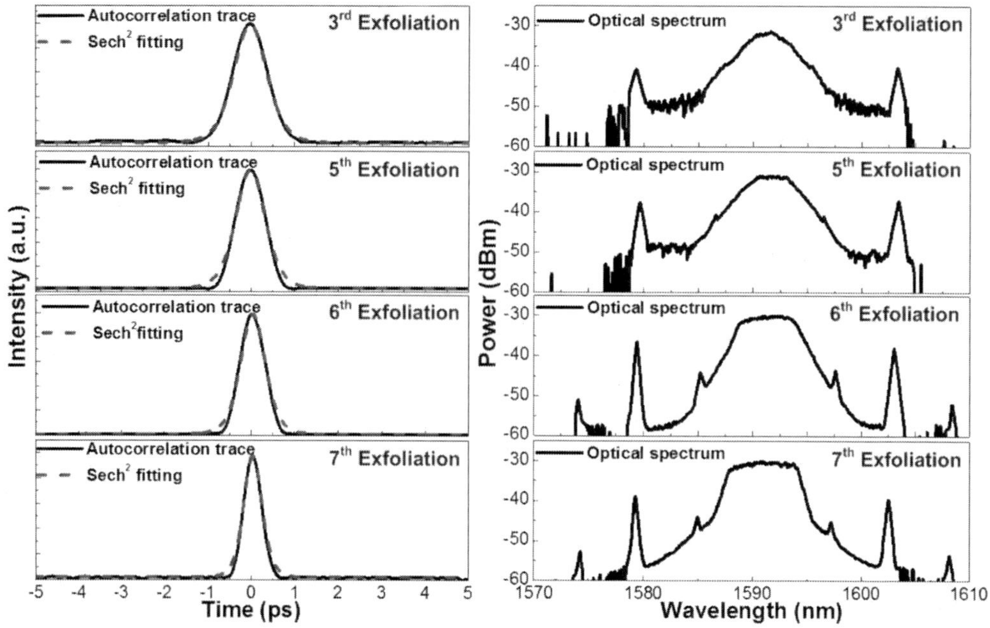

Figure 14. The passively mode-locked EDFL performances with different time-exfoliated graphite nano-particles [56].

For an ultrafast mode-locked laser, the chromatic dispersion caused by the passive components has be to compensated for because the pulse would be broadened due to the different phase velocities of each cavity mode. The cavity dispersion at zero is expected to form a chirp-free resonator, or it should be at least anomalous to generate a soliton pulse [84]. Typically, the dispersion of EDF at 1.55 μm is positive, which can be compensated by the negative dispersion of SMF. In the authors' case, the dispersion coefficients of the EDF ($\beta_{2,EDF}$) and the SMF ($\beta_{2,SMF}$) are +26 ps^2/km and -21 ps^2/km [56]. By using a 12-m SMF to compensate the normal dispersion caused by a 9-m SMF, the group delay dispersion (GDD) is about -0.186 ps^2. The passively mode-locked laser performance is promoted after the imprinting-exfoliation-wiping process. Figure 14 shows the autocorrelation traces and the optical spectra of the passively mode-locked EDFLs with differently exfoliated graphite nano-particles.

By increasing the imprinting-exfoliation-wiping process from 3 to 7 times, the pulsewidth shortens from 620 to 375 fs, the spectral FWHM broadens from 4.17 to 6.85 nm, and the TBP of all the EDFLs are around 0.31. This phenomenon indicates that the passively mode-locked EDFL pulsewidth can be controlled by detuning the layer number of graphite nano-particle. A graphene-saturable absorber can generate an fs-degree passively mode-locked EDFL easily; even a sub-200 fs pulsewidth has been obtained [20, 24]. In comparison with graphene, the exfoliated graphite nano-particle obtained by using a convenient imprinting-exfoliation-wiping method also shows its potential to produce the comparable fs-degree passively mode-locked EDFL.

The Kelly sidebands are observed on the shoulder of the optical spectrum to represent the EDFL in the soliton mode-locking operation [75, 85]. In the resonant cavity, the combined perturbation of the linear anomalous GDD and the nonlinear self-phase modulation (SPM) periodically form a nonlinear-index grating [78, 86], which gradually enlarges the specific dispersive waves after several round-trips.

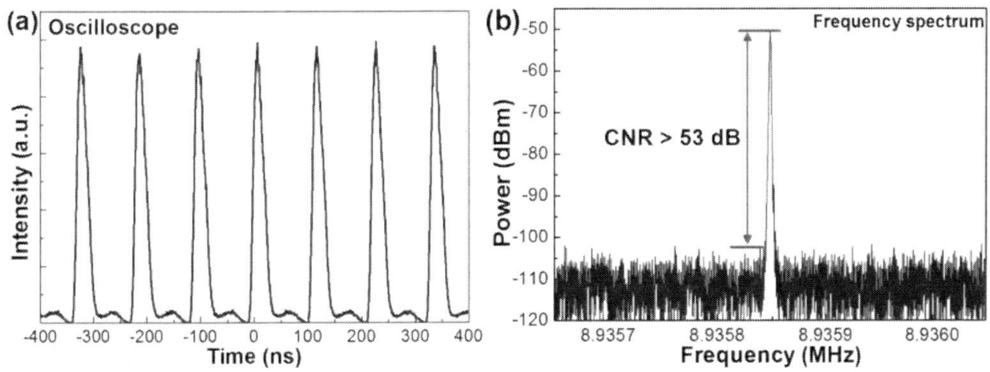

Figure 15. The oscilloscope trace and the frequency spectrum of the passively mode-locked EDFL with the 7th-time exfoliated graphite nano-particle [56].

The enhanced specific dispersive waves occur as the symmetric sidebands,the so-called Kelly sidebands, on the optical spectrum. According to Figure 14, the 1st-order Kelly sidebands obviously appear on the optical spectrum by using the 3rd-exfoliated graphite nano-particle.

The frequency spacing $\Delta\upsilon$ shrinks from ±1.45 to ±1.35 THz by increasing the imprinting-exfoliation-wiping process from 1 to 7 times. After exfoliating6 times, the passively mode-locked EDFL produces the outer pair as the 2nd-order Kelly sideband. It is caused by a higher peak intensity circulated in the resonant cavity and the sideband spacing decreasesfrom±2.04 to ±2.01 THz by increasing the imprinting-exfoliation-wiping process from 6 to 7 times.

The oscilloscope trace and the frequency spectrum of the passively mode-locked EDFL with the 7th-time exfoliated graphite nano-particle are demonstrated in Figures15(a) and (b).The repetition time of 112 ns corresponds with the cavity length. The fluctuation of peak amplitude is less than 5% to indicate the stability. Simultaneously, the stabilized passively mode-locked pulse-train shows a repetition rate of 8.936 MHz with the carrier-to-noise ratio (CNR) of >53 dB.

7. Theoretical Analysis of Passively Mode-Locked EDFL with Graphite Nano-Particles Exfoliated for Different Amounts of Time

The master equation shown in Eq. (5) can simulate the dynamic of the passively mode-locked EDFL. The EDFL is operated with an EDFA as the gain medium; The Gain of the EDFA is >20 dB with a saturation power of 1.1 mW. The cavity loss, from the cavity insertion loss and the saturable absorber, is decreased from 0.82 to 0.19 with the increasing imprinting-exfoliation-wiping process from 1 to 7 times. Figure 16 demonstrates the simulated passively mode-locked EDFL performances with different time exfoliated graphite nano-particles, and Table 2 includes the parameters used in the simulation.

Same as the experimental results shown in Figure 14, the simulated EDFL pulsewidth shortens from 757 to 370 fs with the an increasing imprinting-exfoliation-wiping process from 1 to 7 times, and the spectral FWHM broadens from 3.5 to 6.87 nm. The frequency spacing of the fundamental soliton slightly varies from ±1.5 to ±1.37 THz. The varied parameters used in the simulation are the absorbance (q_{non}+q_{lin}) and saturation power (P_{sat}) of the saturable absorber. The modulation coefficient (defined as q_{non}/P_{sat}) is thus increased from 0.064 to 0.103 with increasing exfoliation from 1to 7 times. According to Eq. (5), the derived pulsewidthτ is found to relate withD_g, $\gamma$and pulse amplitude $|A_{peak}|^2$.

$$\tau = \sqrt{\frac{2D_g}{\gamma \left| A_{peak} \right|^2}} \tag{6}$$

In the above discussion, the authors mention that a saturable absorber with a low linear loss and a high modulation depth can form an intense and sharp pulse. This point of view can be realized from Eq. (6) that the pulsewidth is inversely proportional with the square roots of γ and $|A_{peak}|^2$. The graphite nano-particle with fewer layers of graphene contributes to the low absorption to enhance the peak amplitude; moreover, the modulation depth is higher than the one with multilayer graphene. Therefore, the graphene layer number of graphite nano-particle can determine the pulsewidth variation. The optimized sub-400-fs pulsewidth occurs with the 7th- exfoliated graphite nano-particle.

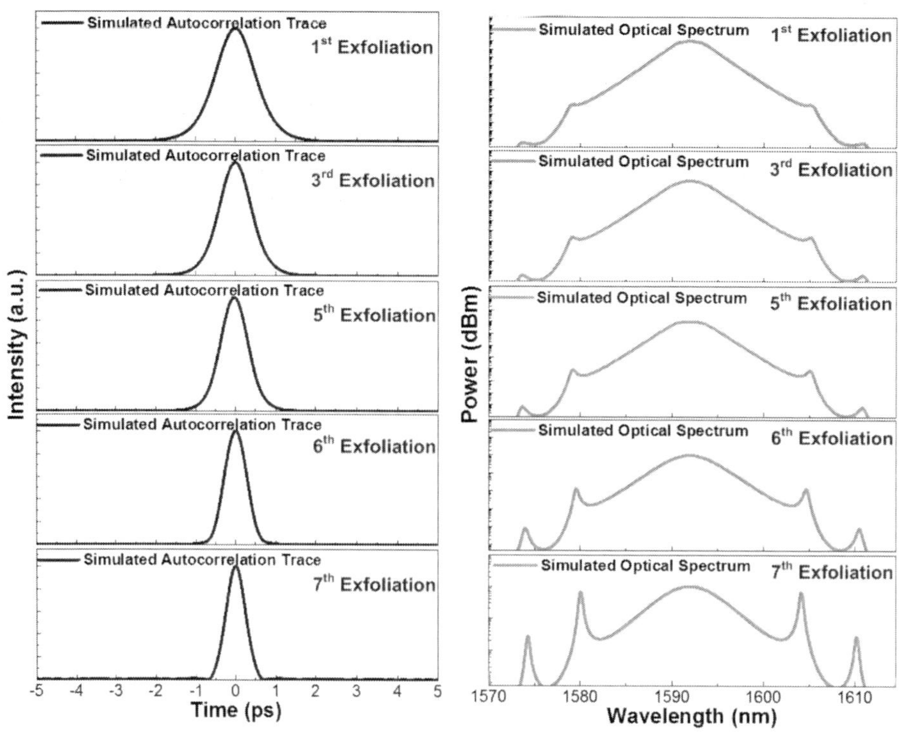

Figure 16. The simulated passively mode-locked EDFL performances with graphite nano-particle exfoliated for different amounts of time [56].

CONCLUSION

The graphite nano-particle, containing just a few layers of graphene and multilayer graphene, has demonstrated to be the saturable absorber for passively mode-locked EDFL. By directly polishing the graphite foil, a triturated graphite nano-particle with an average size of 500±200 nm can be obtained.

The structural information of the graphite nano-particle has been investigated by XRD and Raman scattering spectroscopy. The defects induced by the polishing process are found outside the graphite nano-particle, which may degrade the optical property as being a saturable absorber. In addition, a large graphite nano-particle with multilayer graphene would induce high-cavity losses and low modulation depths. Therefore, the authors invented an imprinting-exfoliation-wiping process to reduce the number of graphene layers and remove the defects existing outside the graphite nano-particle.

According to the Raman spectrum, the G' band of the graphite nano-particle gradually increases the intensity with the linewidth shortening with increasing the imprinting-exfoliation-wiping process from 1 to 7 times. It indicates that the decreasing graphene layer number of the graphite nano-particle, and the property of the graphite nano-particle approaches to the single layer graphene. Moreover, the D band slightly shrinks in intensity with increases in exfoliation, which means that the defects are removed during the imprinting-exfoliation-wiping process.

The transmittance of the graphite nano-particle enlarges from 0.44 to 0.9, and the normalized absorbance decreases from 0.78 to 0.43 with the modulation depth promoting from 22% to 57% by increasing the imprinting-exfoliation-wiping process from 1 to 7 times. The graphite nano-particle with few-layer graphene is more suitable to be the saturable absorber than the one with multilayer graphene. Because the property of low linear loss can reduce the mode-locking threshold, a large modulation depth is expected to form a sharp pulse.

By increasing the exfoliation from 1 to 7 times, the pulsewidth shortens from 755 to 375 fs, the spectral FWHM broadens from 3.55 to 6.85 nm, and the TBP of all the EDFLs are around 0.31. This phenomenon indicates that the passively mode-locked EDFL pulsewidth can be controlled by detuning the layer number of the graphite nano-particle. The Kelly sidebands are observed on the shoulder of the optical spectrum to represent that the EDFL is in the soliton mode-locking operation. Authors used the Haus master equation to simulate the passively mode-locked EDFL dynamic. The simulated pulsewidth and optical spectrum are well correlated with the experimental results.

It is valuable to develop this saturable absorber by direct polishing and direct brushing methods. In comparison with graphene grown by chemical methods, the graphite nano-particle obtained by these physical methods can not only simplify the fabrication, but also generate the fs-degree passively mode-locked EDFL.

The defects induced by the polishing process can be released by an imprinting-exfoliation-wiping process to improve the laser performance. In addition, the pulsewidth can be adjusted by controlling the graphene layer number. The graphite nano-particle has shown its potential to be the mode-locker, which can make a contribution to the development of passively mode-locked lasers.

ACKNOWLEDGMENTS

The book writing and the research works are supported by National Taiwan University and National Science Council of Taiwan, Republic of China, under grants of 102R89083, NSC99-2622-E-002-024-CC2, and NSC101-2622-E-002-009-CC2.

REFERENCES

[1] Novoselov, K. S; Geim, A. K.; Morozov, S. V.; Jiang, D.; Zhang, Y.; Dubonos, S. V.; Grigorieva, I. V.; Firsov, A. A. *Science* 2004, 306 (5696), 666-669.
[2] Geim, A. K.; Novoselov, K. S. *Nature Mater.* 2007, 6, 183–191.
[3] Soldano, C.; Mahmood, A.; Dujardin, E. *Carbon* 2010, 48, 2127-2150.
[4] Avouris, P. *Nano Lett.* 2010, 10, 4285–4294.
[5] Bonaccorso, F.; Sun, Z.; Hasan T.; Ferrari, A. C. *Nature Photonics* 2010, 4, 611-622.
[6] Geim, A. K. *Science* 209, 324, 1530-1534.
[7] Kumar, S.; Anija, M.; Kamaraju, N.; Vasu, K. S.; Subrahmanyam, K. S.; Sood, A. K.; Rao, C. N. R. *Appl. Phys. Lett.* 2009, 95, 191911.
[8] Hasan, T.; Sun, Z.; Wang, F.; Bonaccorso, F.; Tan, P. H.; Rozhin, A. G.; Ferrari, A. C. *Adv. Mater.* 2009, 21, 3874–3899.
[9] Bao, Q. L.; Zhang, H.; Wang, Y.; Ni, Z.; Yan, Y.; Shen, Z. X.; Loh, K. P.; and Tang, D. Y. *Adv. Funct. Mater.* 2009, 19, 3077–3083.
[10] Zhang, H.; Bao, Q.; Tang, D. Y.; Zhao, L.; Loh, K. P. *Appl. Phys. Lett.* 2009, 95, 141103.
[11] Zhang, H.; Tang, D. Y.; Zhao, L. M.; Bao, Q.; Loh, K. P. *Opt. Express* 2009, 17, 17630-17635.
[12] Sun, Z.; Hasan, T.; Torrisi, F.; Popa, D.; Privitera, G.; Wang, F.; Bonaccorso, F.; Basko, D. M.; Ferrari, A. C. *ACS Nano* 2010, 4, 803–810.
[13] Chang, Y.M.; Kim, H.; Lee, J. H.; Song, Y. W. *Appl. Phys. Lett.* 2010, 97, 211102.
[14] Zhao, L. M.; Tang, D. Y.; Zhang, H.; Wu, X.; Bao, Q. L.; Loh, K. P. *Opt. Lett.* 2010, 35,3622-3624.
[15] Martinez, A.; Fuse, K.; Xu, B.; Yamashita, S. *Opt. Express* 2010, 18, 23054-23061.
[16] Sun, Z. P.; Popa, D.; Hasan, T.; Torrisi, F.; Wang, F. Q.; Kelleher, E. J. R.; Travers, J. C.; Nicolosi, V.; Ferrari, A. C. *Nano Research* 2010, 3, 653-660.
[17] Zhang, H.; Tang, D. Y.; Zhao, L. M.; Bao, Q. L.; Loh, K. P. 2010, *Opt. Commun.* 2010, 283, 3334-3338.
[18] Tan, W. D.; Su, C. Y.; Knize, R. J.; Xie, G. Q.; Li, L. J.; Tang, D. Y. *Appl. Phys. Lett.* 2010, 96, 031106.
[19] Bao, Q. L.; Zhang, H.; Yang, J.; Wang, S.; Tang, D. Y.; Jose, R.; Ramakrishna, S.; Lim, C. T.; Loh, K. P. Adv. *Funct. Mater.* 2010, 20, 782-791.
[20] Popa, D.; Sun, Z.; Torrisi, F.; Hasan, T.; Wang, F.; Ferrari, A. C. *Appl. Phys. Lett.* 2010, 97, 203106.
[21] Zhang, H.; Tang, D. Y.; Zhao, L. M.; Bao, Q.; Loh, K. P.; Lin, B.; Tjin, S. C. *Laser Phys. Lett.* 2010, 7, 591–596.

[22] Zhang, H.; Tang, D. Y.; Knize, R. J.; Zhao, L.; Bao, Q.; Loh, K. P. *Appl. Phys. Lett.* 2010, 96, 111112.

[23] Martinez, A.; Fuse, K.; Yamashita, S. *Appl. Phys. Lett.* 2011, 99, 121107.

[24] Cunning, B. V.; Brown, C. L.; Kielpinski, D. *Appl. Phys. Lett.* 2011, 99, 261109.

[25] Xu, J. L.; Li, X. L.; He, J. L.; Hao, X. P.; Wu, Y. Z.; Yang, Y.; Yang, K. *J. Appl. Phys. Lett.* 2011, 99, 261107.

[26] Cho, W. B.; Kim, J. W.; Lee, H. W.; Bae, S.; Hong, B. H.; Choi, S. Y.; Baek, I. H.; Kim, K.; Yeom, D. I.; Rotermund, F. *Opt. Lett.* 2011, 36, 4089-4091

[27] Zhang, L.; Wang, Y. G.; Yu, H. J.; Zhang, S. B.; Hou, W.; Lin X. C.; Li, J. M.; *Laser Phys.* 2011, 12, 2072-2075.

[28] Liu, Z. B.; He, X.; Wang, D. N. *Opt. Lett.* 2011, 36, 3024-3026.

[29] Xu, J. L.; Li, X. L.; Wu, Y. Z.; Hao, X. P.; He, J. L.; Yang, K. J. *Opt. Lett.* 2011, 36, 1948-1950.

[30] Wang, L. R.; Liu, X. M.; Gong, Y. K.; Mao, D.; Feng, H. *Laser Phys. Lett.* 2011, 8, 376-381.

[31] Bao, Q. L.; Zhang, H.; Ni, Z. H.; Wang, Y.; Polavarapu, L.; Shen, Z. X.; Xu, Q. H.; Tang, D. Y.; Loh, K. P. *Nano Research* 2011, 4, 297-307.

[32] Kim, H.; Cho, J.; Jang, S. Y.; Song, Y. W. *Appl. Phys. Lett.* 2011, 98, 021104.

[33] Sobon, G.; Sotor, J.; Abramski, K. M. *Laser Phys. Lett.* 2012, 9, 581–586.

[34] Huang, P. L.; Kuo, H. H.; Dong, R. X.; Hsieh, B. Z.; Huang, S. H.; Hong, S. F.; Shih, P. T.; Yeh, C. Y.; Lin, G.-R.; Lin, J. J.; Cheng, W. H. *IEEE J. Lightwave Technol.* 2012, 30, 3413-3419.

[35] Ugolotti, E.; Schmidt, A.; Petrov, V.; Kim, J. W.; Yeom, D. I.; Rotermund, F.; Bae, S.; Hong, B. H.; Agnesi, A.; Fiebig, C.; Erbert, G.; Mateos, X.; Aguilo, M.; Diaz, F.; Griebner, U. *Appl. Phys. Lett.* 2012, 101, 161112.

[36] Luo, Z. Q.; Wang, J. Z.; Zhou, M.; Xu, H. Y.; Cai, Z. P.; Ye, C. C. Laser *Phys. Lett.* 2012, 9, 229-233.

[37] Bao, Q. L.; Loh, K. P. *ACS Nano* 2012, 6, 3677-3694.

[38] Huang, P. L.; Lin, S. C.; Yeh, C. Y.; Kuo, H,-H.; Huang, S. H.; Lin, G.-R.; Li, L. J.; Su, C. Y.; Cheng, W.-H. *Opt. Express* 2012, 20, 2460-2465.

[39] Xu, J.; Liu, J.; Wu, S.; Yang, Q. H.; Wang, P. *Opt. Express* 2012, 20, 15474-15480.

[40] Martinez A.; Yamashita, S. *Appl. Phys. Lett.* 2012, 101, 041118.

[41] Baek, I. H.; Lee, H. W.; Bae, S.; Hong, B. H.; Ahn, Y. H.; Yeom, D.-I.; Rotermund, F. *Appl. Phys. Express* 2012, 5, 032701.

[42] Liu, J.; Wang, Y. G.; Qu, Z. S.; Zheng, L. H.; Su, L. B.; Xu, *J. Laser Phys. Lett.* 2012, 9, 15–19.

[43] Wang, Y. G.; Chen, H. R.; Hsieh, W. F.; Tsang, Y. H. *Opt. Commun.* 2013, 289, 119-122.

[44] Keller, U. *Nature* 2003, 424, 831-838.

[45] Jacobovitz-Veselka, G. R.; Keller, U.; Asom, T. *Opt. Lett.* 1992, 17, 1791-1793.

[46] Keller, U.; Miller, D. A. B.; Boyd, G. D.; Chiu, T. H.; Ferguson, J. F.; Asom, M. T. *Opt. Lett.* 1992, 17, 505-507.

[47] Keller, U.; Weingarten, K. J.; K¨artner, F. X.; Kopf, D.; Braun, B.; Jung, I. D.; Fluck, R.; Honninger, C.; Matuschek, N.; Aus der Au, J. *IEEE J. Sel. Top. Quantum Electron 1996, 2, 435-453.*

[48] Steinmeyer, G.; Sutter, D. H.; Gallmann, L.; Matuschek, N.; Keller, U. *Science* 1999, 286, 1507-1512.
[49] Set, S. Y.; Yaguchi, H.; Tanaka, Y.; Jablonski, M. *J. Lightwave Technol.*2004, 22, 51-56.
[50] Schibli, T.; Minoshima, K.; Kataura, H.; Itoga, E.; Minami, N.; Kazaoui, S.; Miyashita, K.; Tokumoto, M.; Sakakibara, Y. *Opt. Express* 2005, 13, 8025-8031.
[51] Wang, F.; Rozhin, A. G.; Scardaci, V.; Sun, Z.; Hennrich, F.; White, I. H.; Milne W. I.; Ferrari, A. C. *Nat. Nanotechnol.* 2008, 3, 738-742.
[52] Chiu, J. C.; Chang, C. M.; Hsieh, B. Z.; Lin, S. C.; Yeh, C. Y.; Lin, G.-R.; Lee, C. K.; Lin, J. J.; Cheng, W. H. *Opt. Express* 19, 4036-4041, 2011.
[53] Cheng, K. N.; Lin, Y. H.; Yamashita, S.; Lin, G.-R. *IEEE photon. J.* 2012, 4, 1542-1552.
[54] Lin, G.-R.; Lin, Y.-C. *Laser Phys. Lett.* 2011, 8, 880–886.
[55] Lin, Y. H.; Lin, G.-R. *Laser Phys. Lett.* 2012, 9, 398-404.
[56] Lin, Y. H.; Lin, G.-R. *Laser Phys. Lett.* 2013, 10, 045109.
[57] Li, Z.Q.; Lu, C.J.; Xia, Z.P.; Zhou, Y.; Luo, Z. *Carbon* 2007. 45, 1686–1695.
[58] Subrahmanyam, K. S.; Vivekchand, S. R. C.; Govindaraj A.; Rao, C. N. R. *J. Mater. Chem.* 2008, 18, 1517–1523.
[59] Tang, L; Wang, Y.; Li, Y.; Feng, H. B.; Lu, J.; Li, J. H. *Adv. Funct. Mater.* 2009, 19, 2782–2789.
[60] Yan, J.; Wei, T.; Shao, B.; Fan, Z.; Qian, W.; Zhang, M.; Wei, F. *Carbon* 2010, 48, 487-493.
[61] Cullity, B.D.; Stock, S.R. Elements of X–Ray Diffraction, 3rd ed.; Prentice Hall, 2001.
[62] Ferrari, A. C.; Meyer, J. C.; Scardaci, V.; Casiraghi, C.; Lazzeri, M.; Mauri, F.; Piscanec, S.; Jiang, D.; Novoselov, K. S.; Roth, S.; Geim, A. K. *Phys. Rev. Lett.* 2006, 97, 187401.
[63] Ferrari, A. C. *Solid State Commun.* 2007, 143, 47–57.
[64] Graf D.; Molitor F.; Ensslin K.; Stampfer C.; Jungen A.; Hierold C.; Wirtz L. *Nano Lett.* 2007, 7, 238-242.
[65] Malard, L. M.; Pimenta, M. A.; Dresselhaus, G.; Dresselhaus, M. S. *Phys. Rep. Rev. Sec. Phys. Lett.* 2009, 473, 51–87.
[66] Allen, M. J.; Tung, V. C.; Kaner, R. B. *Chem. Rev.* 2010, 110, 132–145.
[67] Lin, G.-R.; Liao, Y. S.; Xia, G. Q. *Opt. Express* 2004, 12, 2017-2026.
[68] Lin, G.-R.; Chiu, I. H. *Opt. Express* 2005, 13, 8772-8780.
[69] Wang, G.; Fafard, S.; Leonard, D.; Bowers, J. E.; Merz, J. L.; Petroff P. M. *Appl. Phys. Lett.* 1994, 64, 2815.
[70] Jain, S. C.; Willander, M.; Narayan, J.; Overstraeten, R. V. J. *Appl. Phys.* 2000, 87, 965-1006.
[71] Dinu, M.; Quochi, F.; Garcia, H. *Appl. Phys. Lett.* 2003, 82, 2954-2956.
[72] He, G. S.; Tan, L. S.; Zheng, Q.; Prasad, P. N. *Chem. Rev.* 2008, 108, 1245-1330.
[73] Xing, G. C.; Guo, H. C.; Zhang, X.; Sum, T. C.; Hon, C.; Huan, A. *Opt. Express* 2010, 18, 4564-4573.
[74] Lin, C. D.; Cheng, C.H.; Lin, Y. H. ; Wu, C. L.; Pai, Y.-H.; Lin, G.-R. *Appl. Phys. Lett.* 2012, 99, 243501.
[75] Haus, H. A. *IEEE J. Sel. Top. Quantum Electron* 2000, 6, 1173-1185.
[76] Fermann, M. E.; Hart, I. *IEEE J. Sel. Top. Quantum Electron* 2009, 15, 191-206.

[77] Tucker, R.S.; Koren, U.; Raybon, G.; Burrus, C. A.; Miller, B. I.; Koch, T. L.; Eisenstein, G. *Electron Lett.* 1989, 25, 621-622.
[78] Lin, Y. C.; Peng, G. H.; Lin, G.-R. *Opt. Express* 2009, 17, 5526-5532.
[79] Lin, G.-R.; Kang, J. J.; Lee, C. K. *Opt. Express* 2010, 18, 9570-9579.
[80] Lee, C. K.; Kang, J. J.; Lin, G.-R. IEEE/OSA *J. Lightwave Technol.* 2009, 27, 851-856.
[81] Ippen, E. P.; Shank, C. V.; Dienes, A. *Appl. Phys. Lett.* 1972, 21, 348.
[82] Lin, Y. H.; Chi, Y. C.; Lin, G.-R. *Laser Phys. Lett.* 2013, 10, 055105.
[83] K¨artner, F. X.; Au, J. A. der; Keller, U. *IEEE J. Sel. Top. Quantum Electron* 1998, 4, 159-168.
[84] K¨artner, F. X.; Jung, I. D.; Keller, U. *IEEE J. Sel. Top. Quantum Electron* 1996, 2, 540-556.
[85] Kelly, S. M. *Electron. Lett.* 1992, 28 (8), 806-807.
[86] Lin, G.-R.; Chiu, I.-H.; Wu, M. C. *Opt. Express* 2005, 13, 1008-1014.

In: Graphite
Editor: Quinton C. Campbell
ISBN: 978-1-62618-576-0

Chapter 8

CLUSTER DEPOSITION AND IMPLANTATION ON/IN GRAPHITE

Vladimir N. Popok*
Department of Physics and Nanotechnology, Aalborg University, Aalborg, Denmark

ABSTRACT

Cluster ion beam technique is a versatile tool which can be used for controllable formation of nanosize objects on a surface as well as modification and processing of surfaces and shallow layers on an atomic scale. In this chapter an overview of research on cluster interaction with graphite is presented. One of the emphases is put on pinning of metal clusters on graphite with a possibility of following selective etching of graphene layers. The other main topic of concern is related to the phenomenon of cluster stopping and the development of scaling law for cluster implantation in graphite. Graphite is chosen for experiments because it is a good model material; it has an atomically smooth surface that makes it easy to resolve very small deposited clusters or damaged areas. Layered structure of graphite with strong covalent bonds in the graphene sheets and very week van der Waals interactions between them is an interesting type of crystalline arrangement for modelling of clusters implantation. Additionally, there is rapidly increasing research activities on graphene which promises a number of applications in nanoelectronics, plasmonics, sensing etc. Therefore, understanding of particle behaviour on soft landing or pinning on graphene is of considerable practical importance.

INTRODUCTION

Ion-beam treatment of materials is one of very widely applied techniques for numeruous research and industrial purposes. Along with traditional monomer ions, atomic or molecular clusters (aggregates of atoms or molecules) have attracted considerable attention during the last two decades [1-6]. A cluster can be formed from atoms of the same chemical element or

* E-mail: vp@nano.aau.dk.

from two or more different species. Depending on composition, these aggregates may exhibit different binding types as well as different geometric and electronic structures [7, 8]. Their sizes can vary from a few up to many thousands of constituents. Medium and large size clusters have diameters on the scale of nanometers. They are often called nanoparticles (NPs).

Interest in clusters comes from various fields. Clusters are used as models to investigate fundamental physical aspects of the transition from the atomic scale to bulk material [9].Clusters deposited on a surface represent systems highly relevant for practical applications. Finite size effects lead to electronic, optical, magnetic, chemical and other properties that can be quite different from those of the same material in the bulk. They are of great interest for practical applications in areas such as electronics and optics, biology and medicine, catalysis and various nanotechnology-related branches [2, 3, 5, 10-15]. Increase of cluster kinetic energy provides a possibility to tune the cluster-surface interaction regime from soft landing towards implantation [4, 16-20]. This paves the way for a number of applications ranging from pinning of size-selected clusters on surface to ultrashallow junction formation [21] or deeping of clusters into soft (polymer) matrices [22]. In this chapter one of emphases is put on pinning of metal clusters on graphite surface with following etching of graphene layers.

High-energy implantation leads to interesting phenomena related to the very large energy density deposited during the cluster-substrate interaction, generating a highly excited region around the projectile path. For instance, nanosize hillocks or pillars can be formed on surfaces by impact of energetic cluster ions [23-25]. Upon MeV implantation of fullerenes into graphite, the relaxation of the deposited energy leads to a phase transition. In particular, the local compression and increase of temperature are enough for the formation of nanocrystalline diamond particles [26].

For successful applications of clusters beams, a theory of cluster stopping in matter is required. Unfortunately, the existing theory for conventional ion implantation can be applied to cluster ions only for very limited number of cases due to the fact that cluster is an aggregate of weakly bonded atoms or molecules. It becomes broken on impact on very short (femtosecond) time scale and its penetrating constituents generate multiple collisions with target atoms [27]. The primary stage of these collisions affects very much both the dynamics of the cluster and the subsequent phenomena. One of them is materials compression on impact leading to rapid heating to very high temperatures [28-30]. Interestingly, the temperature and pressure values are comparable to those known for meteorite-planet collisions [31]. The cluster impact leads to shock wave propagating from the surface towards bulk [32, 33] and to the development of thermal spikes. Since the deposited energy is rather high, it can cause melting of the material around the collision cascades originated by the penetrating cluster constituents [34]. This description of the cluster stopping is brief and very qualitative but it can already give a feeling about complexity of the phenomenon.

Nowadays, there is no commonly accepted theory satisfactory describing the cluster implantation process, even though cluster-solid interactions have been extensively studies for the past 15-20 years. Therefore, another emphasis of this paper is put on the development of scaling laws for cluster implantation. And the main material under the discussion is graphite or very often highly ordered pyrolytic graphite (HOPG). Graphite is chosen for modelling and experiments because it has an atomically smooth surface that makes it easy to resolve very small features on the sub-nm scale. The surface is also relatively chemically inert, i.e. properties and structure of the deposited clusters are not much disturbed. Layered structure of graphite with strong covalent bonds in the graphene layers and very week van der Waals interactions between them is an interesting type of crystalline arrangement for modelling of clusters implantation. On the other hand, the study of cluster deposition or embedding on/in a few layers of graphene has significant practical importance for the rapidly developing field which is aimed in applications of this material in nanoelectronics, plasmonics, sensing etc. [35].

CLUSTER-SURFACE INTERACTION: ENERGY REGIMES

Clusters are typically produced by special sources and then collimated into a beam. Descriptions of cluster formation and size-selection are beyond the scopes of the current chapter and can be found elsewhere [7, 20, 36, 37]. Depending on source type and its constructional details, velocities (kinetic energies) of clusters can vary from rather low (so-called thermal) values up to supersonic ones. Clusters coming out from a source can be either neutral or charged. Neutral clusters can be charged using electron impact or photoionisation. From the application point of view an advantage of cluster ions is in possibility of simple control of their kinetic energies by deceleration or acceleration in electrostatic fields. Thus, one is able to tune the cluster ion energies within very wide interval providing the appropriate regime of the cluster-surface interaction. In this paper, the following regimes are considered: (i) soft landing, (ii) pinning and (iii) implantation. Soft landing and implantation regimes are schematically shown in Figure 1.

The case corresponds to soft landing if the kinetic energy per atom E_{at} is much below the binding (cohesive) energy of the cluster constituents E_{coh}(which is typically below the level of eV/atom). The soft landing does not lead to cluster fragmentation, i.e. the clusters preserve their composition.

If E_{at} is close to or slightly overcomes E_{coh} the cluster undergoes a plastic deformation, it can lose a few constituents but the most of atoms are still bonded to each other. At the same time, the energy transferred to the substrate atoms upon collision can already be high enough to displace one or a few of them. Thus, some point defects are formed and the residual cluster becomes trapped at these sites, i.e. pinned [38]. Pinning is a boundary case between the soft landing and implantation of clusters.

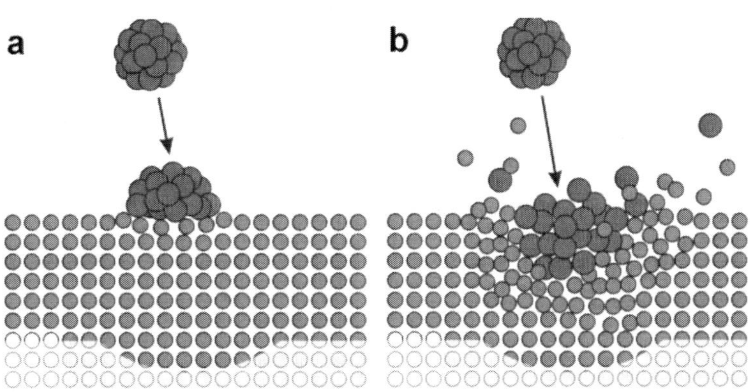

Figure 1. Schematic of cluster-surface interaction: (a) cluster soft landing; (b) cluster implantation.

Pinning suppresses the cluster surface diffusion, thus, making advantages for a number of practical applications in which immobile particles on surface are required.

With further increase of kinetic energy the clusters are decomposed on impact and the constituents become implanted producing collision (radiation) defects in the matrix. Penetration depth of cluster constituents which is called projected range as well as depth and mechanisms of radiation damage are functions of cluster energy and size. These questions are discussed below in more detail.

SOFT LANDING OF CLUSTERS

Study of soft landed clusters is driven by the efforts to utilise various unique properties of supported nanosize objects for applications. From practical point of view, it is important to know what happens to the clusters after deposition. A number of studies were carried out and it was shown that surface diffusion of clusters must be considered. Since the activation barrier strongly depends on the cluster-surface bonding, cluster mobility is favored on weakly interacting materials, such as graphite or amorphous carbon. If the lattice mismatch between the cluster and the substrate is significant, vibrational coupling between these can overcome the small energy barrier of the surface potential, thus, leading to Brownian-like motion of the cluster [39]. For instance, if one compare a hexagonal structure of graphite (with a lattice parameter of 0.246 nm) to a structure of gold cluster (let's assume that the cluster is large enough to follow fcc lattice with a constant of 0.408 nm) it is clear that the lattice mismatch is over 50%. Therefore, interaction potential of HOPG will be very low for the cluster.

Diffusion, Coalescence and Agglomeration of Clusters

Particles moving on the surface can meet and interact yielding two main possible scenarios: coalescence and agglomeration. Coalescence leads to merging of two (or more) particles and the formation of a single larger particle with potentially different shape. It was suggested by Perez et al. [11] that the transition from coalescence to agglomeration of diffusing clusters occurs at a critical size. For instance, very small antimony clusters (n = 4) on HOPG and clusters with n = 4 and 90 on amorphous carbon prefer to coagulate while larger ones (n = 500 and 2200) on HOPG agglomerate with formation of 2D ramified islands as one can see in Figure 2 [40]. Island formation was also observed for Au_n [39] and Pt_n [41] (see Figure 3). In the case of Ag_n clusters soft landed on HOPG, smaller ones (n = 50) were found to coalesce while larger clusters (which can also coalesce) prefer to form ramified islands [42]. Nucleation and growth of agglomerates depend on the surface mobility and sticking coefficient of the clusters. Modeling showed that a combination of rotations and translations can be the dominant mechanism for the motion of Au_n on graphite [43]. Regarding more general properties, though, the motion of large clusters shows strong similarities to adatom diffusion: based on MD simulations it was found that the diffusion rate for Au_{249} on graphite is comparable to that for single adatoms [44].

The versatility of sample preparations based on diffusion is generally limited by its statistical nature. Cluster agglomeration typically results in structures without preferential directional growth. But the diffusion can be guided by the presence of steps or other types of defects on the surface. Clusters tend to collect at natural defects or imperfections. For example, condensation of deposited particles is very often observed at step edges of graphite. It was demonstrated for different metal clusters: Al_{70} [45], Ag_{400} [46], Cu_n [47] etc. If there is a need to reduce the agglomeration, number of surface defects can be artificially increased. For instance, using focused ion beams (FIB) one can "drill" periodic arrays of nanoscale holes; deposited clusters diffuse to the nearest hole where they become trapped, therefore, large islands are not formed [48]. This type of approach was used to prepare a HOPG surface with periodic arrangements of gold clusters collected around the produced defect spots as one can see by the atomic force microscopy (AMF) image in Figure 4 [48]. For a review on different aspects of cluster diffusion, coalescence and island formation see, for example [49].

Applications of Soft Landed Clusters

Soft landed clusters (supported NPs) have a numerous number of practical applications. Clusters of noble metals attract significant attention due to the phenomenon of localised surface plasmon resonance (LSPR) [50]. Position, shape and intensity of the plasmon band strongly depend on the cluster species, size distribution, shape and surface or volumetric fraction of the NPs. Thus, one can control LSPR and utilise this phenomenon for various optical purposes as well as for detection or marking.

Size-controlled magnetic NPs deposited or embedded on/in various matrices can be fabricated using cluster beams of transition metals (Co, Fe, Ni, etc.). Precise control of the metal filling factor and a choice of material forming the matrix for encapsulation of NPs allow tuning the cluster electronic structure and controlling the magnetic properties of the composites [13, 51, 52] as well as their spin-dependent transport [53]. It is worth noting that electronic and magnetic behaviour of small supported or embedded metal clusters is not fully understood so far [54, 55].

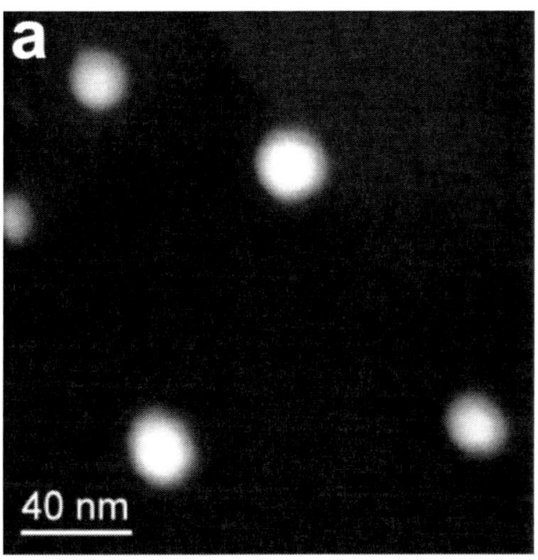

Figure 2. (Continued).

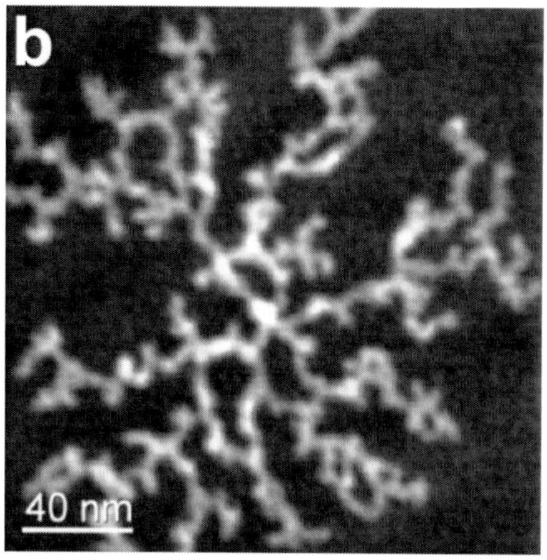

Figure 2. Scanning transmission electron microscopy images of Sb islands grown on graphite surface for different sizes of the incident clusters: (a) <n> = 4; (b) <n> = 500. Reprinted from [40] with permission from Elsevier.

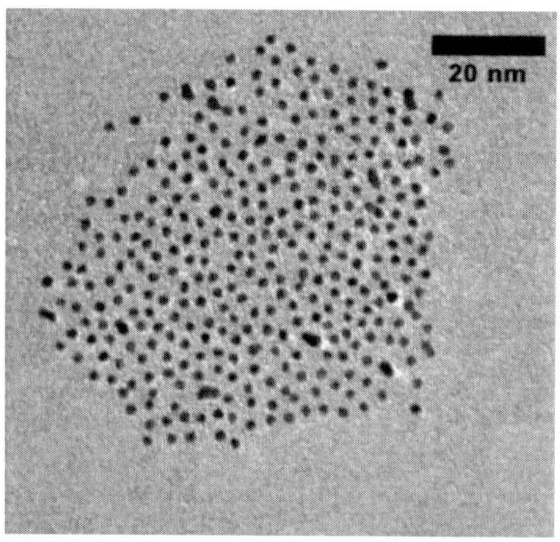

Figure 3. Transmission electron microscopy image of typical morphology of an island formed by the deposition of size-selected (1.6 nm in diameter) platinum clusters on graphite. Reprinted with permission from [41]. Copyright 2008, American Chemical Society.

A number of fundamental physical questions still have to be answered. Research in this direction is continuously bringing surprises and magnetic clusters are of significant interest for formation of novel magnetic storage media as well as magneto-electronic and magneto-optical devices or sensors.

Small clusters have high fraction of surface atoms. Since only the atoms present at surface are chemically active, small NPs are attractive objects for applications in catalysis providing very high efficiency and yield [56].

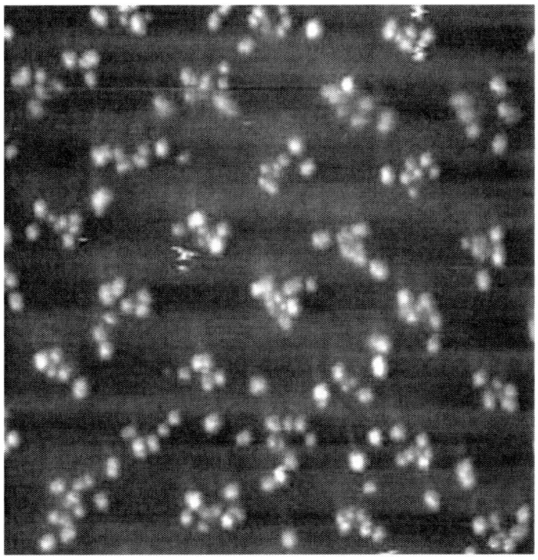

Figure 4. AFM image (1.5x1.5 μm) of the gold clusters deposited with the coverage of 0.02 ML on graphite surface preliminary functionalized using FIB. Periodicity of the defects created by FIB is 300 nm. The clusters are deposited without size selection having initial size distribution centered at around 750 atoms. Reprinted from [48] with permission from Elsevier.

CLUSTER PINNING

As mentioned above, with increase of cluster kinetic energy one can make transition from soft landing to pinning. Due to the energetic impact the structure of the atomic aggregate significantly changes compared to the soft landing case. Good visualization of this transition is shown in Figure 5 by the example of molecular dynamics (MD) modeling of the Ar_{147} impact on graphite [57]. One can see that with the energy increase the cluster produces damage in top graphene layers and becomes slightly embedded. Impacts of a number of different cluster species have been investigated (Ag_n, Au_n, Pd_n, Ni_n and Co_n) [38, 57-62]. Experiments complemented by MD simulations demonstrated a certain threshold (minimum) energy value E_{pin} required for the cluster pinning which was found to be dependent on cluster species, size and substrate composition. A semi-empirical model gave the following relationship between E_{pin} and the transferred energy E_T required to displace a target atom (or produce a recoil one) [63]

$$E_{pin} = \frac{nmE_T}{4NM}, \tag{1}$$

where n is the number of atoms in the cluster, m is the atomic mass of the cluster constituents, M is the mass of the target atoms and N is the number of recoil atoms.

To obtain pinning of small clusters, one can assume that only one carbon atom should be set into motion and this approach agreed rather well with the experiments, showing a linear dependence of E_{pin} on cluster size. In other words, the pinning energy per cluster constituent was found to be a constant value which varies between ca. 6 and 19 eV/atom from Ni to Au [38, 58, 59, 62, 63] for case of HOPG. However, the shape and structure of the pinned clusters is seriously disturbed. For instance, it was shown by MD simulations that large gold clusters are more spread out after the impact and can partially fragment while pinned nickel clusters are more compact [64].

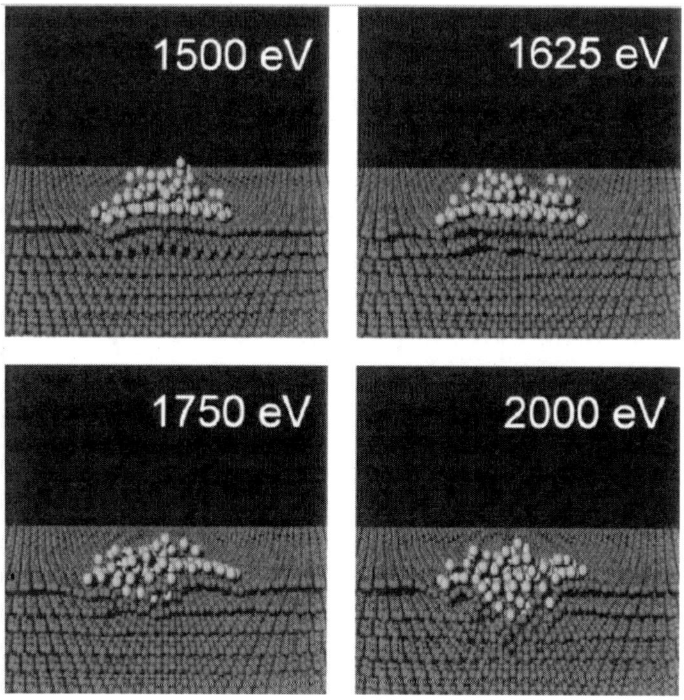

Figure 5. MD simulations, viewed in cross section, of the impact of Ag_{147} cluster at various energies (indicated in the panels) with graphite. Reprinted with permission from [57]. Copyright 2000, American Institute of Physics.

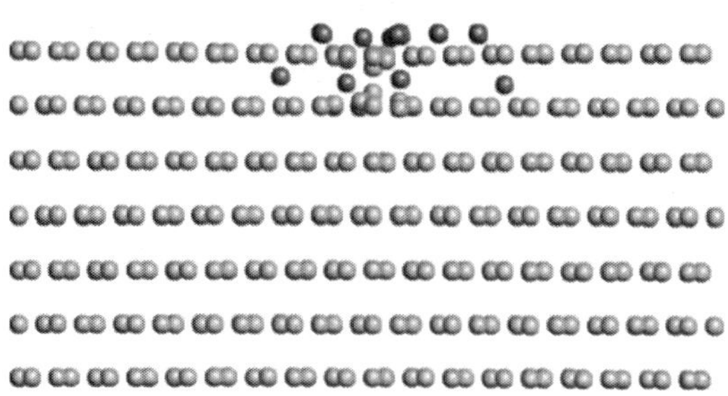

Figure 6. MD simulation, viewed in cross section, of Co_{30} cluster 5 ps after the impact with energy of 300 eV/cluster (10 eV/atom) on graphite. Width of the frame is 4 nm. Co atoms are shown in the darker shade. Reprinted with permission from [62].

Cobalt clusters of a few tens of atoms in size prefer to form single adatom layers on interaction with graphite with several atoms embedded below the top graphene layer (Figure 6) [62]. This picture obtained in modeling was confirmed by the scanning tunneling microscopy (STM) measurements showing the formation of one atomic layer thick small islands of cobalt on the HOPG surface after the impact of clusters with "pinning" energies [61, 62].

It was shown that implantation of Ag_n, Ar_n, $Co_n$and C_{60}clusters[62, 65, 66] into graphite followed by annealing at around 600 °C in presence of oxygen led to the formation of pits with depth corresponding to the introduced radiation damage. This phenomenon is often called oxidative etching. The pits had hexagonal or circular shape. For the case of Co_nclusters impacting HOPG with energies corresponding to the pinning threshold (5-9 eV/atom) or slightly higher the formation of worm-like structures was found [61]. MD simulations predicted that for the energy around E_{pin} the cluster introduces damage to very top graphene layer. With energy increase more layers beneath the surface become damaged [67]. Depth of damage reaches 4-5 monolayers (ML) of graphite if the energy is increased to ca. 50 eV/atom (Figure 7) [62]. For the energies higher than E_{pin} the cluster is broken on impact but the constituents are implanted in close vicinity and their final location in HOPG is close to each other. One can assume that Co-Cobonds are restored to some extent, in other words, there is a residual cluster located in the shallow graphite layer. High temperature increases the diffusive mobility of these residual clusters and at the same time they catalyse the reaction of atmospheric oxygen with carbon, thus, favouring the formation of planar surface channels of random shape as those shown in Figure 8. For the impact energies around E_{pin} the depth of channels is 1 ML and they can be 200-300 nm long after a few minutes of the thermal treatment. For the higher energies (up to 50 eV/atom) the channels are deeper (up to 5 ML) but the length is reduced to 10-15 nm. Further energy increase leads to deep implantation and there are no surface channels produced by the thermal treatment; only pits.

Similar phenomenon of channel formation was reported elsewhere on the deposition of various chemical substances on graphite surfaces followed by heating [68].The heating-induced etching was recently reported for Fe NPs deposited on few-layer graphene [69].While exposed to 900 °C, the nanoparticlesetched away the graphene sheets producing channels. Trench channelling of graphene was also demonstrated by silver NPs under heating up to 650 °C in ambient atmosphere [70]. Schematic picture of this process is shown in Figure 9. Trenching by NPs was suggested as a "catalytic pen" or prospect for high-precision lithography on graphenes. Apart from etching, pinned clusters can have similar applications as those soft landed. But there are also specific directions of interest.

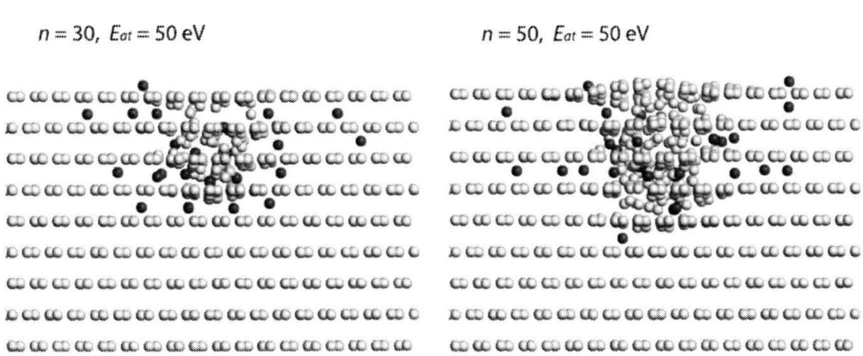

Figure 7. MD simulations, viewed in cross section, of the damage created by impact of 50 eV/atom Co_{30} (left panel) and Co_{50}(right panel)clusters on graphite. Width of the frames is 4 nm. Implanted Co atoms are shown in the darker shade. Reprinted with permission from [62].

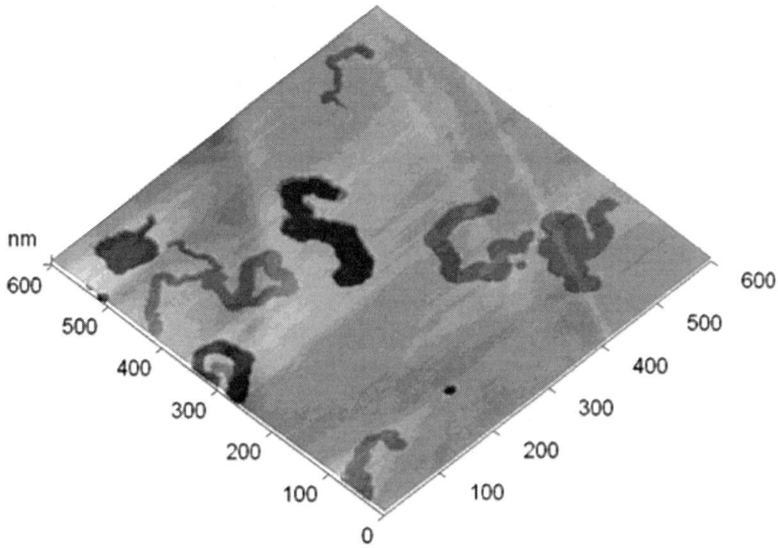

Figure 8. STM image of the graphite surface after Co_{50} cluster impact with energy of 13 eV/atom and following thermal-induced oxidative etching (600 °C, 3 min in ambient atmosphere). Channels depth is 1 (lighter in shade) or 2 (darker in shade) graphene layers.

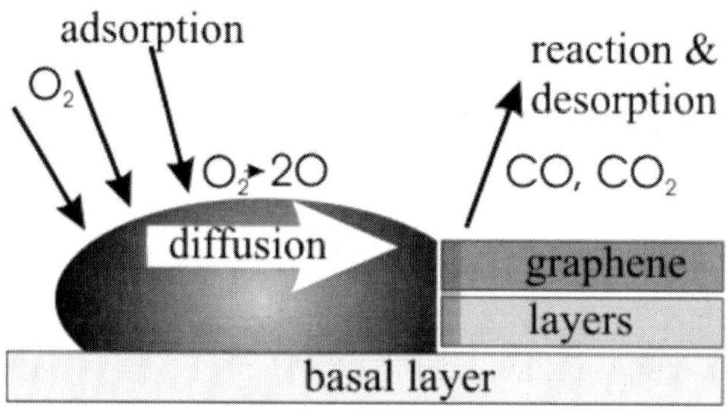

Figure 9. Schematic of a particle channeling a trench into graphene layers: side view of a catalytic particle. Particle diffuses in the plain and catalyzes chemical reaction of chemisorbed oxygen molecules with carbon atoms contacting the particle. Gaseous products of the reaction desorb. Reprinted with permission from [70]. Copyright 2009, American Chemical Society.

For instance, one of them is immobilization of protein molecules for their further investigation or fabrication of biochips [12, 71]. Pinned metal clusters can also be considered as a main component of nanoscale gas sensors [72]. Adsorption of gas molecules on the cluster surface can significantly change its electronic properties. These changes can be monitored either by optical methods (utilizing LSPR) or through measurements of the conductance of the cluster ensembles. The above-mentioned applications are only a very brief overview. More detailed study of these topics is beyond the scopes of this chapter and can be found elsewhere [19].

CLUSTER IMPLANTATION

Crater Formation

Elastic collision of an accelerated monoatomic ion with lattice atoms causes a knock-on effect: the displaced atoms can be sputtered from the surface or become so-called recoils contributing into the formation of a displacement cascade in the target [73, 74]. It was experimentally observed that impact of heavy ions can lead to formation of small craters on the surface [75-78]. The phenomenon was ascribed to the combination of sputtering and development of displacement cascades. Energetic cluster ions were found to be much more efficient in crater formation. This phenomenon was studied both experimentally and by MD modelling [65, 79-88]. The multiple collision effect leads to local high-energy transfer: the target atoms possess high kinetic energy.

Those in the core part of the impact area contribute into the development of inward displacement cascades forming a depression on the surface, i.e. a crater. Large fraction of atoms – those present at periphery of the impact area – obtain lateral momenta or momenta directed away from the surface (Figure 10a) [29]. These atoms either form an atomistic flow that contribute into the formation of the crater rim or become sputtered (Figure 10b).

The efficiency of crater formation is dependent on cluster size and kinetic energy. Detailed analysis of these parameters can be found for instance in [19]. Substrate material also plays an important role. Higher density, melting (or sublimation) point and larger atomic displacement energies provide less favourable conditions for crater formation. Diamond is a good example of such material on which it is difficult to produce craters [89, 90]. Graphite is another allotropic form of carbon having layered structure with strong covalent bonds (significant displacement energies of ca. 12-20 eV along *c*-axis are required) in graphene plains and week van der Waals interactions between the layers.

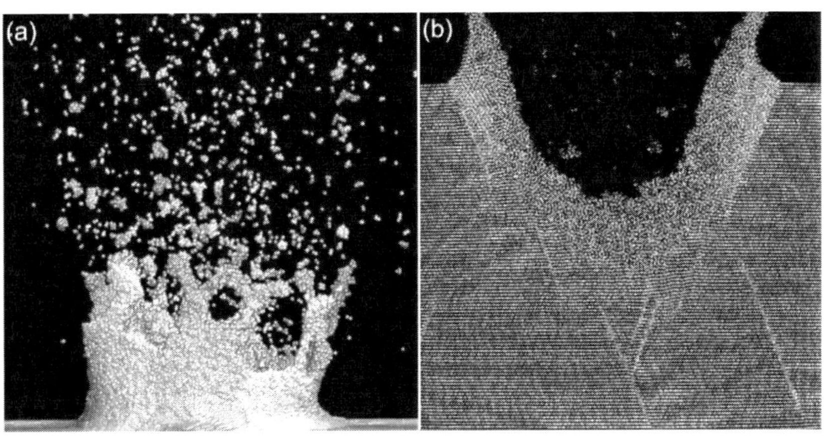

Figure 10. Snapshots of MD simulations of 650 keV Au_{13} cluster impact on gold: (a) sputtering of surface atoms in the beginning of collision and (b) crater formation after 16 ps. Reprinted from [19] with permission from Elsevier.

Therefore, the graphite structure responds very elastically to cluster impact: the collision induces oscillations of the graphene planes [67]. For the case of relatively small Ar_n (n = 16 and 41) colliding HOPG with energies up to 16 keV/cluster, the oscillations have very little influence on the structure outside the immediate impact region with primary displacement cascades, although their amplitude could be as large as the distance between two neighbouring planes.

Thus, a crater can be formed only at the initial stage of impact. The elastic behaviour of graphene sheets at a later stage causes efficient closure of the craters and only disordered areas are finally formed. These damaged areas were found by STM as tiny bumps (Figure 11a). Figure 11b shows the crater closure predicted by MD modelling. It is worth noting that craters can be formed on HOPG but the clusters should be larger and they should have high kinetic energies in order to be able to provide high energy density transfer to the graphite target. One of such examples can be found in [91].

Projected Range and Radiation Damage

Due to the fact that cluster is an aggregate of atoms or molecules with weak bonding between the constituents, implantation of cluster ion is very different from that of monatomic ion [27]. Cluster consisting of many atoms generates multiple collisions with target atoms. Every cluster constituent transfers some amount of energy to the target.

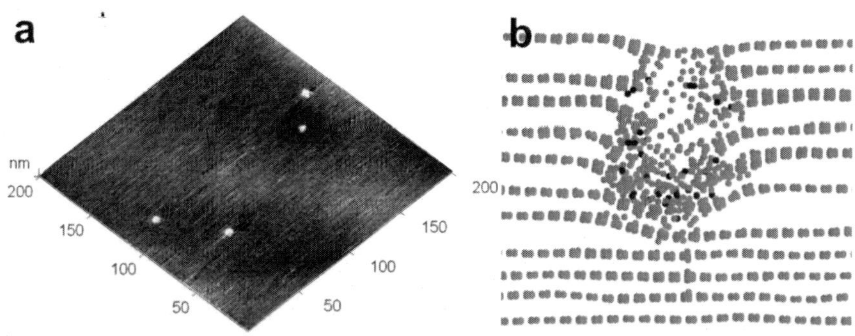

Figure 11. (a) STM image of HOPG surface implanted by Ar_{41} clusters with energy of 4.1 keV (100 eV/atom) and (b) snapshot of MD simulation (after 20 ps) for the same impact conditions. Width of the frame is 5 nm. Implanted argon atoms are shown in the darker shade. Reprinted from [89] with permission from Elsevier.

Since clusters are very compact aggregates, the energy is transferred very locally. The surface interaction area is on the nanometer scale, i.e. the energy density is high. At the beginning of the impact, energy and momentum brought to the target cause its compression and an increase of temperature. Local temperature and pressure can rise up to 10^4-10^5 K and the GPa level, respectively, for the first 10^{-13}-10^{-12} s [28-30]. Shock wave is formed and it propagates from the surface towards bulk [32, 33]. Thermal spikes are produced. In the case of keV-energy cluster implantation the thermal spikes originate via nuclear stopping of the projectiles [92, 93].Closely penetrating cluster constituents produce overlapping displacement cascades leading to high radiation damage [94]. These cascades have complex nature due to the non-linear effects related to interaction of cluster constituents not only with the target atoms but also with each other.

MD and Monte Carlo simulations showed that the penetration depth of clusters is larger than for the ions of the same chemical element at the same incident velocity [95, 96]. This phenomenon was called clearing-the-way effect and explained in a way that the "front" atoms of the cluster push target atoms out of the way of the "rest" atoms. Therefore, the stopping power of the cluster is reduced and the projected range is increased. This phenomenon was confirmed by a number of simulations and experiments. One can find an overview of these results for example in [19]. In this chapter, mostly the data on cluster stopping in graphite are analysed.

It was already mentioned above that cluster ion implantation followed by heat treatment in the presence of oxygen leads to the formation of pits with the depth corresponding to the depth of the radiation damage cascades developed by the projectiles [62, 65, 97]. Etching removes only the damaged volume if the treatment time is on the scale of a few minutes.

To attack the underlying pristine graphene planes the time must be significantly increased. Hence, one can experimentally and quite precisely (using AFM or STM) measure the depth of the radiation damage for the implanted graphite samples. Moreover, MD simulations showed that for keV-energy clusters penetration depth of cluster constituents (mean projected range R_p) is very close to the depth of radiation damage: the secondary (recoil) displacement cascades give only a minor contribution to the defect formation in graphite [97].

For C_{60}^+ ion implantation into graphite in a rather wide range of kinetic energies E from 500 eV to 23 keV the penetration depth was found to follow $E^{1/2}$[80]. The same type of dependence was demonstrated for the case of graphite bombarded by keV-energy Ag_n, Au_n and Si_n (n = 7) clusters [66].

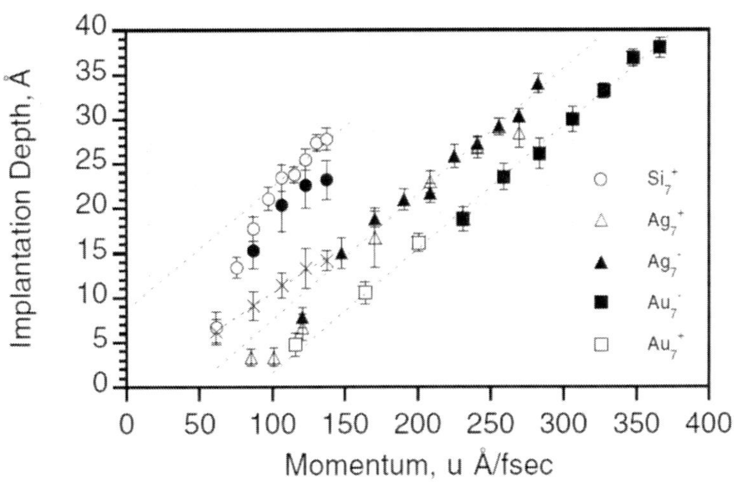

Figure 12. The implantation depth for Au_7, Ag_7 and Si_7 clusters in graphite as a function of momentum. The filled symbols correspond to MD simulations, the open ones to experimental measurements. The dashed lines are the best fit straight lines. The crosses are the calculated centre-of-mass cluster implantation depths. Reprinted with permission from [98].

Since cluster momentum $p{\sim}E^{1/2}$, it was suggested to scale R_p with momentum that leaded to the linear scaling law (Figure12). The use of momentum allows considering the cluster size (through its mass) together with the energy. Thus, the projected range can be defined using only one physical quantity – momentum.

$$R_p = a + bp = a + bmv \quad , \tag{2}$$

where m and v are the cluster mass and velocity, respectively, a and b are the variables. This equation can also be represented using the size-dependent variables a_n and b_n

$$R_p = a_n + b_n v \quad . \tag{3}$$

In this case the mass is included into the variables.

One of the important parameters defining the projected range and radiation damage is stopping power which can be found by considering the main mechanisms of projectile stopping in matter. For the case of clusters with keV-energies the dominant mechanism is nuclear stopping $(dE/dx)_{nucl}$. To find it one needs to take into account that E is the kinetic energy equal to $mv^2/2$ and x represents R_p. Then by substituting v in eq. (3) one can get

$$\left(\frac{dE}{dx}\right)_{nucl} = \frac{mv}{b_n} = \frac{p}{b_n} \tag{4}$$

Thus, both the projected range and nuclear stopping power are found to be linear functions of the cluster momentum.

In Figure12, the best fit lines are different for various cluster species that leads to infer about some implantation parameter which is not considered. Seminara et al. [66] took into account a cross-sectional area of the cluster projected on the surface and divided the cluster momentum by this area, thus, introducing the scaled momentum. A similar approach was used in [62, 97] for Co_n and Ar_n clusters. It was found that the experimental values of depth of radiation damage (which is approximately equal to R_p) for different cluster sizes fall on the same fit straight line vs scaled (or projected) cluster momentum as shown in Figure13a. MD simulations for the same cluster species, sizes and energies demonstrated good agreement with the experiments (Figure13b).

Considering the cross-sectional area of cluster-surface interactions allows to extend the scaling law for different cluster species. As one can see in Figure 13a the same line fits to both argon and cobalt. This provides very strong support for the assertion that the cluster implantation depth is a linear function of momentum per unit surface area.

The generality of this statement can also be seen in Figure14where the straight line fit from Figure13 is plotted together with the implantation data taken from the literature [66, 98] for Si_7, Au_7, and Ag_n (n=7 and 13). Thus, the same universal plot is obtained for implantation in graphite, regardless of the cluster constituent. Recently, it was shown that the scaling law in which both depth of radiation damage and mean projected range of cluster constituents linearly depend on momentum is also valid for keV-energy argon cluster ions implanted in diamond [90]. Hence, this universal scaling law is an important contribution into the development of general theory of cluster stopping in matter.

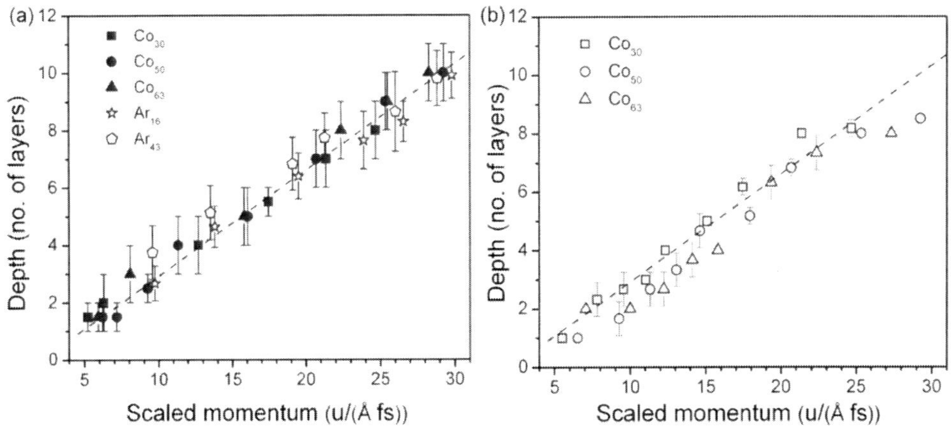

Figure 13. (a) Experimentally found and (b) theoretically simulated dependences of depth of etched pits (in number of graphene layers) on scaled momentum for implantation of size-selected cobalt and argon clusters. Dashed line shows the best fit straight line for the experimental data. Reprinted from [20] with permission from Elsevier.

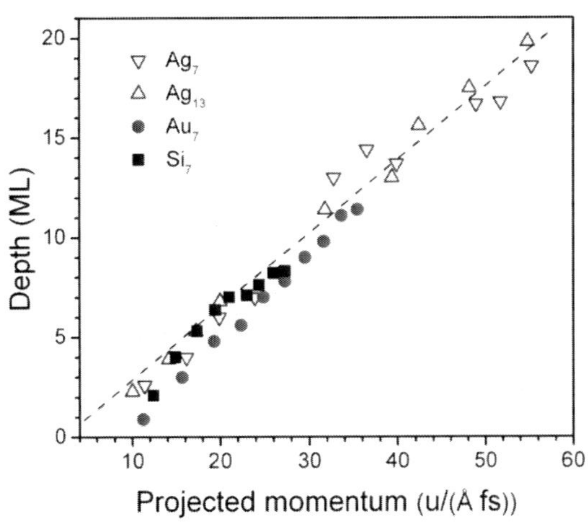

Figure 14. Experimentally found dependences of depth of etched pits (in number of graphene layers) on projected momentum for implantation of size-selected silver, gold and silicon clusters into graphite. Dashed line is the same best fit as in Figure 13. Reprinted from [19] with permission from Elsevier.

CONCLUSION

Cluster beams is a novel approach which is widely used for engineering the physical and chemical properties of materials on the nanoscale. This technique has a number of practical applications which are of increasing interest in both academic and industrial domains. In this chapter the research related to cluster interactions with graphite is overviewed.

For low cluster kinetic energies the bonds between the soft landed particles and graphite surface are very week and the clusters can easily diffuse that leads to either coalescence or agglomeration. These processes are briefly analyzed. The focus is put on pinning of metal clusters where the impacting particle produces a small damaged site in the top graphene layer and becomes bound to it (immobilized).

A semi-empirical model for the cluster pinning is discussed and so-called threshold energies for the pinning are estimated. It is found that the pined clusters can be of different shapes depending on the species and they can even form monolayer islands (example of cobalt clusters with a few tens of atoms in size).

Pinned clusters can produce random in shape, narrow and shallow (1-2 ML deep) trenches under particular heating conditions due to the induced cluster mobility and simultaneous catalysis of the reactions between oxygen and carbon. Thus, small imbedded nanoparticles can be used as a processing tool for graphene.

Cluster implantation in HOPG is another focus of this chapter. A number of effects related to the energetic interaction and penetration of cluster constituents into material is discussed. In particular, a phenomenon of crater formation on graphite is analysed.

Then, it is shown that heating-induced etching of the radiation damage introduced by the cluster impacts provides a measure of the depth to which the collision cascades are developed and allows a comparison of these data with MD simulations. Very good quantitative agreement between the experiments and simulations was demonstrated for the cases of Ar_n and Co_n cluster implantation. Comparison of these studies with the results on implantation of other cluster species in HOPG leads to the conclusion about universal stopping law for cluster constituents in graphite where both the projected range and depth of radiation damage linearly scale with cluster momentum. This scaling law is an important contribution into the general theory of cluster implantation in matter.

REFERENCES

[1] Milani, P.; Iannotta,S. Cluster Beam Synthesis of Nanostructured Materials;Springer: Berlin, 1999.

[2] Meiwes-Broer, K.-H.; Ed.; Metal Clusters at Surfaces; Springer: Berlin, 2000.

[3] Binns,C. *Surf. Sci. Rep.* 2001, 44, 1-49.

[4] Popok, V.N., Campbell, E.E.B. Rev. Adv. Mater. Sci. 2006, 11, 19-45.

[5] Wegner, K., Piseri, P., Tafreshi, H.V., Milani, P. *J. Phys. D.* 2006, 39, R439-R459.

[6] Toyoda, N., Yamada, I. IEEE Trans. Plasma Sci. 2008, 36, 1471-1488.

[7] Haberland, H.; Ed.; Clusters of Atoms and Molecules; Springer-Verlag: Berlin, 1994.

[8] Johnston, R.L.; Atomic and Molecular Clusters; TaylorandFrancis: London, 2002.

[9] Jena, P., Castleman, A.W. *Jr. Proc. Nation. Acad. Sci.* 2006, 103, 10560-10569.

[10] Kreibig,U.; Vollmer, M. Optical Properties of Metal Clusters; Springer: Berlin, 1995.

[11] Perez, A., Melinon, P., Dupius, V., Jensen, P., Prevel, B., Tuaillon, J., Bardotti, L., Marlet, C., Treilleux, M., Broyer, M., Pellarin, M., Vaille, J.L., Palpant, B., Lerme, J. *J. Phys. D: Appl. Phys.* 1997, 30, 709-721.

[12] Palmer, R. E., Pratontep, S. H., Boyen, G. *Nature Mater.*2003, 2, 443-448.

[13] Bansmann, J., Baker, S.H., Binns, C., Blackman, J.A., Bucher, J.-P., Dorantes-Davila, J., Dupius, V., Favre, L., Kechrakos, D., Kleibert, A., Meiwes-Broer, K.-H., Pastor, G.M., Perez, A., Toulemonde, O., Trohidou, K.N., Tuaillon, J., Xie, Y. *Surf. Sci. Rep.* 2005, 56, 189-275.

[14] Woodruff, D.P.; Ed.; Atomic Clusters: From Gas Phase to Deposition; The Chemical Physics of Solid Surfaces; Elsevier: Amsterdam, 2007, Vol. 12.

[15] Kaden, W. E., Wu, T., Kunkel, W. A., Anderson, S. L. *Science*.2009, 326, 826-829.

[16] Gspann, J. In Large Clusters of Atoms and Molecules; Martin, T.P.;Ed.; Kluwer: Amsterdam, 1997; pp 443-461.

[17] Yamada, I., Matsuo, J., Toyoda, N., Kirkpatrick, A.*Mater. Sci. Eng. R*.2001, 34, 231-295.

[18] Popok, V. In Handbook of Nanophysics: Clusters and Fullerenes; Sattler, K.D.; Ed.; *CRC Press: Boca Raton*, 2010; pp 19-1 – 19-19.

[19] Popok, V.N. *Mater. Sci. Eng. R*.2011, 72, 137-157.

[20] Popok, V.N., Barke, I., Campbell, E.E.B., Meiwes-Broer, K.-H. *Surf. Sci. Rep.* 2011, 66, 347-377.

[21] Borland, J., Hautala, J., Gwinn, M., Tetreault, T. G., Skinner, W.*Sol. State Technol.* 2004, 47, 64-67.

[22] Corbelli, G., Ghisleri, C., Marelli, M., Milani, P., Ravagnan, L. *Adv. Mater.* 2011, 23, 4504-4508.

[23] Thevenard, P. A., Dupin, J. P., Vu Thien, B., Purcell, S. T., Semet, V.*Surf. Coat. Technol.* 1999, 128-129, 59-65.

[24] Toulemonde, M., Trautmann, C., Balanzat, E., Hjort, K., Weidinger, A.*Nucl. Instrum. Meth. Phys. Res.B.* 2004, 216, 1-8.

[25] Popok, V.N., Prasalovich, V. S., Campbell, E.E.B. *Nucl. Instrum. Meth. Phys. Res. B.* 2003, 207, 145-153.

[26] Dunlop, A., Jaskierowicz, G., Ossi, P. M., Della-Negra, S.*Phys. Rev. B.* 2007, 76, 155403(1-14).

[27] Takagi, T.*Mater. Sci. Eng. A.* 1998, 253, 30-41.

[28] Cleveland, Ch.L., Landman, U.*Science.* 1992,257, 355-361.

[29] Colla, T.J., Aderjan, R., Kissel, R., Urbassek, H.M.*Phys. Rev. B.* 2000,62, 8487-8493.

[30] Allen, L.P., Fenner, D.B., Santeufemio, C., Brooks, W., Hautala, J., Shao, Y.*Proc. SPIE.* 2002, 4806, 225-232.

[31] Melosh, H.J.; Impact Cratering: A Geologic Process; Oxford University Press: New York, 1989.

[32] Averback, R.S., Ghaly, M. *Nucl. Instrum. Meth. Phys. Res.* B. 1994, 90, 191-201.

[33] Insepov, Z., Yamada, I.*Nucl. Instrum. Meth. Phys. Res. B.* 1996, 112, 16-22.

[34] Peltola, J., Nordlund, K., *Phys. Rev. B.* 2003, 68, 035419(1-5).

[35] *MRS Bulletin.* 2012, 37, 1119-1321.

[36] de Heer, W.A., *Rev. Mod. Phys.* 1993, 65, 611-676.

[37] Pauly, H.; Atom, Molecule and Cluster Beams; Springer: Berlin, 2000; Vol. 2.

[38] Carrol, S.J., Weibel, P, von Issendorf, B., Kuipers, L., Palmer, R.E. J. *Phys.: Condens. Mater.* 1996, 8, L617-L624.

[39] Vandamme, N., Janssens, E., Vanhoutte, F., Lievens, P., Van Haesendonck, C. *J.Phys.: Condens. Mater.* 2003, 15, S2983-S2999.

[40] Yoon, B., Akulin, V. M., Cahuzac, P., Carlier, F., de Frutos, M., Masson, A., Mory, C., Colliex, C., Brechignac, C.*Surf. Sci.*1999, 443, 76-88.

[41] Tainoff, D., Bardotti, L., Tournus, F., Guiraud, G., Boisron, O., Melinon, P. *J. Phys. Chem. C.* 2008, 112, 6842-6849.

[42] Goldby, I.M., Kuipers, L., von Issendorf, B., Palmer, R.E. *Appl. Phys. Lett.* 1996, 69, 2819-2821.

[43] Guerra, R., Tartaglino, U., Vanossi, A., Tosatti, E. *Nature Mater.* 2010, 9, 634-637.

[44] Lewis, L.J., Jensen, P., Combe, N., Barrat, J.-L.*Phys. Rev. B.* 2000, 61, 16084-16090.

[45] Klipp, B., Grass, M., Müller, J., Stolcic, D., Lutz, U., Ganteför, G., Schlenker, T., Boneberg, J., Leiderer, P. *Appl. Phys. A.* 2001, 73, 547-554.

[46] Carroll, S.J., Palmer, R.E., Mulheran, P.A., Hobday, S., Smith, R. *Appl. Phys. A.* 1998, 67, 613-619.

[47] Hartmann, H., Popok, V.N., Barke, I., von Oeynhausen, V., Meiwes-Broer, K.-H. *Rev. Sci. Instrum.*2012, 83, 073304(1-6).

[48] Bardotti, L., Prevel, B., Jensen, P., Treilleux, M., Melinon, P., Perez, A., Gierak, J., Fainin, G., Mailly, D. *Appl. Surf. Sci.* 2002, 191, 205-210.

[49] Jensen, P. *Rev. Mod. Phys.* 1999, 71, 1695-1735.

[50] Quinten, M.; Optical Properties of Nanoparticle Systems;Wiley-VCH:Weinheim, 2011.

[51] Peng, D.L., Sumiyama, K., Hihara, T., Yamamuro, S., Konno, T.*J. Phys. Rev. B.* 2000, 61, 3103-3109.

[52] Binns, C., Trohidou, K.N., Bansmann, J., Baker, S.H., Blackman, J.A., Bucher, J.-P., Kechrakos, D., Kleibert, A., Louch, S., Meiwes-Broer, K.-H., Pastor, G.M., Perez, A., Xie, Y. *J. Phys. D: Appl. Phys.* 2005, 38, R357-R379.

[53] Lius, F., Bartolome, F., Petroff, F., Bartolome, J., Garcıa, L.M., Deranlot, C., Jaffres, H., Martınez, M.J., Bencok, P., Wilhelm, F., Rogalev, A., Brookes, N.B.*Europhys. Lett.* 2006, 76, 142-148.

[54] Lau, J.T., Föhlisch, A., Nietubyc, R., Reif, M., Wurth, W.*Phys. Rev. Lett.* 2002, 89, 057201(1-4).

[55] Brune, H., Gambardella, P.*Surf. Sci.* 2009, 603, 1812-1830.

[56] Vajda, S., Pellin, M.J., Greeley, J.P., Marshall, C.L., Curtiss, L.A., Ballentine, G.A., Elam, J.W., Catillon-Mucherie, S., Redfern, P.C., Mehmood, F., Zapol, P. *Nature Mater.* 2009, 8, 213-216.

[57] Carroll, S.J., Pratontep, S., Streun, M., Palmer, R.E., Hobday, S., Smith, R.*J. Chem. Phys.*2000, 113, 7723-7727.

[58] Di Vece, M., Palomba, S., Palmer, R.E.*Phys. Rev. B.* 2005, 72, 73407(1-4).

[59] Gibilisco, S., Di Vece, M., Palomba, S., Faraci, G., Palmer, R.E.*J. Chem. Phys.* 2006, 125,084704(1-3).

[60] Vučković, S., Svanqvist, M., Popok, V.N. *Rev. Sci. Instrum.* 2008, 79, 073303(1-8).

[61] Vučković, S., Samela, J., Nordlund, K., Popok, V.N. *Eur. Phys. J. D.* 2009, 52, 107-110.

[62] Popok, V.N., Vučković, S., Samela, J., Järvi, T., Nordlund, K., *Campbell, E.E.B. Phys. Rev. B.*2009, 80, 205419(1-12).

[63] Smith, R., Nock, C., Kenny, S.D., Belbruno, J.J., Di Vece, M., Palomba, S., Palmer, R.E.*Phys. Rev. B.* 2006, 73, 125429(1-5).

[64] Smith, R.; Nock, C.; Kenny, S.D.; Belbruno, J.J.; Palmer, R.E.; In Atomic Clusters: From Gas Phase to Deposition; Woodruff, D.P.; Ed.; The Chemical Physics of Solid Surfaces; Elsevier: Amsterdam, 2007; Vol. 12; pp 589-616.
[65] Bräuchle, G., Richard-Schneider, S., Illig, D., Rockenberger, J., Beck, R.D., Kappes, M.M.*Appl. Phys. Lett.*1995, 67, 52-54.
[66] Seminara, L., Convers, P., Monot, R., Harbich, W.*Eur. Phys. J. D.* 2004, 29, 49-56.
[67] Samela, J., Nordlund, K., Keinonen, J., Popok, V.N., Campbell, E.E.B. *Eur. Phys. J. D.* 2007, 43, 181-184.
[68] Chang, H., Bard, A. *J. Am. Chem. Soc.* 1991, 113, 5588-5596.
[69] Datta, S.S. Strachan, D.R., Khamis, S.M., Johnson, A.T.C. *Nano Lett.* 2008, 8, 1912-1915.
[70] Severin, N., Kirstein, S., Sokolov, I.M., Rabe, J.P. *Nano Lett.* 2009, 9, 457-461.
[71] Palmer, R.E., Leung, C., *Trends Biotechnol.* 2007, 25, 48-55.
[72] Watson, T.F., Belic, D., Convers, P.Y., Boyd, E.J., Brown, S.A.*Eur. Phys. J. D.* 2011,61, 81-85.
[73] Sigmund, P. *Nucl. Instrum. Meth. Phys. Res. B.* 1987, 27, 1-20.
[74] Nastasi, M.; Mayer, J.W. Ion Implantation and Synthesis of Materials; Springer: Berlin, 2006.
[75] Merke, K.L., Jäger, W.*Philos. Mag. A.* 1981, 44, 741-762.
[76] Wilson, I.H., Zheng, N.J., Knipping, U., Tsong, I.S.T.*Phys. Rev. B.* 1988, 38, 8444-8450.
[77] Neumann, R.*Nucl. Instrum. Meth. Phys. Res. B.* 1999, 151, 42-55.
[78] Teichert, C., Hohage, M., Michely, T., Comsa, G. *Phys. Rev. Lett.*1994, 72, 1682-1685.
[79] Seki, T, Aoki, T, Tanomura, M., Matsuo, J., Yamada, I.*Mater. Chem. Phys.*1998, 54,143-146.
[80] Webb, R.P., Kerford, M., Kappes, M., Bräuchle, G.*Nucl. Instrum. Meth. Phys. Res. B.* 1997, 122, 318-321.
[81] Allen, L.P., Insepov, Z., Fenner, D.B., Santeufemio, C., Brooks, W., Jones, K.S., Yamada, I.*J. Appl. Phys.*2002, 92, 3671-3678.
[82] Popok, V.N., Prasalovich, S.V., Campbell, *E.E.B. Vacuum.* 2004, 76, 265-272.
[83] Popok, V.N., Prasalovich, S.V., Campbell, *E.E.B. Surf. Sci.*2004, 566-568, 1179-1184.
[84] Samela, J., Nordlund, K., Keinonen, J., Popok, V.N., *Nucl. Instr. Meth. Phys. Res. B* 2007, 255, 253-258.
[85] Samela, J., Nordlund, K., Popok, V.N., Campbell, E.E.B.*Phys. Rev. B.* 2008, 77, 075309(1-15).
[86] Popok, V.N., Jensen, J., Vučković, S., Mackova, A., Trautmann, C. J. *Phys. D: Appl. Phys.* 2009, 42, 205303(1-6).
[87] Yamaguchi, Y., Gspann, J.*Phys. Rev. B.* 2002, 66, 155408(1-10).
[88] Nordlund, K., Samela, *J. Nucl. Instrum. Meth. Phys. Res. B.* 2009, 267, 1420-1423.
[89] Popok, V.N., Samela, J., Nordlund, K., Popov, V.P.*Nucl. Instrum. Meth. Phys. Res. B* 2012, 282, 112-115.
[90] Popok, V.N., Samela, J., Nordlund, K., Popov, V.P. *Phys. Rev. B.* 2012, 85, 033405(1-4).
[91] Houzumi, S., Mochiji, K., Toyoda, N., Yamada, I. *Jpn. J. Appl. Phys.* 2005, 44, 6252 6254.
[92] Kelly, R.*Rad. Eff.*1977, 32, 91-100.

[93] Miotello, A., Kelly, R.*Nucl. Instrum. Meth. Phys. Res. B*. 1997,122, 458-469.
[94] Shao, L., Nastasi, M., Wang, X., Liu, J., Chu, W.-K. *Nucl. Instrum. Meth. Phys. Res.* B. 2006, 242, 503-505.
[95] Shulga, V.I., Vicanek, M., Sigmund, P. *Phys. Rev. A*. 1989, 39, 3360-3372.
[96] Yamamura, Y., Muramoto, T. *Rad. Eff. Def. Sol.* 1994,130-131, 225-231.
[97] Popok, V.N., Samela, J., Nordlund, K., Campbell, E.E.B. *Phys. Rev. B* 2010, 82, 201403(R1-R4).
[98] Pratontep, S., Preece, P., Xirouchaki, C., Palmer, R.E., Sanz-Navarro, C.F., Kenny, S.D., Smith,R. *Phys. Rev. Lett.*2003, 90, 055503(1-4).

In: Graphite
Editor: Quinton C. Campbell
ISBN: 978-1-62618-576-0

Chapter 9

THERMAL STABILITY OF FULLY LITHIATED GRAPHITE

Yoon-Soo Park and Sung-Man Lee*
Department of Advanced Materials Science and Engineering,
Kangwon National University, Chuncheon Kangwon-Do, Republic of Korea

ABSTRACT

The safety-related performance of lithium ion batteries is a major concern in their potential applications. The safety of lithium ion batteries is closely related to the thermal stability of the cell materials, in which appears that the thermal stability of anode can be critical to the thermal runaway of lithium-ion batteries. Here is an overview of the studies on the thermal stability of fully lithiated natural graphite investigated using differential scanning calorimetry (DSC), which deal with the effect of particle size and carbon coating on thermal stability of lithiated graphite anode. The initial heat generation is due to the solid electrolyte interface (SEI) decomposition. The onset temperatures for exothermic reactions after initial SEI decomposition, which is associated with the release of intercalated lithium from graphite, appear to be affected by particle sizes and surface carbon coating. A thermal induced delithiation is facilitated by reduced diffusion path and higher surface area in smaller graphites, but alleviated by carbon coating. These structural changes in graphites during DSC scan are confirmed by ex situ X-ray diffraction analysis.

INTRODUCTION

The safety-related performance of lithium ion batteries is a major concern in their potential applications. The safety of lithiumion batteries is closely related to the thermal stability of the cell materials. It is also known that the thermal stability of anode is critical to the thermal runaway of lithium-ion batteries [1–4]. According to previous studies using DSC [3, 5–16] and accelerated rate calometry (ARC) [17–23], the thermal reaction of graphite as a

* E-mail: smlee@kangwon.ac.kr.

representative anode material is dependent on many factors such as the carbon type, state of charge, binder level, the presence of electrolyte and so on. However, little attention has been paid to the effect of particle size of graphite powders on the thermal stability of graphite anodes. Moreover, in previous studies on the thermal stability of graphite electrodes using DSC, the shapes and onset temperature of exotherms seem to be not consistent.

Surface modification by carbon-coating the graphite surface has been reported as an effective approach to suppress the irreversible intercalation of the solvated species and the side-reactions leading to the formation of the SEI layer [4, 24–27]. Spherical natural graphite is more favorable than flaky natural graphite as an anode material because it can provide an improved energy density and rate capability. Accordingly, carbon-coated spherical natural graphite is considered to be a very promising anode material for lithium-ion batteries. Understanding the thermal stability of carbon-coated natural graphite is also important for its application in Li-ion battery. Nevertheless, only a few studies have investigated the thermal stability of the carbon-coated natural graphite [15]. The activity and the enthalpy of the exothermic reaction of the lithiated carbon anode in the presence of electrolyte/PVDF binder are dependent on the morphology and the surface chemistry of graphite, as well as on the degree of lithiation [3, 5-7].

The work is focused on the thermal stability of lithiated flake-type graphite electrodes with different particle sizes [28], and the effect of carbon coating on the thermal stability of spherical natural graphite [29]. The main results are as follows.

1. Particle Size Effect [28]

In order to investigate the effect of particle size, four natural graphites with different sizes were used. Figure 1 shows SEM images of four graphites used in this study. The graphite particles had a flaky appearance. These graphite samples had particle size distributions that peak at 3.6, 5.7, 25.4, and 138.4 μm, as shown in Figure 2. We call these samples G1, G2, G3 and G4, respectively. The BET surface areas of these samples were 13.4, 12.9, 2.95 and 0.95 m^2/g, respectively. XRD patterns of the graphites show typical characteristics of 'highly' crystallized graphite (see Figure 3). The d_{002} spacing, which was determined by X-ray diffraction analysis, is 3.35Å and was essentially the same for all of the samples.

The first charge and discharge potential profiles of four graphite samples are presented in Figure 4 and show similar intercalation/deintercalation of lithium in graphite. However, it is evident that during the first intercalation, the plateau at 0.7 V, which is related to the decomposition of the electrolyte and formation of solid electrolyte interface (SEI) film on graphite [30], decreases with increasing the particle size of graphite as expected from the corresponding BET surface areas.

The thermal behavior of the graphite samples have been studied by means of DSC. For DSC measurements, the cells were pre-cycled three times to reach a stable capacity level and the cycling was interrupted when the cells were charged to a fully intercalated state. The charged cells were disassembled in a glove-box. Discs with a diameter of 0.5cm and a copper current-collector were cut from the electrode sheet without removing the electrolyte and transferred to a high-pressure stainless-steel pan with a gold-plated copper seal. The weight of each DSC pan with sample was measured before and after the experiment. The weight was constant in all cases, indicating that there were no leaks during the experiments.

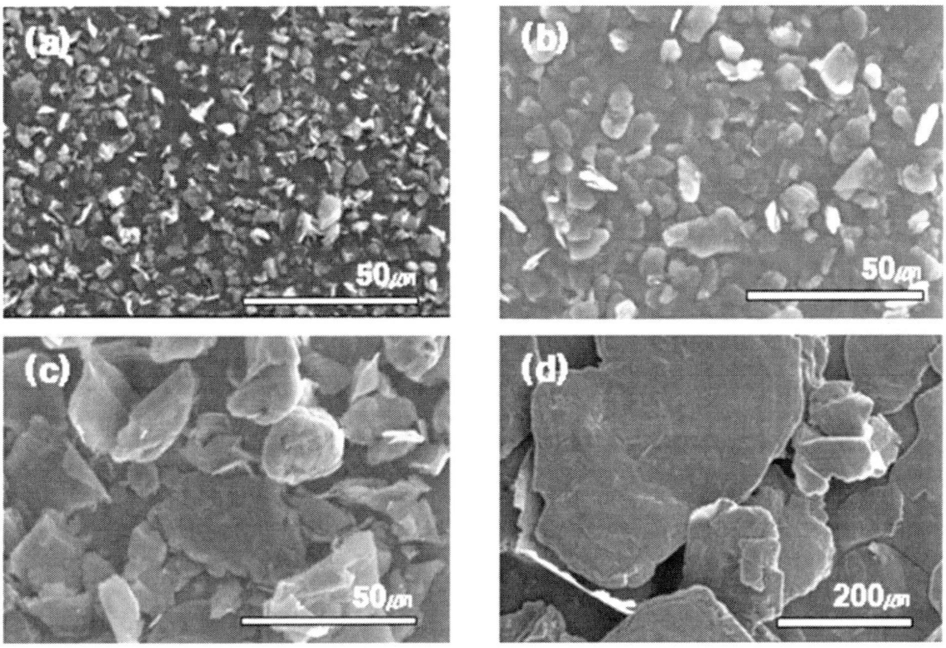

Figure 1. FE-SEM images of graphites used in this study: (a) G1, (b) G2, (c) G3, (d) G4.

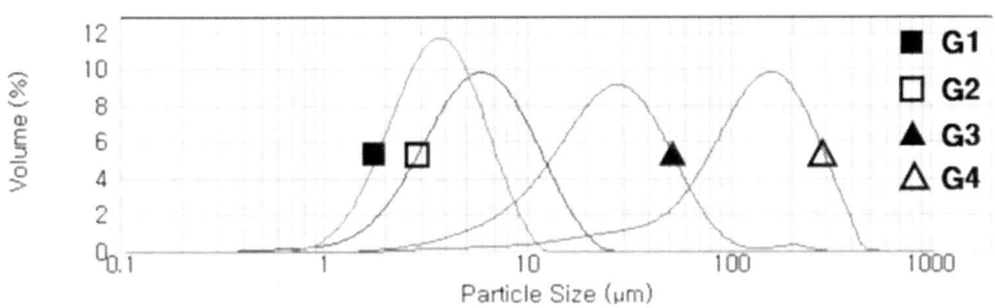

Figure 2. Particle size distributions of the graphites.

Figure 5 shows DSC profiles of graphite anodes in fully charged state at a heating rate of 10°C•min^{-1} from 50 to 420°C, in which DSC profiles were determined by subtracting the base line, obtained by subsequent heating, from the initial profile. All samples show similar heat evolution characteristics. It typically consists of three heat evolution peaks designated as A, B, and C for all of the samples. However, it appears that there are some distinct differences in the amounts of heat evolved and the positions of exothermic peaks, in which the peak B and C shift to higher temperature with increasing particle size. Judging from previous studies [12, 15–18], the heat evolution at around 150°C is mainly attributed to the thermal decomposition of SEI film. The amount of heat evolved at around 150°C was determined from the dotted line curves in inset of Figure 5. in which it was assumed that the maximum height measured in the profile was the peak height and the peak is symmetrical. The peak area increases with the irreversible capacity loss in the first cycle, and is closely related with BET surface area as illustrated in Figure 6.

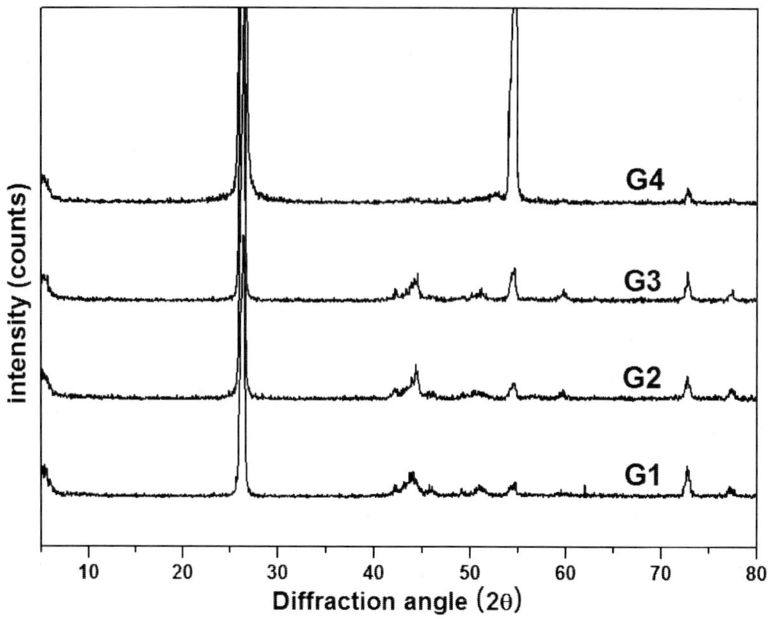

Figure 3. The XRD profiles of the graphites.

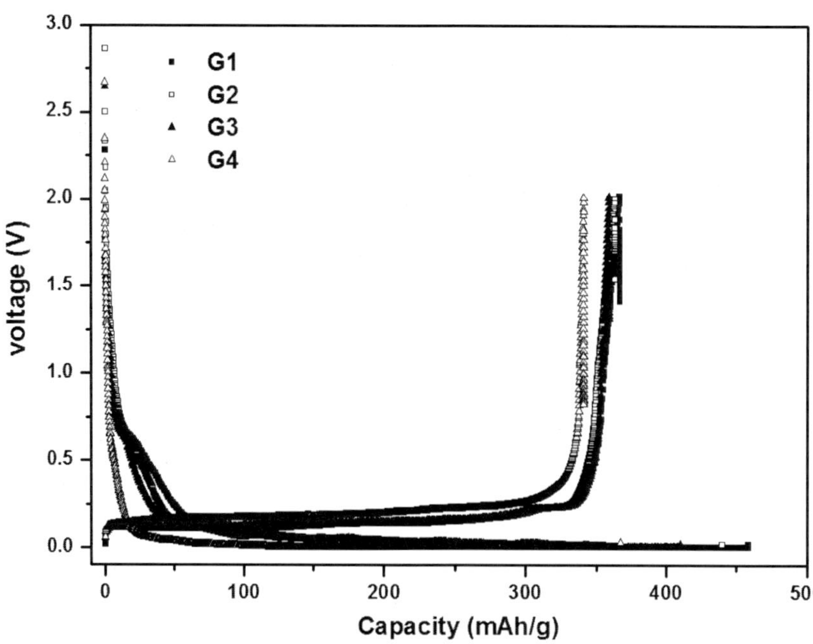

Figure 4. Charge and discharge potential profiles of four graphite samples.

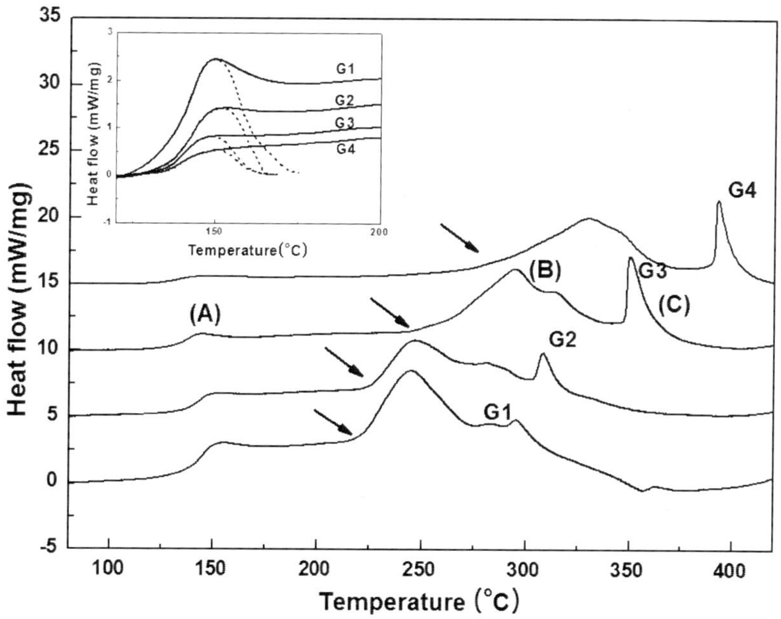

Figure 5. DSC profiles of four graphite electrodes in fully lithiated state with electrolyte.

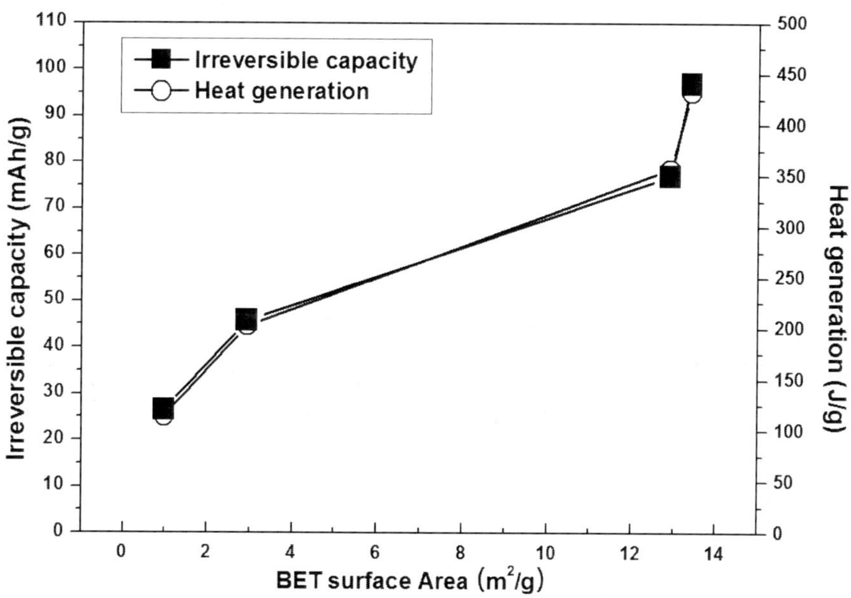

Figure 6. Irreversible capacity and heat generation of graphite samples with different BET surface areas. Note: The heat generation was measured from the dotted line curves in inset of Figure 5.

A mild heat evolution continues after the exothermic peak around 150°C until a large exothermic process (the reaction peak B) appears. This broad peak is ascribed to the reaction of the intercalated lithium and the electrolyte after SEI layer breakdown [5, 12, 15, 16, 18]. As a result of SEI layer breakdown, the lithiated graphite surface is directly contact with the electrolyte and the intercalated Li ions move from the inner structure to the edge of graphite particles during heating [12, 31]. The amount of lithium that leaks from the graphite during

DSC scan was monitored by ex situ XRD measurements, in which the samples for XRD analysis were heated to a given temperature in the DSC run and then rapidly cooled. The structural change of a graphite anode (sample G3) was checked at different step of DSC scans (points 1, 2, 3, 4 and 5 in Figure 7(a)) as illustrated in Figure 7(b). After the exothermic peak at around 150°C (point 1 in Figure 7), it is still in fully intercalated state, LiC_6 (stage 1). However, upon further heating, the lithium content changes, which leads to the formation of LiC_{12} phase (stage 2). Figure 8 shows the lithiated state of the graphite electrodes, when heated to a temperature at which the exothermic peak, B begins (marked with an arrow in Figure 5). It is revealed that a smaller amount of lithium has been deintercalated for a sample with larger particle size of graphite. Sample G1 with the smallest particle among graphites used shows only a small amount of stage 2, LiC_{12} remaining, while in the case of sample G4 with the largest graphite a substantial amount of stage 1, LiC_6 still exists though it has been heated to higher temperature. This indicates that for smaller graphites, substantially more lithium has been consumed in the exothermic reaction in the range of peak A due to the reduced Li ion diffusion path and higher surface area of the graphites. This leads to a larger heat generations due to the reaction of the intercalated lithium and the electrolyte after the SEI breakdown, as shown in Figure 5. Thus, the particle size of graphite affects significantly the heat evolution below 250°C.

On heating above 250°C, the exothermic reactions, designated as peak B and C in Figure 5 also proceed with the loss of intercalated lithium as shown in Figure 7. Interestingly enough, the onset temperature of the peak B and C increases with increasing the particle size of graphite. It is reported that these exothermic peaks are mainly attributed to the reaction of lithiated graphite with PVDF [5, 10] and/or with electrolyte [15, 16]. Yamaki et al. [13] suggested that the sharp exothermic reaction (corresponding to the peak C in this work) is due to a direct reaction between lithiated graphite and the electrolyte accompanied by a breakdown of SEI. Therefore, for the thermal reaction of peak B and C, the intercalated lithium should move from the inner structure of lithiated graphite particles to the surface region of graphites. This can be rationalized by considering that the intercalated lithium is very active at high temperatures and thus delithiation is facilitated by heating the lithiated graphite, and moreover the Lewis acid PF_5 produced from the salt $LiPF_6$ existing in electrolyte system also accelerates the lithium movement by removing electrons from graphite [12, 15, 16]. Since lithium must diffuse though graphite to reach the surface of the edge planes, the reaction of the lithiated graphite with PVDF and/or electrolyte can be limited by the diffusion. As the particle size of graphite decreases, more lithium becomes available for the reaction at lower temperatures due to the reduced diffusion path and higher surface area. As a result, smaller graphites have a greater heat generation at lower temperatures compared with larger ones.

As the particle size of graphite decreases, more lithium becomes available for the reaction at lower temperatures due to the reduced diffusion path and higher surface area. As a result, smaller graphites have a greater heat generation at lower temperatures compared with larger ones. This seems likely to explain the reason why the heat evolution of peak C of sample (G4) with the largest particle size is the largest.

Figure 9 shows DSC profiles of various graphite (G3) electrodes in fully lithiates state with electrolyte, fully lithiated state without electrolyte and fully delithiated state with electrolyte. This indicates that the thermal evolution of peak B and C is mainly related with a reaction between the intercalated lithium and electrolyte.

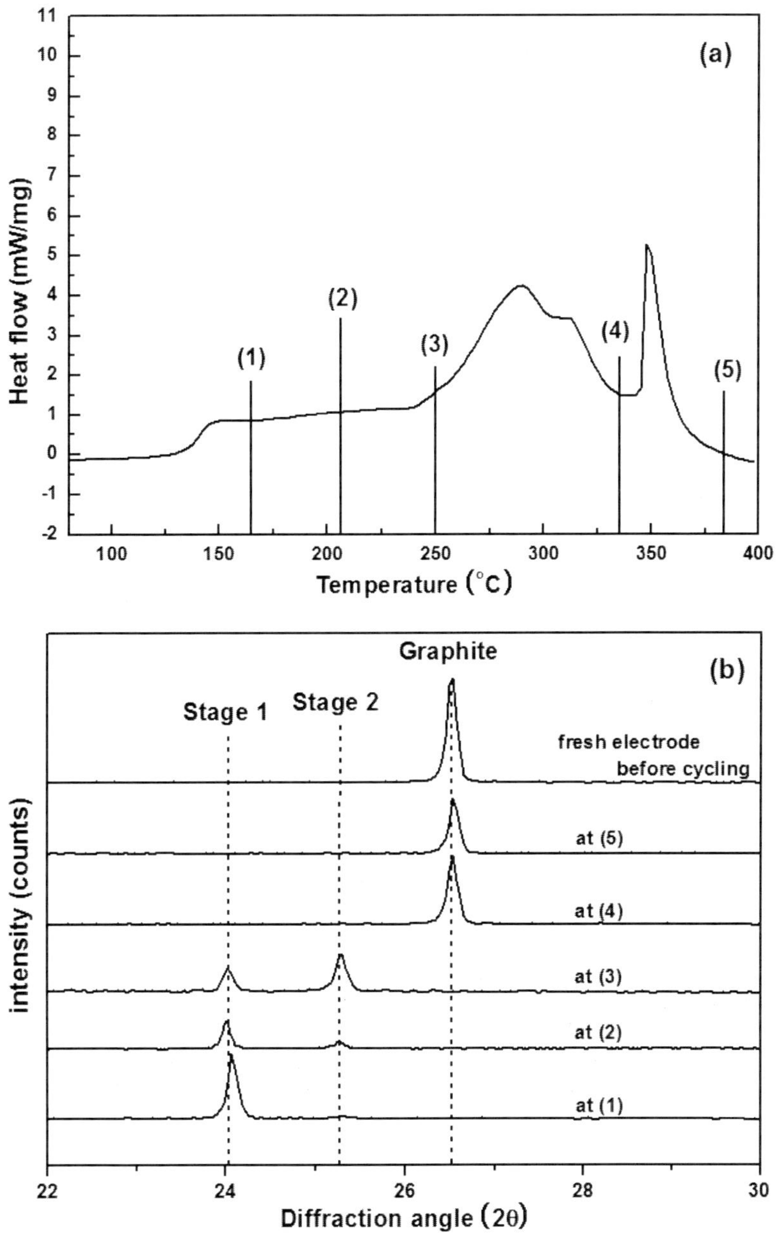

Figure 7. (a) DSC scan for a lithiated graphite electrode (G3) and (b) the XRD patterns of the G3 electrode heated to a temperature marked as 1, 2, 3, 4, and 5 in (a).

On the other hand, at this point it is noteworthy that the XRD peak corresponding to the point (4) in DSC profile (Figure 7(a)) was observed at the position similar to that of fresh electrode as shown in Figure 7(b). This indicates that considerable proportion of intercalated lithium was released before the thermal reaction of peak C, though a quantitative analysis of lithium content through XRD pattern would be limited. Therefore it seems to be possible to suppose that the heat evolution at peak C may be attributed to a mechanism other than a direct reaction between lithiated graphite and electrolyte as mentioned above. However, another possible reaction mechanism for the peak C remains unknown at present.

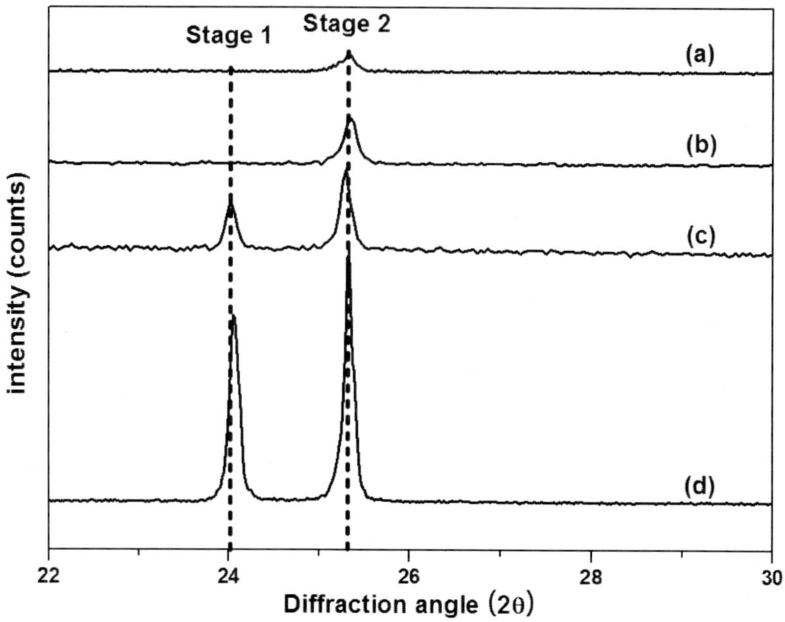

Figure 8. The XRD patterns of lithiated electrodes heated to a temperature at which the exothermic peak, B begins, marked which an arrow in Figure 5: (a) G1 electrode, (b) G2 electrode, (c) G3 electrode and (d) G4 electrode.

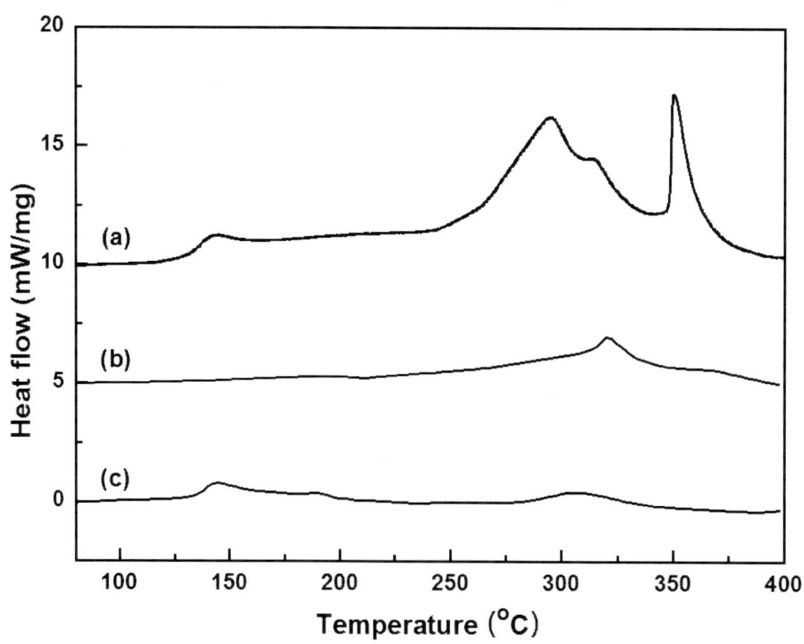

Figure 9. DSC profiles of graphite (G3) electrodes in (a) fully lithiated state with electrolyte, (b) fully lithiated state without electrolyte and (c) fully delithiated state with electrolyte.

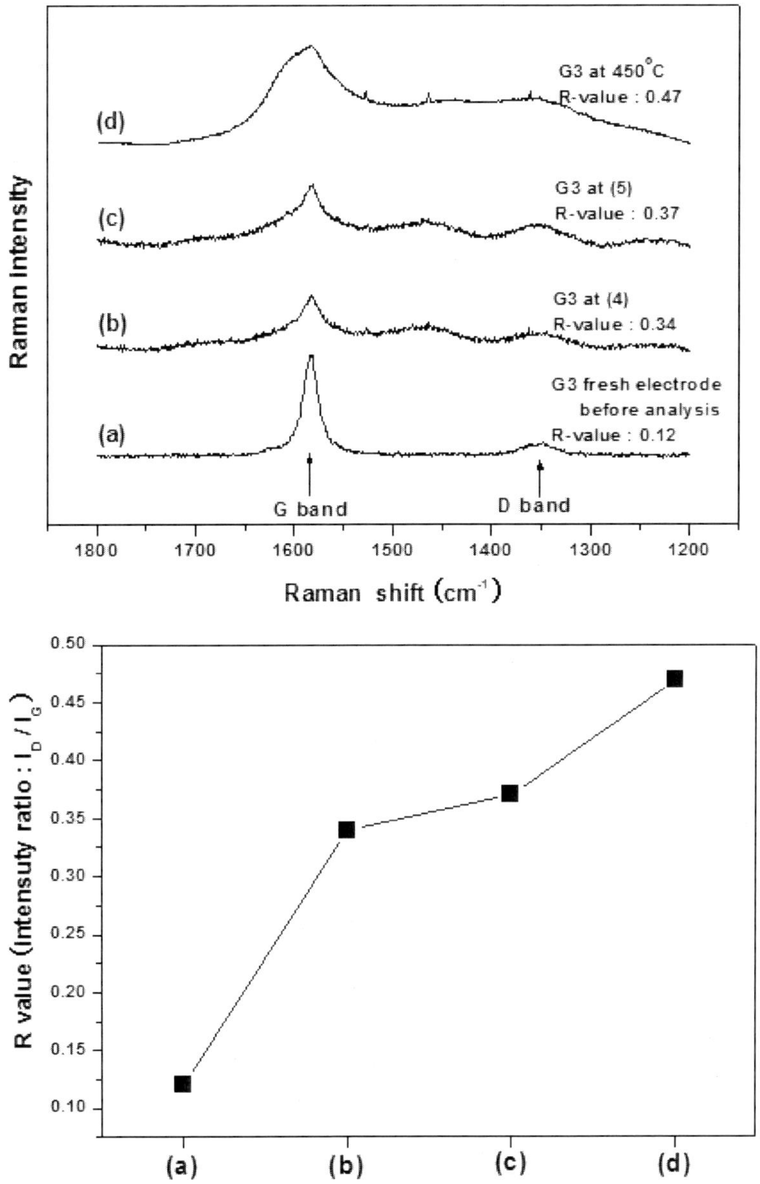

Figure 10. Raman spectra and R value of graphite electrode with G3 sample: (a) fresh electrode, lithiated electrode heated at: (b) (4), (c) (5), and (d) temperature of 450°C.

The XRD patterns of the graphite anodes, when heated to an ending temperature of peak C and rapidly cooled to room temperature, exhibited a sharp diffraction peak at the position of a graphite carbon as shown in Figure 7. However, peak intensity seems to be slightly reduced compared with that of fresh electrode before cycling. The Raman spectrum and the R-value of the graphite anode (sample G3) after DSC measurement are compared with those of the fresh electrode (Figure 10). The R-value (R = ID/IG), defined as the ratio of intensity of D-band (at 1360 cm-1) to that of G-band (at 1580 cm-1) exhibits the degree of disordering of near surface region of carbon materials [32, 33]. After the DSC scan was conducted up to the temperature which the reaction of peak C finishes, the peak at about 1580 cm-1 broadened and the peak height at 1360 cm-1 increased, resulting in the increases of R-value. This means an increase in the surface disordering. When DSC scan was performed up to a higher temperature, the R-value further increases, which indicates that the surface structure of the graphite particle becomes more disordered. On the other hand, Yang et al. [15] reported that the sharp exothermic peak reaction (corresponding to peak C in our experiment) results in the structural collapse into an amorphous structure. However, it is worthwhile to note that they have heat treated the lithiated graphite at 400°C for 10 min to examine a structural change occurring during the sharp exothermic peak reaction, which was much higher than the sharp exothermal peak temperature (283°C) in their experiment. Nevertheless, it appears to be consistent that the changes in the surface structure occur at high temperatures. This structural change at high temperatures may be due to the exfoliation of the surface layers of graphite caused by the entrance of EC and LiPF6 into the graphite sheets to react with intercalated lithium as suggested by Aurbach et al. [34] and Chung et al. [35]. Yang et al. [15] reported that the remaining intercalated lithium in the graphite structure at high temperatures will be released as a result of graphite structure collapse and react with PVDF. It is considered, therefore, that the decrease of the graphite particle size leads to the decrease in the height of the exothermic peak C, as shown in Figure 5, because more intercalated lithium can be consumed in exothermic reactions related to peak A and B due to more enhanced delithiation in smaller graphites.

2. Carbon Coating Effect [29]

Scanning electron micrographs of the unmodified NG and the carbon-coated NG are presented in Figure 11. Both graphite samples have similar shapes and particle sizes. It can be seen from the inset in Figure 11(a) that the unmodified NG has rough and craggy surface, whereas the surface of the carbon-coated NG is smooth and shows no clear cracks (see the inset in Figure 11(b)). To investigate the surface structure and confirm the existence of the coated-carbon layer on the surface of NG, Raman spectroscopy has been employed to evaluate the degree of surface disorder of the carbon materials [32, 33]. Raman spectra of the unmodified and the carbon-coated NG samples are shown in Figure 12. A pair of peaks in the range of 1200 - 1800 cm^{-1} is observed for both samples. The band at approximately 1580 cm^{-1} is a G-band corresponding to the E_{2g2} vibration mode in the graphitic structure of carbon materials, whereas the band at 1360cm^{-1} (D-band) represents the A_{1g} vibration mode caused by the disordered structure in the carbon materials. The intensity ratio of the two peaks, R (R = I_D/I_G), can be used to estimate the degree of disorder on the surface of the unmodified NG and the carbon-coated NG. As shown in Figure 12, the spectrum for unmodified NG exhibits a distinct G-band and a small D-band. In case of carbon-coated NG, the G-band becomes broader and smaller, and the D-band becomes broader and higher. As a result, the intensity ratio R of the unmodified NG is 0.26 and increases to 0.5 for the carbon-coated NG. This implies that a layer of non-graphitic carbon exists on the surface of the original NG after carbon coating.

Figure 13 shows the initial discharge and charge (intercalation and deintercalation) curves for both the unmodified NG and the carbon-coated NG in the voltage range of 0.005–

2.0 V. For electrochemical measurements, the working electrode was prepared by coating a slurry of graphite (90 wt.%) and poly(vinylidene fluoride) (PVDF) binder (10 wt.%) dissolved in N-methyl pyrrolidone (NMP) on a copper foil. The slurry-coated electrodewas dried at 120°C for 12 h. The electrolyte was 1M LiPF6 in a mixture of ethylene carbonate (EC) and diethyl carbonate (DEC) (1:1 by volume, provided by Techno Semichem Co., Ltd., Korea). Coin cells were assembled in an argon-filled glove-box with a working electrode and lithium-foil counter electrode. The cells were galvanostatically cycled at 0.2 mA•cm-2 between 0.005 and 2.0 V at 30°C.

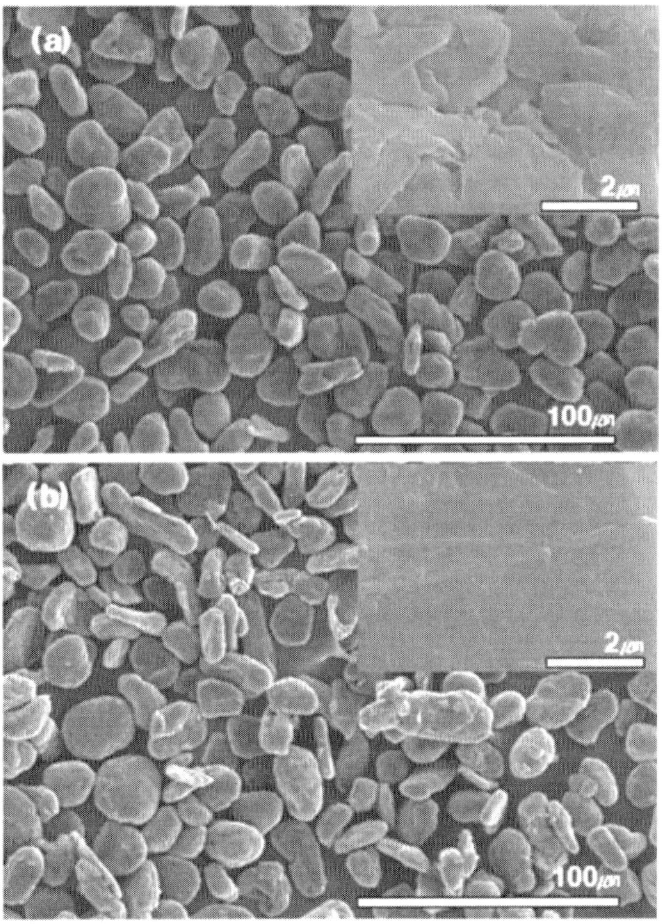

Figure 11. FE-SEM images of (a) unmodified and (b) carbon-coated NG.

The initial discharge capacity (intercalation) of the unmodified NG is 415 mAh•g^{-1} and the charge capacity (deintercalation) is 362 mAh•g^{-1}, which results in a coulombic efficiency of 87.2% on at the first cycle. The carbon-coated NG delivers 374 mAh•g^{-1} of discharge capacity and 348 mAh•g^{-1} of charge capacity with a coulombic efficiency of 93.0%. The reversible capacity of the carbon-coated NG slightly decreased, compared to that of the unmodified NG (362 mAh•g^{-1}). However, the irreversible capacity of the carbon-coated NG drastically decreases from 54 to 26 mAh•g^{-1} with improved coulombic efficiency. The enhanced coulombic efficiency and the reduced irreversible capacity of the carbon-coated NG on the first cycle can be attributed to a decrease in BET specific surface area and the existence of non-graphitic carbon on the surface structure. The measurement of the BET specific surface area for the unmodified NG shows 5.67 m^2•g^{-1}, whereas that for the carbon-coated NG is only 0.6 m^2•g^{-1}.

The DSC curves of unmodified NG and carbon-coated NG electrodes at a fully-lithiated state are given in Figure 14. For DSC measurements, the cells were pre-cycled three times to reach a stable capacity level and the cycling was interrupted when the cells were charged to a fully intercalated state. The charged cells were disassembled in a glove-box. Discs with a diameter of 0.5cm and a copper current-collector were cut from the electrode sheet without removing the electrolyte and transferred to a high-pressure stainless-steel pan with a gold-plated copper seal. The DSC scans were performed with a DSC 200 F3 instrument (NETZSCH, Germany) at various scanning rates in the range from 2.5 to 20°C•min^{-1}. The weight of each DSC pan with the sample was constant before and after DSC measurements, indicating that there was no leakage during the experiments.

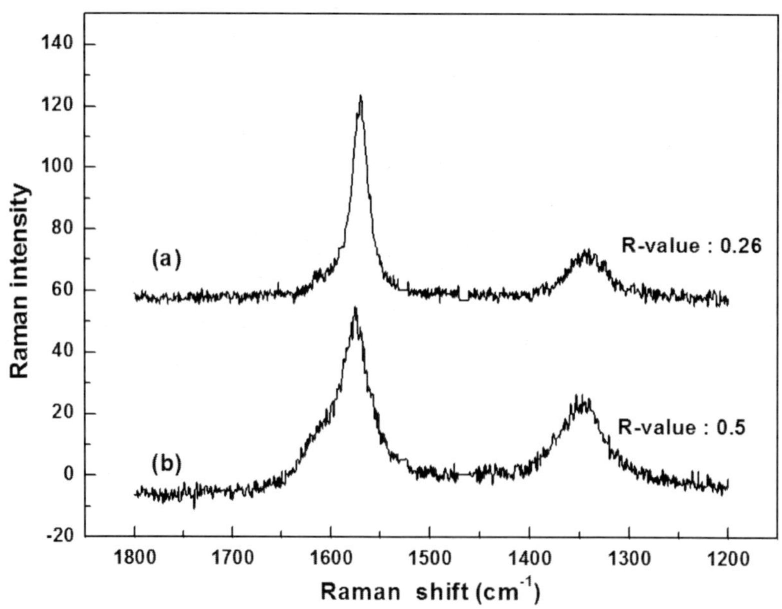

Figure 12. Raman spectra and R-values of (a) unmodified and (b) carbon-coated NG: R-value defined as intensity ratio of two peaks ($R = I_D/I_G$).

It is well known that the SEI film formed during the initial cycle at room temperature is decomposed by temperature-induced reactions, followed by continuous reformation and decomposition as a function of temperature, which is responsible for the exothermic reaction observed below approximately 225°C on the DSC curves [12, 15–18]. The DSC curve of the unmodified NG shows an exothermic reaction starting from 125°C and then continues to 250 °C with a similar magnitude of heat generation. Two major exothermic reactions are observed after 250°C which are related to the structural collapse of carbon material and inter-reaction with electrolyte [15]. The exothermic peak due to the initial SEI film decomposition is illustrated as the dotted line curves in the inset of Figure 14. The estimated heat generation of the unmodified NG is 240 J•g^{-1}. Surprisingly, the carbon-coated NG shows a lower heat generation of 144 J•g^{-1}, which is 40% less than from the unmodified NG. Considering the reduction ratio of the irreversible capacity (0.48 = 26 mAh•g^{-1} / 54 mAh•g^{-1}), the heat generation for the initial SEI decomposition is proportionally reduced as the initial irreversible capacity is reduced. This is consistent with the previous finding of Yang et al. [15]. This is worth noting, however, that a drastic decrease in the BET specific surface area of the carbon-coated NG does not result in the same reduction ratio of the irreversible capacity and exothermic heat generation of the initial SEI film decomposition.

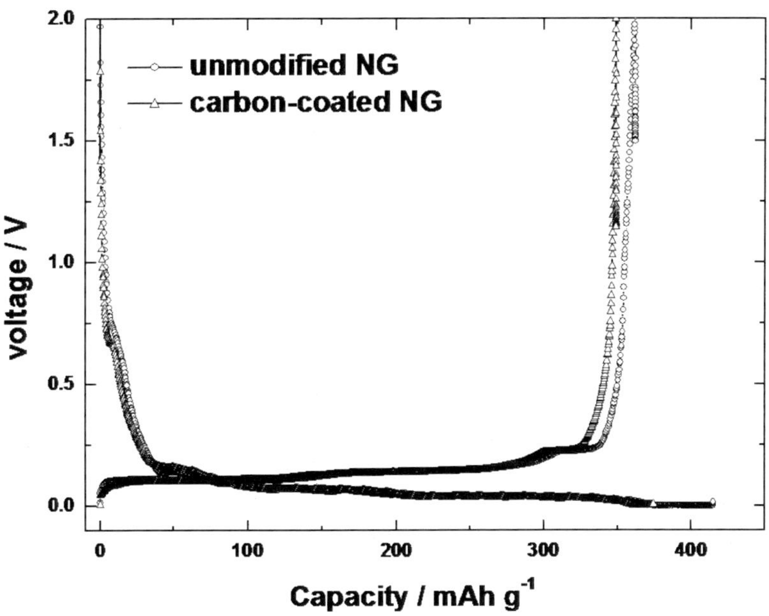

Figure 13. Initial discharge and charge curves of (a) unmodified and (b) carbon-coated NG in voltage range of 0.005–2.0V at current density of 0.2 mA·cm^2.

It has been reported that after SEI decomposition, the reactions of the lithiated graphite with electrolyte and a fluorinated binder, such as PVDF, occur at elevated temperatures [5, 12, 15, 18]. Moreover, these exothermic reactions might lead to the collapse of the graphite structure [15]. The consecutive reactions after SEI film decomposition result from deintercalation of the lithium in the graphite whereby Li ions diffuse from the inner structure to the surface region of the graphite particles at an elevated temperature [12, 25]. To evaluate the effect of carbon coating on this phenomenon, the structural characteristics and the lithiated state of graphite electrodes as a function of temperature were examined using XRD. The XRD patterns of the unmodified NG and the carbon-coated NG electrodes heated at various temperatures are given in Figure 15, in which the sample for XRD analysis was prepared by heating fully-lithiated samples to the designated temperature in the DSC system, followed by rapid cooling to room temperature.

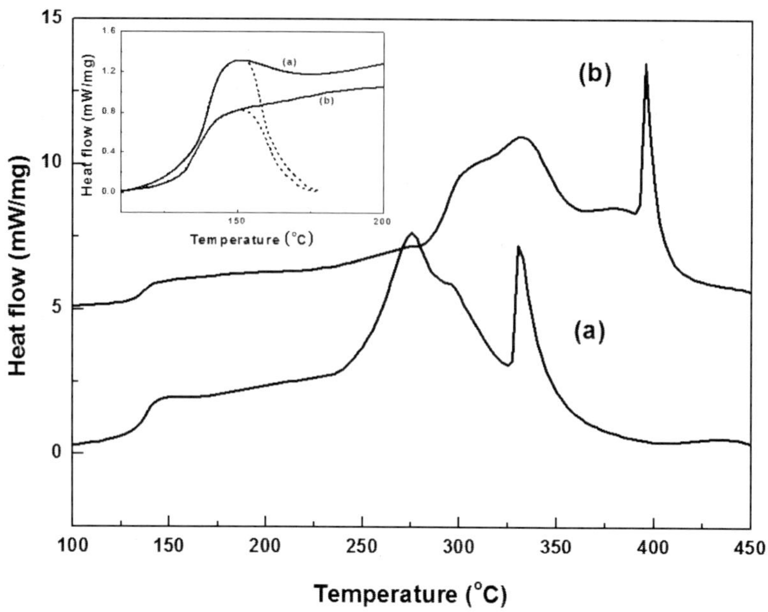

Figure 14. DSC curves of (a) unmodified and (b) carbon-coated NG electrodes in fully lithiated state at scan rate of 10°C min^{-1}.

Based on the assumption that the intensity of the XRD peaks is proportional to the amount of the phase present, it can be inferred from Figure 15 that more lithium is extracted from the unmodified NG than from the carbon-coated NG when heated to a given temperature. For example, when the samples are heated to 300°C, the Bragg peaks corresponding to Stage-2 (LiC_{12}) and Stage-3 still remain in the carbon-coated NG electrode, Furthermore, the peak corresponding to pure graphite phase is hardly observed. By contrast, the peak corresponding to the phase of the fully-delithiated graphite is observed in unmodified NG. In addition, the peaks for Stage-2 and Stage-3 have almost disappeared. The XRD results imply that less heat may generate at the temperature where the structural collapse occurs because intercalated Li^+ still remains in the graphite structure. The remaining lithium in the structure could, however, be detrimental if the release of intercalated lithium occurs and causes further reactions with the SEI and the other components in the electrode. To examine closely the phase variation for both samples, the intensity ratio of the peak for the Stage-1 phase (LiC_6) (I_T/I_0) as a function of temperature is presented in Figure 16. The I_T/I_0 value was defined as the intensity ratio of the Stage-1 phase on the XRD patterns obtained at room temperature and at the designated temperatures. The peak intensity ratio of the unmodified NG electrode more rapidly decreased than that of the carbon-coated NG electrode. Therefore, it can be reasonably assumed that the carbon coating on the surface of NG suppresses the delithiation process from the inner structure to the surface of graphite at an elevated temperature.

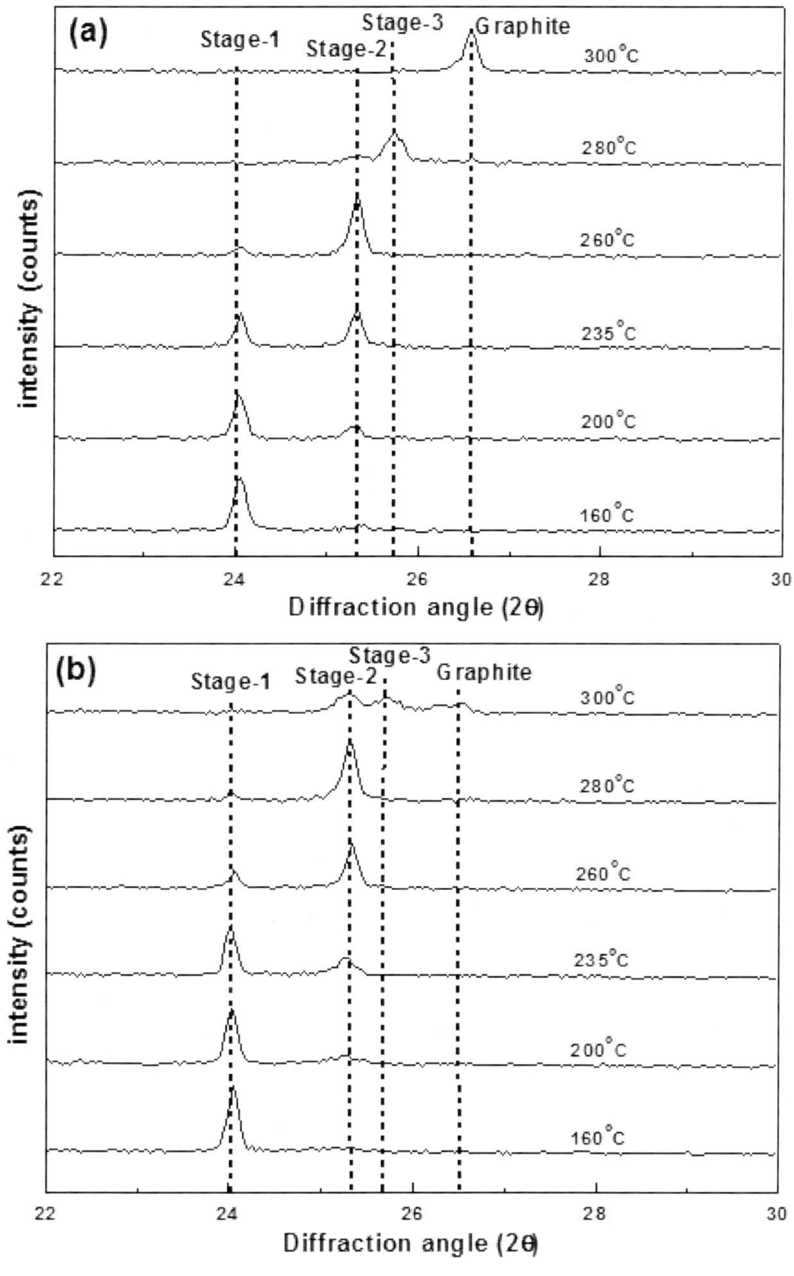

Figure 15. XRD patterns of (a) unmodified and (b) carbon-coated NG electrodes heated to indicated temperatures.

Figure 17 shows the comparison of the heat generation of both samples calculated from the DSC curves as a function of temperature. As assumed above, mild deintercalation of the carbon-coated NG at elevated temperatures leads to a lower heat evolution than that of the unmodified NG. It can be inferred that the carbon-coated NG has better thermal stability than the unmodified NG. This is consistent with results reported by Lee et al. [24] in which the less capacity loss after storage at an elevated temperature was observed for the fully-lithiated carbon-coated graphite than an untreated graphite electrode.

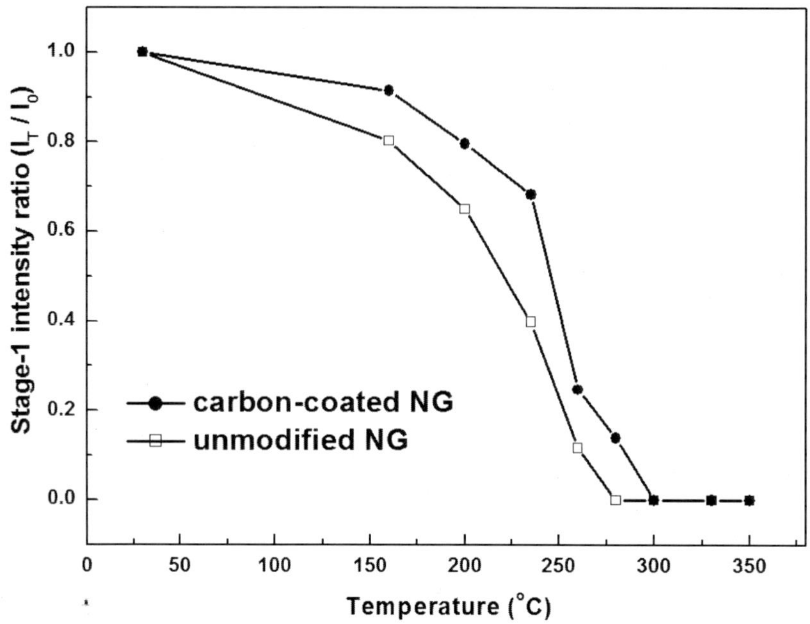

Figure 16. Peak intensity ratio (I_T/I_0) of Stage-1 phase (LiC_6) as function of temperature: I_T/I_0 value was defined as ratio of intensity of Stage-1 phase on XRD patterns of samples heated to designated temperatures.

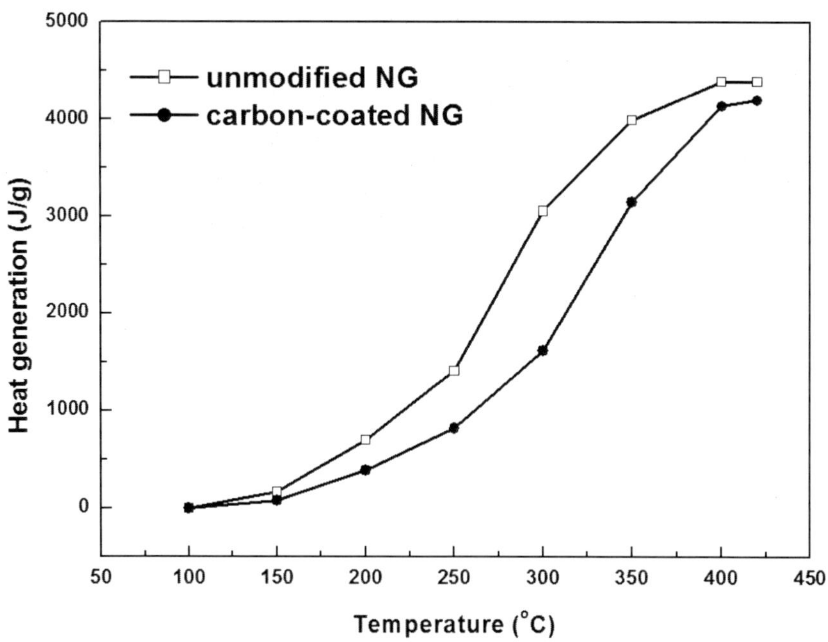

Figure 17. Overall heat evolutions of unmodified and carbon-coated NG electrodes in fully-lithiated state as function of temperature.

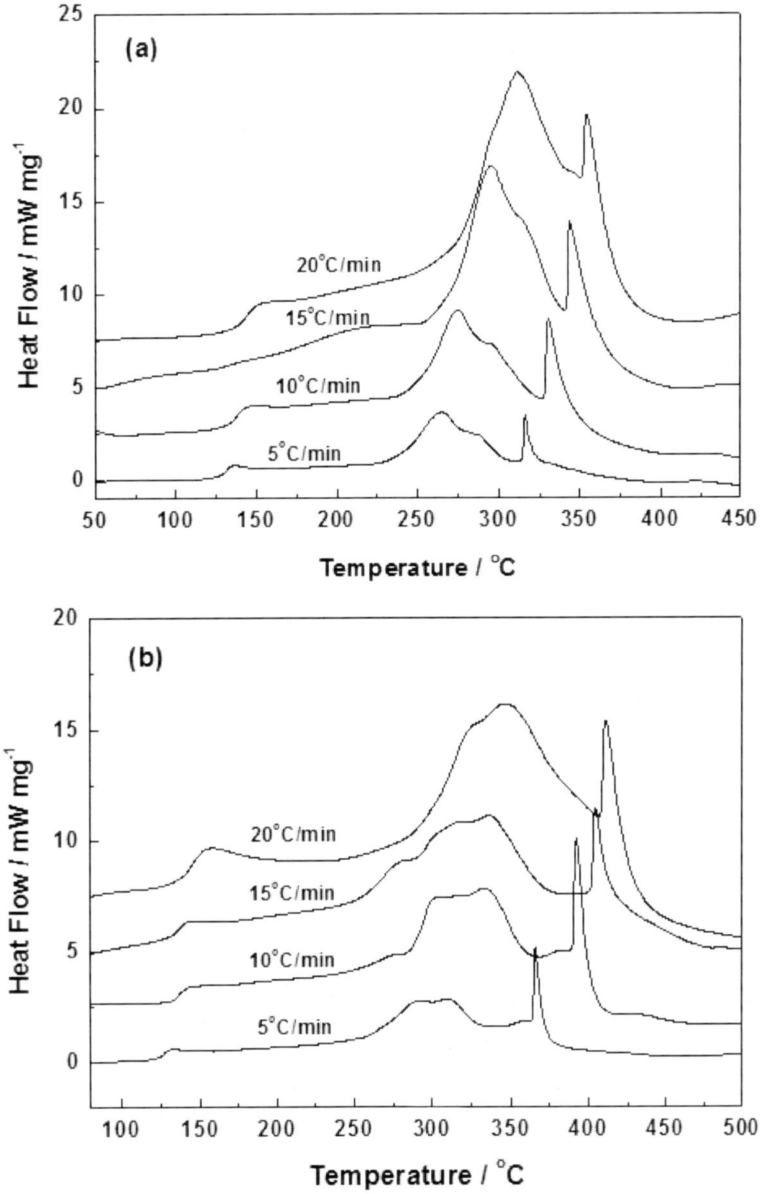

Figure 18. DSC curves of (a) unmodified and (b) carbon-coated NG electrodes in fully lithiated state at different scan rates from 5, 10, 15 and 20°C•min^{-1}.

As shown in Figure 14, a sharp exothermic peak was detected at 330 and 392°C for the unmodified NG and carbon-coated NG sample, respectively, which was attributed to the collapse of the graphite structure [15]. A possible reaction mechanism for the sharp exothermic peak is that EC and $LiPF_6$ may have entered between the graphite sheets and reacted with the intercalated lithium to form lithium carbonate, that, in turn, results in the exfoliation of graphite [35, 36]. Consequently, the lithium-ions released as a result of the structural collapse react with the PVDF binder and the remaining electrolyte to form producing by-products [15, 16]. Surface modification by coating with a non-graphitic carbon layer suppresses the release of the intercalated lithium from the graphite and also protects the graphite structure from electrolyte attack. Therefore, it can be expected that the structural collapse reaction for the carbon-coated NG electrodes will be delayed to a higher temperature.

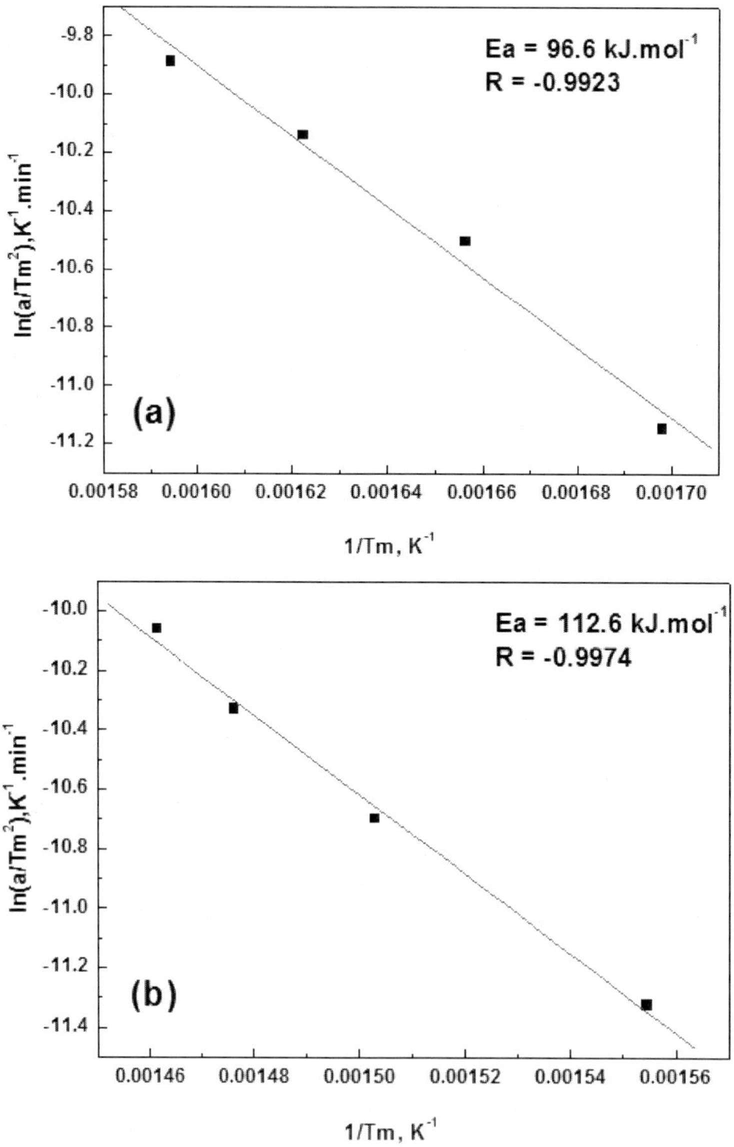

Figure 19. Arrhenius plots of (a) unmodified and (b) carbon-coated NG electrodes.

The activation energies associated with the structural collapse reaction for the unmodified NG and the carbon-coated NG were determined by performing several DSC scans at different rates. Figure 18 shows the DSC traces of both unmodified and carbon-coated NG at different scan rates. It can be observed that as the scan rate increases, the sharp exothermic peak moves to a higher temperature with an increase in intensity. The activation energy for the structural collapse reaction was calculated by the Kissinger method [37] from DSC curves and is presented in Figure 19. The carbon-coated NG sample shows higher activation energy than the unmodified NG sample for the structural collapse reaction. This supports the conclusion that a carbon coating on NG improves the thermal stability of unmodified NG.

CONCLUSION

The DSC study on fully lithiated graphite anodes reveals that the thermal stability of the graphite anode is affected by the particle size of graphite and the surface modification by coating with a non-graphitic carbon layer on natural graphite.

The heat evolved at temperatures below 200°C, mainly due to the SEI decomposition and the reaction of the intercalated lithium with the electrolyte, increases with decreasing the particle size. This is attributed to the amount of SEI film formed during precycling, as inferred from the initial irreversible capacity, and a thermal induced delithiation facilitated by the reduced diffusion path and higher surface area in smaller graphites. As a result of the letter, the onset temperature for main heat generation above 200°C, due to the reaction of the lithiated graphite with PVDF and/or electrolyte, also moves to a higher temperature with an increase in particle size.

Carbon-coated natural graphite shows less irreversible capacity and improved coulombic efficiency than unmodified natural graphite. The BET specific surface area of unmodified NG drastically decreases after carbon coating. The thermal stability of spherical natural graphite is improved by surface coating with a non-graphitic carbon layer. It is confirmed by XRD analysis that the carbon coating layer suppresses the release of intercalated lithium from NG at high temperatures and protects the graphite structure from electrolyte attack.

ACKNOWLEDGMENTS

This study was supported by the National Research Foundation of Korea Grant funded by the Korean Government (MEST) (NRF-2009-C1AAA001-0093307).

REFERENCES

[1] Maleki H., Deng G., Anani A., Howard J., *J. Electrochem. Soc.* 146 (1999) 3224.

[2] Aurbach D., *J. Power Sources* 89 (2000) 206.

[3] Edstrom K., Andersson A. M., Bishop A., Fransson L., Lindgren J., Hussenius A., *J. Power Sources* 97–98 (2001) 87.

[4] Yoshio M., Wang H. Y., Fukuda K., Hara Y., Adachi Y., *J. Electrochem. Soc.* 147 (2000) 1245.

[5] Pasquier A. D., Disma F., Bowmer T., Gozdz A. S., Amatucci G., Tarascon J. M., *J. Electrochem. Soc.* 145 (1998) 472.

[6] Andersson A. M., Edstrom K., Rao N., Wendsjo A., *J. Power Sources* 81–82 (1999) 286.

[7] Joho F., Novak P., Spahr M. E., *J. Electrochem. Soc.* 149 (2002) A1020.

[8] Yamaki J., Baba Y., Katayama N., Takatsuji H., Egashira M., Okada S., *J. Power Sources* 119–121 (2003) 789.
[9] Spotnitz R., Franklin J., *J. Power Sources* 113 (2003) 81.
[10] Maleki H., Deng G., Haller I. K., Anani A., Howard J. N., *J. Electrochem. Soc.* 147 (2000) 4470.
[11] Roth E. P., Doughty D. H., Franklin J., *J. Power Sources* 134 (2004) 222.
[12] Lee H. H., Wan C. C., Wang Y. Y., *J. Electrochem. Soc.* 151 (2004) A542.
[13] Yamaki J., Takatsuji H., Kawamura T., Egashira M., *Solid State Ionics* 148 (2002) 241.
[14] Holzapfel M., Alloin P., Yazami R., *Electrochim. Acta* 49 (2004) 581.
[15] Yang H., Bang H., Amine K., Prakash J., *J. Electrochem. Soc.* 152 (2005) A73.
[16] Wang Q., Sun J., Yao X., Chen C., *J. Electrochem. Soc.* 153 (2006) A3 29.
[17] Richard M. N., Dahn J. R., *J. Electrochem. Soc.* 83 (1999) 71.
[18] Richard M. N., Dahn J. R., *J. Electrochem. Soc.* 146 (1999) 2068.
[19] Richard M. N., Dahn J. R., *J. Electrochem. Soc.* 146 (1999) 2078.
[20] MacNeil D. D., Larcher D., Dahn J. R., *J. Electrochem. Soc.* 146 (1999) 3596.
[21] Jiang J., Dahn J. R., *Electrochem. Solid-State Lett.* 6 (2003) A180.
[22] Jiang J., Chen J., Dahn J. R., *J. Electrochem. Soc.* 151 (2004) A2082.
[23] Maleki H., Howard J. N., *J. Power Sources* 137 (2004) 117.
[24] Lee Y.-T., Yoon C. S., Prakash J., Sun Y.-K., *J. Electrochem. Soc.* 151 (2004) A1728.
[25] Kuribayashi I., Yokoyama M., Yamashita M., *J. Power Sources* 54 (1995) 1.
[26] Qiu W., Zhang Q., Lu S., Liu Q., *Solid State Ionics* 121 (1999) 73.
[27] Lee H. Y., Baek J. K., Jang S. W., Lee S. M., Hong S. T., Lee K. Y., Kim M. H., *J. Power Sources* 101 (2001) 206.
[28] Park Y. S., Lee S. M., *Electrochimica Acta* 54 (2009) 3339.
[29] Park Y. S., Bang H. J., Oh S. M., Sun Y. K., Lee S. M., *Journal of Power Sources* 190 (2009) 553.
[30] Winter M., Novak P., Monnier A., *J. Electrochem. Soc.* 145 (1998) 428.
[31] Andersson A. M., Edstrom K., *J. Electrochem. Soc.* 148 (2001) A1100.
[32] Suzuki K., Hamada T., Sugiura T., *J. Electrochem. Soc.* 146 (1999) 890.
[33] Tsumura T., Katanosaka A., Souma I., Ono T., Aihara Y., Kuratomi J., Inagaki M., *Solid State Ionics* 135 (2000) 209.
[34] Aurbach D., Zaban A., Ein-Eli Y., Weissman I., Chusid O., Markovsky B., Levi M., Levi E., Schechter A., Granot E., *J. Power Sources* 68 (1997) 91.
[35] Chung G. C., Kim H. S., Yu S. I., Jun S. H., Choi J. W., Kim M. H., *J. Electrochem. Soc.* 147 (2000) 4391.
[36] Aurbach D., Zaban A., Ein-Eli Y., Weissman I., Chusid O., Markovsky B., *J. Power Sources* 68 (1997) 91.
[37] Kissinger H. E., *Anal. Chem.* 29 (1957) 1702.

In: Graphite
Editor: Quinton C. Campbell

ISBN: 978-1-62618-576-0

Chapter 10

ETHYLENE-OCTENE COPOLYMER/GRAPHITE COMPOSITES: ELECTRICAL AND THERMAL CONDUCTIVITIES AND FLAME RESISTANCE

Petr Svoboda[1,2*]

[1] Department of Polymer Engineering, Faculty of Technology,
Tomas Bata University in Zlin, Zlin, Czech Republic
[2] Centre of Polymer Systems, University Institute,
Tomas Bata University in Zlin, Zlin, Czech Republic.

ABSTRACT

A series of ethylene-octene copolymer (EOC) composites have been prepared by melt-mixing with different weight ratios of expandable graphite filler (0-50% by weight). Electrical conductivity [both alternating current (AC) and direct current (DC)] and thermal conductivity studies were carried out.

Effect of filler loading and frequency on electrical conductivity was studied. DC conductivity has increased from $1.51\text{x}10^{-13}$ Scm^{-1} to $1.17\text{x}10^{-1}$ Scm^{-1}. Percolation threshold by DC and also AC methods was observed at about 16 vol. % of the filler. Real part of permittivity was found to be decreasing with increase in frequency while conductivity was increasing.

Thermal conductivity was also found to be increasing gradually from 0.196 to 0.676 $Wm^{-1}K^{-1}$ which is about 245% increase. Graphite not only increases the electrical and thermal conductivities but at and above 40 wt. %, also acts as a halogen-free, environmental friendly flame retardant. Horizontal burn rate was studied in detail with interesting results.

Initially the addition of graphite increases the horizontal burn rate but above 35 wt. % of graphite the burning speed decreases. Shore-A hardness of EOC/graphite composites shows that even with high graphite loading, the hardness is increased from about 50 to 68 only so that the rubbery nature of the composite is not affected very much.

* E-mail: svoboda@ft.utb.cz.

Introduction

Polymer composites with conducting fillers serve as electrical or thermal conductors and have wide range of applications in the electric and electronic fields. These composites have an advantage of conductivity of a filler and favorable properties of polymer material together. They are widely used in many areas like electrodes, separators, electromagnetic shielding and anti-static coating [1, 2]. The fillers generally subjected for conductivity studies are carbon fibers [3, 4], multi-wall carbon nanotubes (MWCNT) [5], natural fibers e.g. sisal fibers [6], expanded graphite [7], carbon black [8] and graphene [9].

Ethylene-octene copolymer (EOC) is a new type of copolymer developed by The Dow Chemical Company through constrained geometry catalyst technology (CGCT) under the trade name ENGAGE®. It is being considered for use in diverse applications such as cushioning agents, gaskets, and is particularly good alternative for sealing application due to its structural regularity and non-toxic composition. Several researchers have studied the conductivity properties of EOC composites using different fillers such as carbon black, polyaniline, and multi-wall carbon nanotubes [10-12]. Conducting polymer composites (CPC) with elastomeric polymer matrices have applications in the field of sensors, e.g. temperature sensors [13] and strain sensors [14].

Charge transport in polymer composite materials has been the object of many researches in the past. Conducting mechanism in composites containing conductive particles can be described by percolation theory [15]. The volume fraction of fillers above which the continuous conducting paths are created is known as percolation threshold. At this level, a continuous chain of particles in the matrix results in a considerable increase in conductivity of several orders. Conductivity above percolation threshold increases sharply and composites start behaving like semiconductors [16-19].

Permittivity is the ability of a material to store charge when used as a dielectric material. Capacitance and conductance as a function of frequency can be measured by dielectric analysis which can be then utilized to calculate real part (ε') and imaginary part (ε'') of permittivity and the dielectric loss factor ($\tan \delta$) [20].

Extensive studies have been performed on conductivity of numerous composites with different polymers as matrix and various fillers. Several researchers prepared composites by the incorporation of halogen-free flame retardant expandable graphite for the improvement of flame retardancy and thermal stability [21-24]. Expandable graphite as a conducting filler has been studied only few times [25, 26]. Only few researchers performed the conductivity studies using EOC as a matrix [10, 12]. In this paper, electrical and thermal conductivities of EOC/expandable graphite composites have been studied. Also, flame retardancy according to UL-94 (Underwriters Laboratories) horizontal test was evaluated.

Materials and Methods

Materials

Ethylene-octene copolymer was Engage 8842 supplied by Dow Chemicals. The octene content was 45 wt. %, density = 0.8595 g cm^{-3} and melt flow index (MFI) = 1.02 dg min^{-1} (at

190°C /2.16 kg). Expandable graphite flakes were Grafguard 160-50N, used as received from Graftech Inc., Ohio, Cleaveland, USA. Mean particle size was 350 μm and density was about 2 g cm^{-3}.

Composite and Sample Preparation

EOC and Grafguard 160-50N were melt-mixed in a BrabenderPlasti-corder PL2000 mixing machine with a mixing volume of 50 cm^3 at 100 °C at mixing speed 50 rpm for 10 minutes. Graphite was preheated in a hot air oven at 100 °C for 12 h to remove any absorbed moisture. Graphite content was: 5, 10, 15, 20, 25, 30, 35, 40, 45, 50 wt. % or 2, 5, 7, 10, 13, 16, 19, 22, 26, 30 vol.%, respectively.

The melt-mixed composites were then compression molded to 2 mm thick sheets at 110 °C. From this sheet two discs were cut out: for the electrical conductivity measurement the disc had diameter 15 mm and for the thermal conductivity measurement it was 50 mm. Samples for flammability tests were cut out of the sheet according to UL-94 standard with sample dimensions being 125x13x2 mm. For shore-A hardness measurements the discs had diameter 50 mm and thickness 6 mm.

Scanning Electron Microscopy (SEM)

Morphology of the expandable graphite filler and composites was studied by using SEM technique. SEM analysis was carried out using Vega II LMU (Tescan, Czech Republic) with a beam acceleration voltage set at 10 kV.

Direct current (DC) Conductivity

The current-voltage dependencies were measured in the two-point setup using electrodes of cylindrical shape (13 mm in diameter and covered by a layer of gold) with a programmable electrometer (Keithley6517 A, USA).

The DC conductivity was calculated from the current–voltage dependence in the range of 1–20 V. Current-Voltage characteristics were measured in DC field. Specific electrical conductivity, σ_{DC}, can be expressed as follows:

$$\sigma_{DC} = \frac{I}{U} \cdot \frac{d}{A} \tag{1}$$

where, I is the electrical current, U is the voltage, A is area of electrodes and d is the thickness of the sample.

Four-point method (Van der Pauw) [27] was used for the measurement of samples with conductivity higher than 10^{-3} S cm^{-1}. A main advantage of four-probe measurements is the elimination of contact resistance [28]. All the properties were measured at room temperature (22-25 °C).

Dielectric Properties and Alternating Current (AC) Conductivity

Dielectric characteristics involving the frequency dependence of the real part (dielectric constant, ε') and imaginary part of the complex permittivity (dielectric loss, ε'') in the frequency range of 1 MHz – 3 GHz were determined with an Radio Frequency (RF) Impedance Analyzer (Agilent E4991A, USA) using capacitive method (on a dielectric material test fixture which comes furnished with the device). After inputting the sample thickness the device automatically calculates and yields frequency dependence of complex permittivity ε^*. In the range of 10 Hz-100 KHz, dielectric characteristics were measured by using Hioky 3522 (LCR Hi Tester, Japan). The dielectric constant was calculated from the measured capacitance.

AC conductivity, σ_{AC}, was calculated from imaginary part of complex permittivity, ε'', which was directly obtained from the instrument, following the equation (2):

$$\sigma_{AC} = 2\pi f \cdot \varepsilon_0 \cdot \varepsilon'' \quad (2)$$

where, ε_0 is permittivity of vacuum and f *is* frequency.

Thermal Conductivity

Thermal conductivity of EOC/graphite composites was measured according to Fitch (1935) [29] where the sample is sandwiched between a heat source insulated on all faces but one and a constant temperature brass cylinder as heat sink, as shown in Figure 1.The advantage of this approach is most of all short time of measurement and also the simplicity of the procedure and also the simplicity of the measuring instrument. During the quasisteady heat conductivity measurement, the temperature at certain place changes with time.

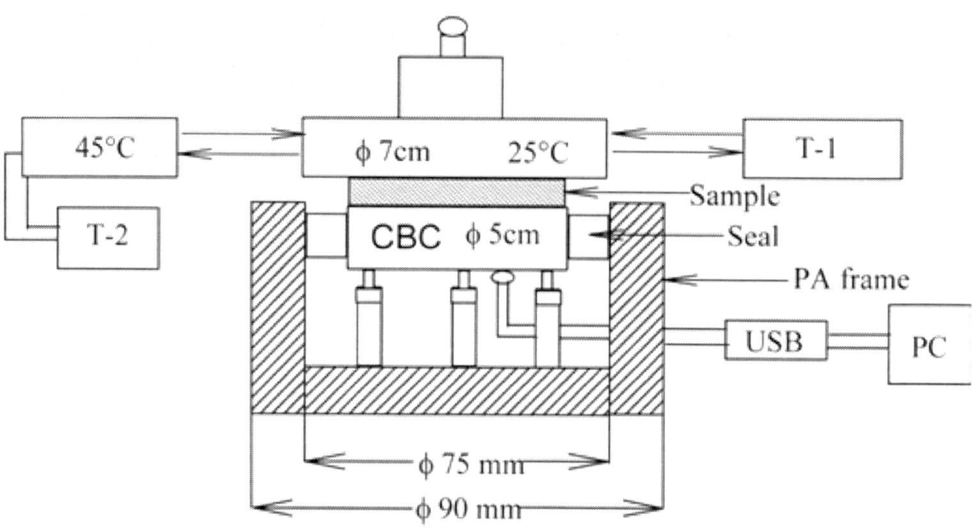

Figure 1. Schematic diagram of the apparatus used for thermal conductivity measurement.

The common procedure involves a case when the heat flows only in one direction. Then temperature depends only on time and one coordinate. This instrument is usually used for the measurement of thermal conductivity of thin sheets or slabs from various plastics and rubber and also for leather. The principle of measuring instrument and the procedure is explained below. Initially the central brass cylinder (CBC) with diameter 5cm is annealed to reach t_2=45°C with the help of another hollow brass cylinder that was connected to a water thermostat by rubber hoses. The water thermostat's accuracy was 0.1°C. This takes about 6 min. Then the 45°C cylinder is quickly removed, measured sample with diameter 5cm and thickness about 2mm is placed on the top of the CBC and on the top of measured sample another hollow brass cylinder connected to another water thermostat with temperature t_1=25°C is placed. On the top a 100g weight was placed. The heat is being transferred from the CBC though the sample to colder brass cylinder (25°C). The temperature of the CBC is decreasing rapidly. The data acquisition of the decreasing temperature of the CBC is being done with the help of a thermocouple (type copper-constantan) that was connected to a National Instruments data acquisition equipment (NI USB-9211A, Portable USB-Based DAQ (Data Acquisition) for Thermocouples) that was connected to a computer by USB. Software LabVIEW Signal Express 2.5 was used for temperature data acquisition. Sampling period was set to 5s. The measurement took about 25 min. The non-linear analysis (exponential decay) of the time-temperature curve was performed.

Mathematical Model for the Measurement

For the derivation of mathematical model describing dependence of the temperature of the CBC on time we start with following heat balance.

$$-K\frac{dt}{d\tau} = \frac{S\lambda(t - t_1)}{\delta} + B(t - t_1) \tag{3}$$

where,

$t(\tau=0) = t_2 = 45$°C
K is the heat capacity of the CBC, central brass cylinder (J K^{-1})
S is the sample area (m^2)
λ is the thermal conductivity ($W.m^{-1}.K^{-1}$)
t is acquired temperature of the CBC (°C)
t_1 is the temperature of hollow brass cylinder 25 (°C)
t_2 is initial temperature of CBC (°C)
δ is the thickness of the sample (m)
B is the coefficient accounting for a heat loss ($J.s^{-1}.\ K^{-1}$)
τ is the time (s)
B can be calculated according to

$$B = \alpha S_z \tag{4}$$

where α is heat transfer coefficient ($W.m^{-2}.K^{-1}$) and S_z is the heat loss area (m^2).

Right side of Equation (3) represents heat flow through the mass of the measured sample. The heat loss caused by natural flow of the air around the measuring instrument is accounted for.

We start with energy balance. Left side of Eq. (3) corresponds to immediate heat output of the central brass cylinder from initial t_2=45°C to the equilibrium temperature t_1=25°C. Thermal energy that is leaving the CBC is realized by two ways. Larger part occurs by conduction heat transfer through the measured sample. Smaller part is heat loss to the surrounding environment that is caused by natural air flow and also by radiation. This part of the equation is determined by so called "blind experiment" when instead of measured sample we place material with very small thermal conductivity. We have used expanded polystyrene with λ=0.035 W/mK. The importantis a good contact of the measured sample with both brasscylinders. In case of hard samples like PP we have to use special the normally conductive paste (alluminum powder in siliconeoil). In our case of EOC/graphite compositeswedid not have to use this special conductive paste because the samples were soft and the contact was perfect.

By solvingEq. (3) weget

$$t = t_1 - (t_1 - t_2) * e^{[-(A_1+A_2)\tau]} \tag{5}$$

where for A_1 and.A_2 constants are given by Eq. (6) and (7).

$$A_1 = \frac{S\lambda}{\delta K} \tag{6}$$

$$A_2 = \frac{B}{K} \tag{7}$$

Practical calculation goes as follows. Eq. (5) can be simplified for the non-linear regression (exponential decay):

$$y = y_0 + ae^{(-bx)} \tag{8}$$

From a non-linear regression we get coefficient b. For example for pure EOC we got b=0.002382 (from Figure 2).

The heat loss from the instrument obtained by blind experiment is A_2 = 0.000368 s^{-1}. Thermal capacity of the central brass cylinder was K_1 = 94.107 J K^{-1}. Area of measured cylinder was S = 1.9635.10^{-3} m^2.

Then the thermal conductivity can be calculated as shown by Eq. (9) and (10).

$$A_1 = b - A_2 = 0.002382 - 0.000368 = 0.002014 \tag{9}$$

$$\lambda = \frac{A_1 \delta K_1}{S} = \frac{0.002014 * 0.002305 * 94.107}{\frac{\pi\, 0.05^2}{4}} = 0.196\, W m^{-1} K^{-1} \quad (10)$$

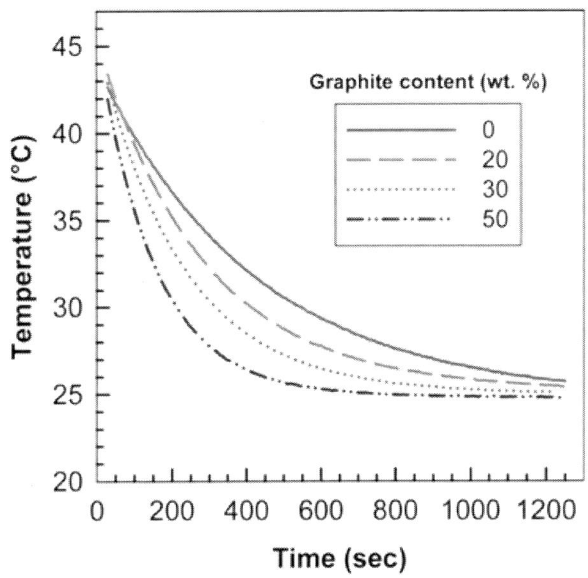

Figure 2. Thermal conductivity measurement: raw data. Temperature as a function of time for EOC/graphite composites with various graphite content. Initial temperature was 45°C and equilibrium temperature was 25°C. Sample had diameter 50 mm and thickness about 2 mm.

Flammability Test

Horizontal flammabilty test was carried out according to the UL-94 standard. Burner with yellow-tipped blue flame of 20 $\pm$ 1 mm height was used for the test. Samples were held horizontally at an angle of 45° and ignited using the flame for 30 $\pm$ 1 seconds. Samples had two marks; one at 25 mm and another at 100 mm. Test was conducted for three specimens in each category.

Shore-A hardness

Hardness of the EOC/expandable graphite composites has been measured using Bareiss Shore-A hardness tester (Model HHP-2001).

Results and Discussion

Morphology of Composites

Flakes-like structure of the expandable graphite filler is clearly shown in Figure 3a. Morphology of the composites, displayed in Figure 3b, 3c and 3d, helps us to understand the distribution of expandable graphite flakes in the polymer matrix. There are three micrographs of composites shown here: one below percolation threshold (3b), one around (3c) and the third one (3d) above the percolation threshold. One can observe increasing amount of graphite particles in the composite (3b→3d), with uniform dispersion of the particles which implies thorough mixing.

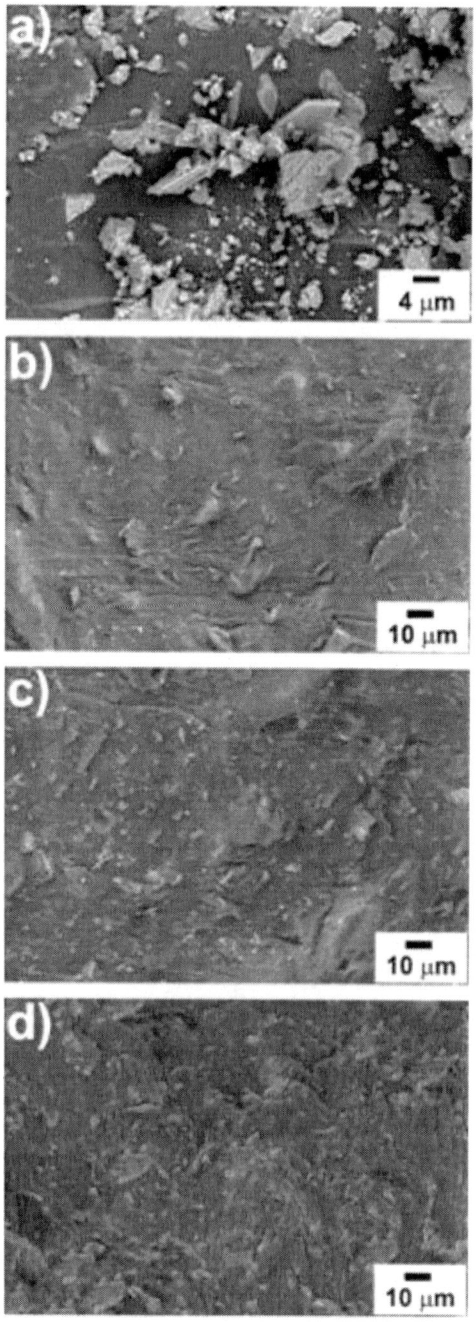

Figure 3. SEM micrographs of (a) Expandable graphite and EOC/expandable graphite composites with (b) 20, (c) 30 and (d) 40 wt. % of filler.

DC Conductivity (σ_{DC})

Conductivity of polymer composites with conducting fillers is greatly affected by filler loading level and distribution or filler alignment in the matrix. Figure 4 shows the dependence of DC conductivity on graphite level. It is very clearly visible from the plot that a sharp increase in the conductivity at particular filler level called percolation threshold. At and above this level, free movement of charge carriers makes the system more conductive or in other words, continuity between the conducting graphite flakes is created. Below percolation threshold, for example, at 5% graphite level, the conductivity is only 1.5 x 10^{-13} S cm^{-1}. Here, conductive paths were not developed and the conductivity may be due to that of ionic conductivity of the polymer composite. From the graph, one can estimate that the percolation threshold lies around 30 wt. % (or 16 vol. %) of graphite loading, where conductivity shows a sharp increase. For a mathematical model system of conducting spheres, randomly distributed in a non-conducting matrix, percolation at 16–17 vol. % was calculated [30]. And our value of percolation level is in good agreement with this model. Above percolation threshold, amount and distribution of filler increases and thus conductivity becomes easier. Small value of the deviations reveals good homogeneity of polymer composite systems, which is also supported by the SEM micrographs of the composites.

The relation of conductivity, σ with volume fraction of the filler can be explained by the power law model of conductivity [15] which is given as:

$$\sigma = \sigma_0 \, (\phi - \phi_c)^t \tag{11}$$

where, $\sigma_0$is the conductivity of the filler (1.33x10^4 Scm^{-1}, in the case of graphite flakes), ϕ is the volume fraction of the filler, ϕ_c is the critical volume fraction and t is the universal exponent determining the power of the electrical conductivity increase above ϕ_c.

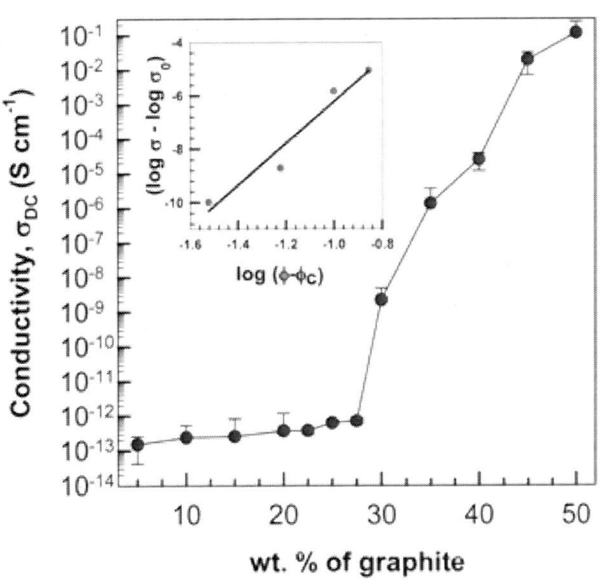

Figure 4. Dependence of DC conductivity on graphite content.

The plot of (log σ- log σ_0) versus log(ϕ-ϕ_c) is shown as an inset in the Figure 4. Slope of this curve gives the value of t. Calculated value of t from previous research was found to be in the range of 1.5 - 2.0 [31], in our case t=1.77, which is in good agreement with the literature.

AC Conductivity (σ_{AC})

Dielectric properties of EOC/graphite composites were measured in an alternating electric field. Dependence of AC conductivity on frequency is shown in Figure 5.

According to Eqn. (1), AC conductivity, σ_{AC}, is proportional to imaginary part of permittivity and frequency. As the frequency increases, AC conductivity also increases, according to Equation (2).

For lower concentrations (20, 25, 30 wt.%) the AC conductivity is increasing with frequency, therefore conductivity is provided by polarization current of fixed dipoles. AC conductivity in the low frequency region have some relations with DC conductivity. Above percolation level (35-40 wt.%), at low frequencies, the plateau relates to the values of DC conductivity, which is in good agreement with results determined by DC conductivity measurements. Similar observations were made by other researchers also [32].

AC conductivities of all compositions at four selected frequencies, 10^3, 10^6, 10^7 and 10^9Hz, are shown in Figure 6. As the frequency increases, percolation behaviour in the 30-35 wt. % graphite range is not so prominently seen like in the case of low frequencies and almost disappears at high frequencies. At high frequencies, rather than percolation, there is a gradual increase in AC conductivity with increasing graphite content. It shows the effect of frequency on the AC conductivity.

In the case of composites with 10 – 50 wt.% of graphite, at low frequency, conductivity increases from 10^{-11} to 10^{-2} Scm^{-1}; we can observe an increment of 9 orders. But at higher frequency, this rise ranges only from 10^{-3} to 1 Scm^{-1}, only of 3 orders.

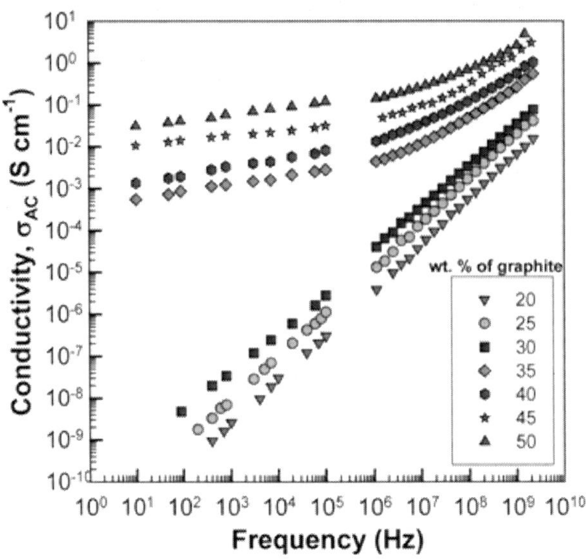

Figure 5. AC conductivity vs. frequency of EOC/graphite composites.

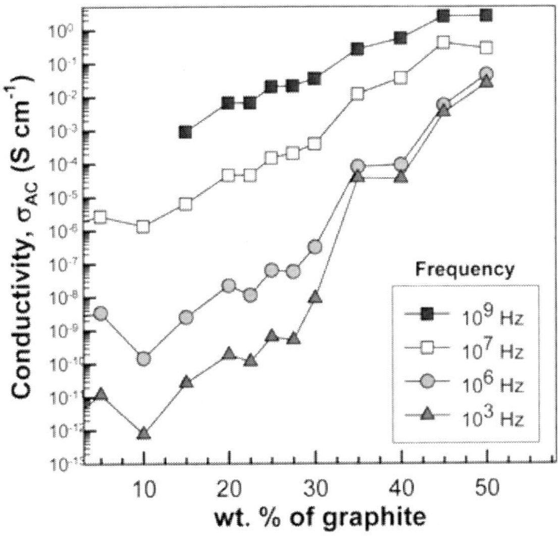

Figure 6. AC conductivity as a function of graphite content at different frequencies.

Dielectric Constant (ε')

As shown in Figure 7, for filler content 0-30 wt. %, ε' value is equal to around 10^1 and for filler content above 30 wt. %, ε' value is around 10^7 at the frequency 10^1 Hz and it is illustrated that as the filler content increases, ε' value also increases. As the frequency increases, permittivity decreases. This can be attributed to the interfacial polarization (IP) of the filler. At low frequencies, all types of polarization contribute to the resulting value of permittivity, in our case predominantly interfacial polarization, typical for systems containing phases of different specific conductivity.

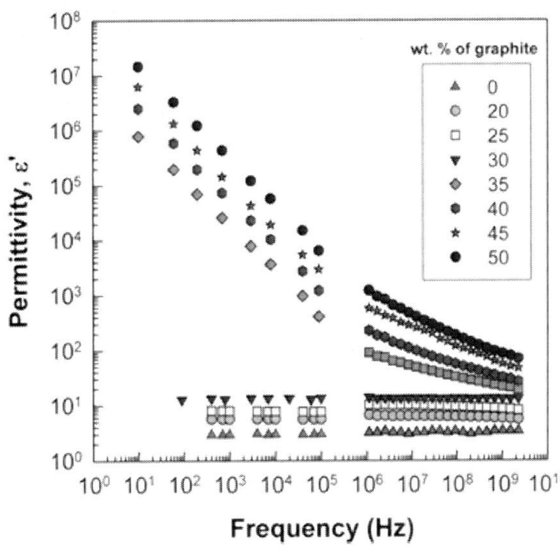

Figure 7. Real part of permittivity as a function of frequency.

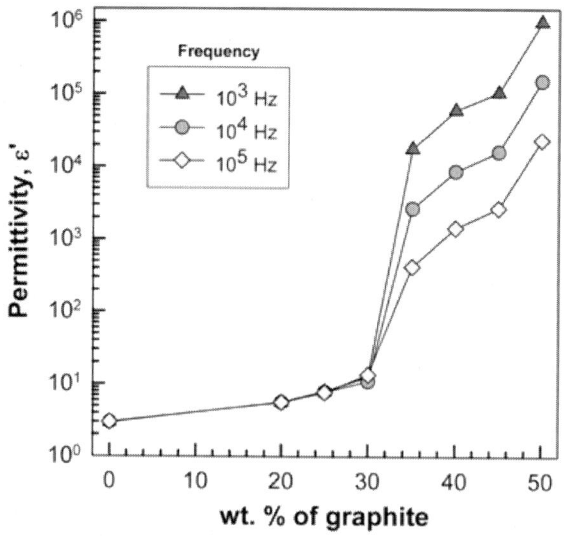

Figure 8. Permittivity as a function of graphite content at different frequencies.

IP arises in electrically heterogeneous materials such as composites and known as the Maxwell-Wagner-Sillars (MWS) effect [20]. The IP leads to an increase in ε'due to the motion of virtual charges, which get trapped at the interface of components of a multiphase material of different conductivity [33]. Similar observations have been made by Paul and Thomas [6]. Permittivity as a function of graphite content is shown in Figure 8. As in the case of conductivity, percolation behavior at low frequencies (10^3 Hz) is more prominent than at higher frequencies.

Thermal Conductivity (λ)

Thermal conductivity as a function of graphite content is given in Figure 9. As it can be seen from the figure, thermal conductivity increases as the filler content increases. Thermal conductivity of pure EOC matrix was found to be 0.196 W $m^{-1}K^{-1}$ while a 30 wt.% graphite loading increased the value almost 100% to 0.393 $Wm^{-1}K^{-1}$ and for 50 wt.%, the λ value was 0.676 W $m^{-1}K^{-1}$ i.e. about 245% increase.

Uniform dispersion of the filler has high influence on the increase in thermal conductivity as in the case of electrical conductivity. Since graphite is good conductor of heat and electricity, the increase in graphite level will result in an increase in electrical and thermal conductivities of the EOC/graphite composites.

Flammability Testing

Flammability test (Horizontal Burning - HB classification) was carried out according to the UL-94 standard. Composites with graphite loading less than 40 wt. % burnt all the way up to the holder. Only samples with filler loading 40, 45 and 50 wt. % did not burn to the holder, the flame self-extinguished.

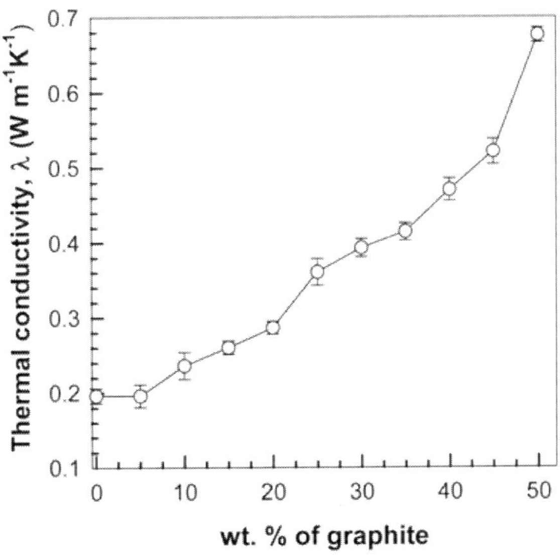

Figure 9. Thermal conductivity as a function of graphite content.

Figure 10. Samples with 40, 45 and 50 wt. % of graphite after horizontal flame testing.

Photograph of these three specimens is given in Figure 10. The EOC/graphite composites can be classified as HB according to the UL-94 standards since their burning rates did not exceed 75 mm min^{-1}, the maximum burning rate was found to be 47 mm min^{-1}.

Shore-A hardness

Composites of soft matrix polymers with conducting fillers are being used as strain or pressure sensors [34]. Thus hardness of these composites has practical importance. Shore-A hardness of the graphite/EOC composites has been measured and it was found to be increasing with increasing graphite content. See Figure 11.

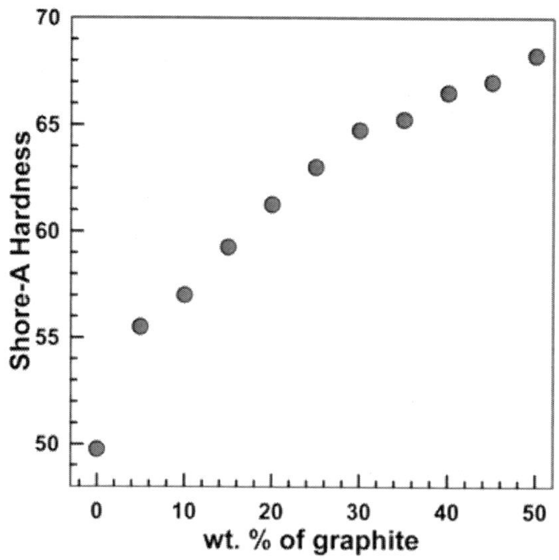

Figure 11. Shore-A hardness as a function of graphite content.

Shore-A hardness of pure EOC matrix was 50 while addition of 5% graphite increased the value to 55. Even the addition of 50 wt. % of graphite resulted in a hardness value of only 68. This shows that hardness of EOC/expandable graphite composites did not increase markedly with the addition of graphite filler, or in other words higher loadings of graphite did not hamper the rubbery nature of the EOC matrix. This is important in case of applications involving sensors when the electric conductivity is changing with pressure [34]. This will be the subject of our future study.

The softness of the composite was important also for the measurement of thermal conductivity. Very hard samples such as polypropylene (PP) might not have flat surface (especially after injection molding). Rough surface creates an air gap between the sample and the conducting metal. When the sample is soft, there is no such air gap, rendering perfect contact which assures proper thermal conductivity measurement.

CONCLUSION

Both AC and DC conductivities of EOC/graphite composites were found to be increasing with increase in graphite content. Percolation threshold has been observed at about 16 vol. % in both cases (DC and AC conductivities). Thermal conductivity of the composites showed a trend of increment with addition of graphite filler, but no percolation threshold was observed. Horizontal flame test according to the UL-94 standard proved the role of expandable graphite filler as an environmentally friendly halogen-free flame retardant. EOC composites with graphite content higher than 40 wt. % (or 22 vol. %) showed self-extinguishing behavior during horizontal flame test. Higher loadings of expandable graphite filler did not cause serious reduction in rubbery nature of the EOC/expandable graphite composites which could be potentially used in applications like pressure or temperature sensors. In summary these composites possess good electrical and thermal conductivities together with significant flame retardancy at higher loadings without extreme increase in hardness values.

APPENDIX

Since there is a possibility to use these composites in automotive industry and in electronics because of acceptable flame resistance the burn rate was examined in detail. Figures 12-15 show the horizontal burn tests with the evaluation of burn rate. When we compare Figures 12 and 13 it is clear that it takes longer to burn the sample with 30 wt.% of graphite.

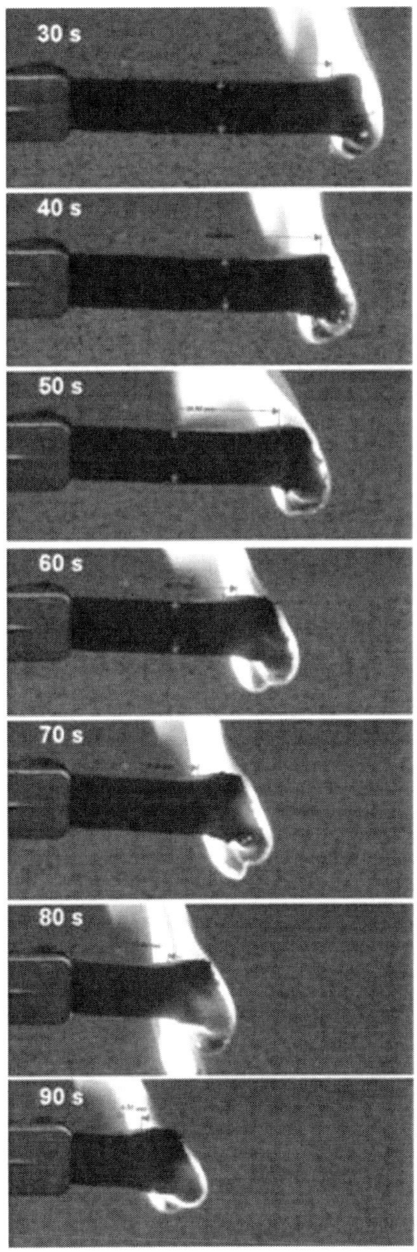

Figure 12. Horizontal burn test of EOC with 20 wt.% of graphite.

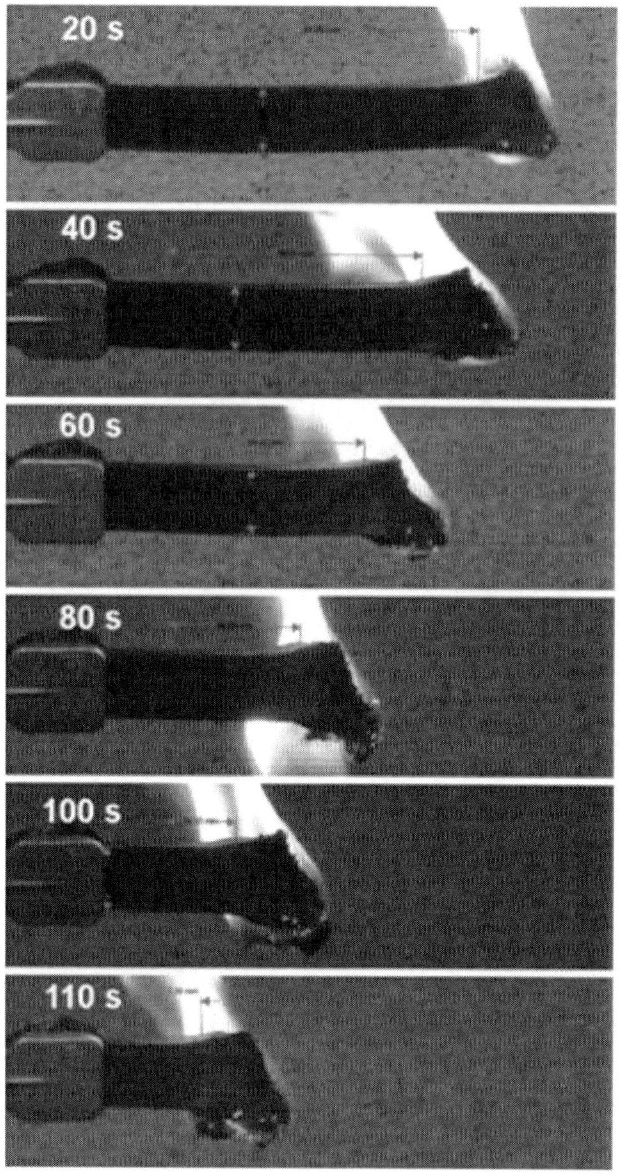

Figure 13. Horizontal burn test of EOC with 30 wt.% of graphite.

When we plot the distance that the flame travels as a function of time we can get a plot that is illustrated on Figure 14. When one evaluates the slope of the line it is possible to get a burn rate. This burn rate was plotted as a function of graphite content on Figure 15. Very interestingly initially the rate increases with increasing graphite content (up to 30 wt. % of graphite).

It is possible to explain this phenomenon by increasing thermal conductivity (shown on Figure 9). The heat travels faster when there is more graphite present. However at concentrations higher than 20 wt.% the influence of graphite as flame retardant prevails over the increased thermal conductivity effect. There is a visible decrease in burning kinetics when the concentration is increased from 20 to 30 wt.% of graphite.

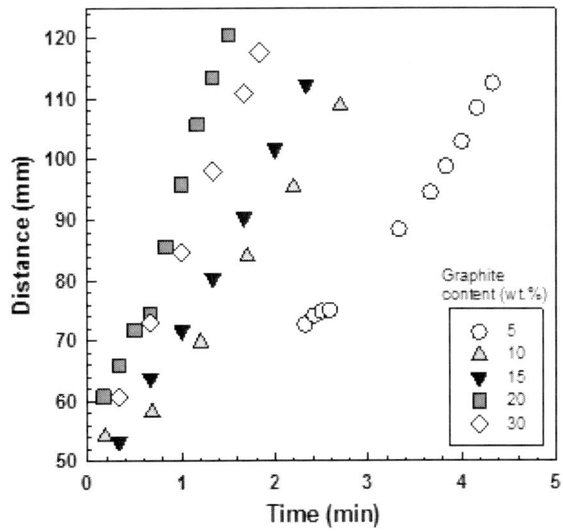

Figure 14. Evaluation of the burn rate during the horizontal burn test.

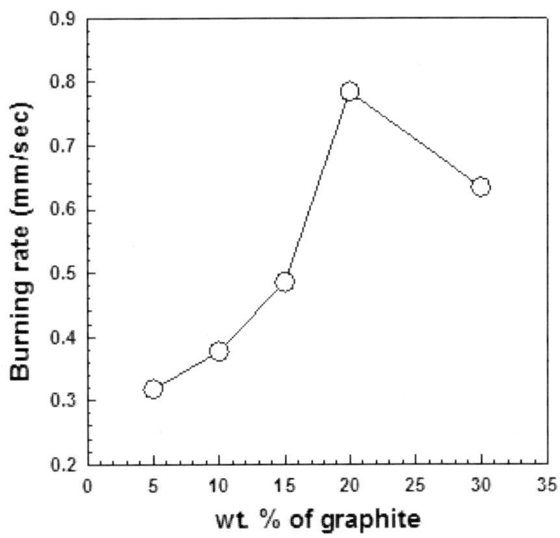

Figure 15. Burning rate as a function of graphite content.

The decrease in burn rate continues even more at concetntrations 40-50 wt.% as was shown on Figure 10 when the samples actually self-extinguished during the tests (therefore we could not evaluate the burn rate).

ACKNOWLEDGMENTS

This work has been supported by Operational Programme Research and Development for Innovations co-funded by the European Regional Development Fund (ERDF) and national budget of Czech Republic within the framework of the Centre of Polymer Systems project (reg.number:CZ.1.05/2.1.00/03.0111).

References

[1] V.S. Mironov, J.K. Kim, M. Park, S. Lim, W.K. Cho, *Polym.Test.*, 26, 547–555 (2007).

[2] Q. Zhang, H. Xiong, W. Yan, D. Chen, M. Zhu, *Polym. Eng. Sci.,* 48, 2090 - 2097 (2008).

[3] Q. Chen, Y. Xi, Y. Bin, M. Matsuo, *J. Polym. Sci. Pol. Phys.,* 46, 359-369 (2008).

[4] J. Vilcakova, P. Saha, O. Quadrat, *Eur. Polym. J.,* 38, 2343-2347 (2002).

[5] W.-y. Wang, G.-h.Luo, F. Wei, J. Luo, *Polym. Eng. Sci.,* 49, 2144-2149 (2009).

[6] A. Paul, S. Thomas, *J. Appl. Polym. Sci.,* 63, 247-266 (1997).

[7] R.K. Goyal, S.D. Samant, A.K. Thakar, A. Kadam, *J. Phys. D. Appl.Phys.*, 43, 1-7 (2010).

[8] N.M. Renukappa, Siddaramaiah, R.D.S. Samuel, J.S. Rajan, J.H. Lee, *J. Mater. Sci.-Mater. El.*, 20, 648-656 (2009).

[9] H. Kim, A.A. Abdala, C.W. Macosko, *Macromolecules,* 43, 6515–6530 (2010).

[10] S. Bhadra, N.K. Singha, D. Khastgir, *Polym. Eng. Sci.,* 48, 995-1006 (2008).

[11] S. Hom, A.R. Bhattacharyya, R.A. Khare, A.R. Kulkarni, M. Saroop, A. Biswas, *Polym. Eng. Sci.,* 49, 1502-1510 (2009).

[12] J.-C. Huang, H.-L. Huang, *J. Polym. Eng.,* 17, 213-229 (1997).

[13] W.-P. Shih, L.-C.Tsao, C.-W.Lee, M.-Y.Cheng, C. Chang, Y.-J.Yang, K.-C. Fan, *Sensors,* 10, 3597-3610 (2010).

[14] S.A. Mansour, *eXPRESSPolym. Lett.,* 2, 836-845 (2008).

[15] S. Kirkpatrik, *Rev. Mod. Phys.,* 45, 574-588 (1973).

[16] Y.R. Hernandez, A. Gryson, F.M. Blighe, M. Cadek, V. Nicolosi, W.J. Blau, Y.K. Gun'ko, J.N. Coleman, *Scripta Mater.,* 58, 69-72 (2008).

[17] M.H. Al-Saleha, U. Sundararaj, *Carbon,* 47, 2-22 (2009).

[18] W. Zhang, A.A. Dehghani-Sanij, R.S. Blackburn, *J. Mater. Sci.,* 42, 3408-3418 (2007).

[19] R. Strumpler, J. Glatz-Reichenbach, *J. Electroceram.*, 3, 329-346 (1999).

[20] H. Hammami, M. Arous, M. Lagache, A. Kallel, *J. Alloy. Compd.,* 430, 1-8 (2007).

[21] X.Y. Meng, L. Ye, X.G. Zhang, P.M. Tang, J.H. Tang, X. Ji, Z.M. Li, *J.Appl. Polym. Sci.,* 114, 853-863 (2009).

[22] X.L. Chen, J. Yu, S.J. Lu, H. Wu, S.Y. Guo, Z. Luo, *J. Macromol. Sci. Part B-Phys.,* 48, 1081-1092 (2009).

[23] C.-H. Chen, W.-H.Yen, H.-C.Kuan, C.-F.Kuan, *Polym. Composite.*, 31, 18-24 (2010).

[24] C.-L. Chiang, S.-W. Hsu, *J. Polym. Res.,* 17, 315-323 (2010).

[25] B.J. Qu, R.C. Xie, *Polym. Int.,* 52, 1415-1422 (2003).

[26] H. Fukushima, L.T. Drzal, B.P. Rook, M.J. Rich, *J. Therm. Anal. Calorim.,*85, 235-238 (2006).

[27] L.J. Van der Pauw, *Philips Res. Rep.,* 13, 1-9 (1958).

[28] F.G.d.S. Jr., B.G. Soares, J.C. Pinto, *Eur. Polym. J.,* 44, 3908-3914 (2008).

[29] J.E. Lozano, *Fruit Manufacturing: Scientific Basis, Engineering Properties and Deteriorative Reactions of Technological Importance,* Springer Science Business Media LLC., New York, (2006).

[30] H. Scher, R. Zallen, *J. Chem. Phys.,* 53, 3759-3761 (1970).

[31] N.K. Srivastava, V.K. Sachdev, R.M. Mehra, *J. Appl. Polym. Sci.,* 104, 2027-2033 (2007).

[32] M. Pelíšková, J. Vilčáková, R. Moučka, P. Sáha, J. Stejskal, O. Quadrat, *J. Mater. Sci.*, 42, 4942-4946 (2007).

[33] F. Kremer, A. Schonhals, *Broadband Dielectric Spectroscopy*, Springer-Verlag, Berlin Heidelberg, (2003).

[34] S. Bhadra, N.K. Singha, D. Khastgir, *Curr. Appl. Phys.*, 9, 396–403 (2009).

INDEX

A

B

C

E

F

N

O

P

Q

R

S

T

U

V

W

X

Y

Z